Code✓Check

Complete

2nd Edition

An Illustrated Guide to the Building, Plumbing, Mechanical, and Electrical Codes

BUILDING

PLUMBING

MECHANICAL

ELECTRICAL

Redwood Kardon / Douglas Hansen / Illustrated by Paddy Morrissey

© 2012 by The Taunton Press, Inc.   ISBN 978-1-60085-493-4   **Douglas Hansen & Redwood Kardon** • Illustrations & Layout: **Paddy Morrissey**

INTRODUCTION

Code Check Complete 2nd Edition is a compilation of the individual Code Check field guides to codes, including Code Check Building 3rd edition, Code Check Plumbing & Mechanical 4th edition, and Code Check Electrical 6th edition.

The basic code referenced in these books is the 2009 International Residential Code®. The book also includes references for the 2011 National Electrical Code®, the 2009 Uniform Plumbing and Mechanical Codes, and several specialized codes. The major codes are updated on a three-year cycle. Building jurisdictions do not always adopt a code during the calendar year of publication, and some actually lag several years behind. This book can still be used in areas where older codes apply, thanks to the "code changes" feature at the end of each section.

Significant changes are highlighted throughout the text and summarized there, and users of older codes are alerted to items where the most current code does not apply.

This book is intended for inspectors, design professionals, plan reviewers, contractors, home inspectors, educators, and do-it-yourselfers. The information contained in this document is believed to be accurate; however, it is provided for informational purposes only and is not intended as a substitute for the full text of the referenced codes. Publication by Taunton Press, ICC, and the authors should not be considered by the user to be a substitute for the advice of a registered design professional. For additional information, or to contact the authors, visit **www.codecheck.com** and **www.taunton.com**.

TABLE OF CONTENTS

Code ☑ Check® Building Third Edition

BY DOUGLAS HANSEN & REDWOOD KARDON
Illustrations & Layout by Paddy Morrissey

Based on Chapters 1 – 10 of the 2009 International Residential Code®

For updates and information related to this book, visit www.codecheck.com

Code Check Building 3rd edition is a condensed guide to the building portions of the *2009 International Residential Code (IRC) for One- & Two-Family Dwellings*. The IRC is the most widely used residential building code in the United States. Significant code changes are highlighted in the text and summarized in the table on pages 62-64. By using that table, the book is also applicable in areas using older editions of the IRC. Check with the local building department to determine which code is used in your area, and for local amendments.

REFERENCE DOCUMENTS

The IRC is part of the suite of codes published by the **International Code Council**. It is limited to one- and two-family dwellings and townhouses not more than three stories above grade. It is a prescriptive document containing minimum rules and instructions for conventional construction. These can be exceeded by using the design-based provisions of the International Building Code (IBC), a more comprehensive document containing engineering regulations for structural design. Aspects of a building that exceed the scope of the IRC must be built to the IBC or one of the following prescriptive codes:

The American Forest and Paper Association publishes the *Wood Frame Construction Manual for One- and Two-Family Dwellings (WFCM)*, which can be used as an alternative to IRC designs for wood framing.

The American Iron and Steel Institute (AISI) publishes the *Standard for Cold-Formed Steel Framing–Prescriptive Method for One- and Two-Family Dwellings (AISI S230)*, which can be used as an alternative to the IRC.

The American Concrete Institute (ACI) publishes two documents that supplement the prescriptive rules of the IRC. These are *ACI 318–Building Codes for Structural Concrete* and *ACI 530–Building Code Requirements for Masonry Structures*.

The Truss Plate Institute (TPI) publishes *TPI 1–National Design Standard for Metal Plate Connected Wood Truss Construction*, which is mandatory for metal-plate-connected truss design. TPI also contributes to *BCSI 1-03–Guide to Good Practice for Handling, Installing & Bracing of Metal Plate Connected Wood Trusses*.

Acknowledgments: Thanks to Sandra Hyde, PE, and the International Code Council.

KEY TO USING THIS BOOK

The line for each code rule starts with a checkbox and ends with the code reference in brackets. Exceptions and lists start with a bullet and also end with the code reference in brackets. Changes to the 2009 code are highlighted by printing the reference in a different color; the superscript number refers to the table which begins on page 62 (**p.62**). The following example is from **p.13**:

☐ Floor or landing min 36 in. deep on each side of door EXC [311.3]
• Balconies < 60 sq. ft. OK for landing to be < 36 in. deep [311.3X]¹⁹

These lines give the basic rule that landings at least 36 inches deep are required on each side of a door, and the code reference in the IRC is section 311.3. (In the IRC, the number is actually R311.3. We omit the letter R at the beginning to save space and include more information on each line.) The line that follows is an exception to the rule, and the code reference is 311.3 Exception. This exception is a new code change, and is explained further on p.62 as code change #19.

Tables and Figures are referenced in the code text lines, as in the following example from **p.29**:

☐ Notching & boring per F30, T14 [502.8.1]

This line says that the rules for notching & boring joists are found in section 502.8.1 and illustrated in Figure 30, with further explanation in Table 14.

SEQUENCE OF THIS BOOK

This book follows the same basic sequence as the IRC. It begins with the planning and administrative sections in the IRC chapter 1, followed by the planning and nonstructural topics in the IRC chapter 3. The structural sections are arranged "from the ground up," beginning with foundations (chapter 4), followed by floors (5), wall construction (6), wall coverings (7), roof-ceiling construction (8), roof assemblies (9), and chimneys and fireplaces (10).

ABBREVIATIONS

L&L	=	listed & labeled
lb.	=	pound(s)
max	=	maximum
min	=	minimum
mph	=	miles per hour
o.c.	=	on center
PL	=	property line (lot line)
PT	=	pressure treated
psf	=	pounds per square foot
psi	=	pounds per square inch
req	=	require
req'd	=	required
req's	=	requires, requirements
SDC	=	Seismic Design Category
SDC D	=	Seismic Design Categories D_0, D_1 & D_2, inclusive
sq.	=	square, as in sq. ft.
UL	=	Underwriters Laboratories
WRB	=	water-resistive barrier
WSP	=	wood structural panel

AAMA	=	American Architectural Manufacturers Association
ACI	=	American Concrete Institute
AMI	=	in accordance with manufacturer's instructions
ASTM	=	American Society for Testing & Materials
BO	=	building official
BWL	=	braced wall line
BWP	=	braced wall panel
BUR	=	built-up roof
cfm	=	cubic feet per minute
CMU	=	concrete masonry unit
EXC	=	exception to rule will follow in the next line
FSD	=	fire separation distance
ft.	=	foot / feet
GB	=	gypsum board
hr.	=	hour
IBC	=	International Building Code
ICF	=	insulating concrete form
in.	=	inch(es)

CODE CHECK BUILDING CONTENTS

Aspect ratio: The ratio of longest to shortest dimensions, or for wall sections, the ratio of height to length.

Attic: The unfinished space between the ceiling assembly of the top story and the roof assembly.

Attic, habitable[1]: A finished or unfinished area meeting minimum room dimension and ceiling height requirements and enclosed by the roof assembly above, knee walls (if applicable) on the sides, and the floor–ceiling assembly below.

Basement: A portion of a building that is partly or completely below grade.

Braced wall line (BWL)[2]: A straight line through the building plan representing the location of lateral resistance provided by wall bracing.

Braced wall panel (BWP): A full-height section of wall constructed to resist shear forces by application of bracing materials.

Building thermal envelope: The basement walls, exterior walls, floor, roof, and other building elements that enclose conditioned space.

Connector: A device such as a joist hanger, post base, hold-down, mudsill anchor, or hurricane tie used to connect structural components–also see *Fastener.*

Cripple wall: Wood-framed wall extending from the foundation to joists below the first floor. Found in the underfloor area.

Dampproofing: A coating intended to protect against the passage of water vapor through walls or other building elements. It is a lesser degree of protection than waterproofing.

Dead load: The weight of all materials of the building and fixed equipment.

Diaphragm: A horizontal or nearly horizontal system, such as a floor, acting to transmit lateral forces to the vertical resisting elements.

Fastener: Generic category that includes nails, screws, bolts, or anchors–also see *Connector.*

Fire separation distance: The distance measured perpendicular from the building face to the closest interior lot line or to the centerline of a street, alley, or public way.

Grade: The finished ground level adjoining the building at all exterior walls.

Habitable space: Space in a building for living, sleeping, eating, or cooking. Bathrooms, bathroom closets, hallways, storage, or utility areas are not considered habitable space.

Live loads: Loads produced by use and occupancy of the building and not including wind, snow, rain, earthquake, flood, or dead loads.

Monolithic: Concrete cast in one continuous operation with no joints, such as a footing and floor slab or a footing and foundation stem wall.

Perm: The unit of measurement of water vapor transmission through a material, based on the number of grains of water vapor at a given pressure differential. Vapor retarders are rated in perms.

Plain concrete or masonry: Structural concrete or masonry with less reinforcement than the minimum amount specified for reinforced concrete or masonry.

Seismic Design Category (SDC): Classification assigned to buildings based on the occupancy category & severity of earthquake ground motion expected at the site.

Story: That portion of a building that is between the upper surface of one floor and the upper surface of the next floor above or the roof.

Story above grade: The parts of the building that are entirely above grade, or basements that are more than 6 feet above grade for more than 50% of the total building perimeter or more than 12 feet above ground at any point.

Townhouse[3]: Single-family dwelling unit constructed in groups of three or more attached units in which each unit extends from foundation to roof and with a yard or public way on at least two sides.

Waterproofing: Materials that protect walls or other building elements from the passage of moisture as either vapor or liquid under hydrostatic pressure.

Wood structural panel (WSP): A panel manufactured from veneers (plywood) or wood strands (OSB) and bonded with waterproof synthetic resins. Wood structural panels must bear a grade stamp (see F35 on p.32) and are used in floors, roof diaphragms, and shear walls.

PLANNING, PERMITS & INSPECTIONS

Before beginning a building or project, plans must be approved by the local building department and must conform to applicable climatic and geographic design criteria. The plans must include setbacks from the property lines and adjacent slopes.

Plans & Permits   09 IRC
- [ ] Scope of code is 1- & 2-family dwellings & townhouses _____ [101.2]
- [ ] Approved plans & permit card on site _____ [105.7 & 106.3.1]
- [ ] Plans to include BWL locations & methods _____ [106.1.1][4]
- [ ] Alternative materials, design & methods OK when approved by BO _ [104.11]
- [ ] Local statutes may req registered design professional to draw plans _ [106.1]
- [ ] Site plan or plot plan to be included in construction documents _____ [106.2]
- [ ] Permits req'd for new work, additions, repairs & alterations _____ [105.1]
- [ ] Permits not req'd for: _____ [105.2]
  - Detached one-story accessory structures (tool sheds) ≤ 200 sq. ft.[5]
  - Fences ≤ 6 ft., sidewalks, driveways, swings & playground equipment
  - Retaining walls ≤ 4 ft. bottom of footing to top of wall & with no surcharge
  - Water tanks on grade ≤ 5,000 gallons & height/width ratio ≤ 2:1
  - Painting, tiling, carpeting, cabinets, counters & similar finish work
  - Awnings projecting ≤ 54 in. from exterior wall & supported from wall
  - Decks ≤ 200 sq. ft. & ≤ 30 in. above grade & not attached to dwelling or serving req'd exit door[6]

Required Inspections   09 IRC
- [ ] Inspection & approval prior to concealing any work _____ [109.4]
- [ ] Foundation forms & steel prior to placing concrete _____ [109.1.1]
- [ ] In flood hazard areas, registered design professional req'd to document lowest floor elevation before construction above it _____ [109.1.3]
- [ ] Rough plumbing, mechanical & electrical before concealment _____ [109.1.2]
- [ ] Frame & masonry after fireblocking & bracing in place _____ [109.1.4]
- [ ] Air barrier & insulation inspection (may be 3rd party) _____ [1102.4.2.2]

Required Inspections (cont.)   09 IRC
- [ ] Drywall nailing of fire resistance–rated walls prior to taping _____ [109.1.5.1]
- [ ] Special inspections as authorized by BO _____ [109.1.5]
- [ ] Final inspection_____ [109.1.6]

DESIGN

The IRC assigns a Seismic Design Category (SDC) from A to E, with A the least likely to experience seismic activity and E the most vulnerable. Category D is broken into three subparts, D_0, D_1 & D_2. Buildings in SDC E must be designed to the IBC. The BO can allow a SDC E to be designated as D_2 in some circumstances, such as buildings of "regular shape" with wall bracing continuous in one plane from the foundation to the uppermost story and no cantilevers.

Design Criteria General   09 IRC
- [ ] Nonconventional elements designed per IBC_____ [301.1.3]
- [ ] Determine climatic & geographic design criteria **T1** _____ [301.2]
- [ ] Complete **T1** from maps & BO _____ [T301.2(1)]
- [ ] In flood hazard areas, determine design flood elevation_____ [322.1.4]
- [ ] BO may req soil tests if expansive, compressible, or questionable ___ [401.4]

TABLE 1		CLIMATIC AND GEOGRAPHIC CONDITIONS [T301.2(1)]				
Ground Snow Load	Wind Speed (mph)[A]	Seismic Design Category[A]	Weathering (Concrete)[B]	Frost Line Depth[A]	Termites Hazard[A]	Flood Hazards[C]

A. To be filled in by local building department.
B. Choose negligible, moderate, or severe–affects strength of concrete & grade of CMUs.
C. To reference entry date into National Flood Insurance Program, date of Flood Insurance Study & numbers & dates of currently effective maps.

Wind Design 09 IRC

☐ Determine basic wind speed from maps _____ [301.2.1.4 & F301.2(4)]

☐ If history of damage due to wind speed-up at hills, modify
map values to consider topographic effects _____ [301.2.1.5][7]

☐ If basic wind speed > 100mph in hurricane-prone region or
> 110mph, design per ICC-600, ASCE-7, WFCM, or AISI S230 _[301.2.1.1]

☐ Cladding, covering & fenestration req design for specified pressure
loads or per T301.2(2&3) & F301.2(7) _____ [301.2.1]

☐ Windows in windborne debris regions to ASTM E 1886 & E 1996
impact-resistance standards EXC _____ [301.2.1.2]

   • 1- & 2-story buildings WSP protection precut & predrilled to
   fit on permanently installed anchors on building _____ [301.2.1.2X]

STRUCTURAL PLANNING

Story Height (also see p.33) 09 IRC

☐ Wood framing 10 ft. **T18** + 16 in. for height of floor framing EXC ____ [301.3]

   • 12 ft. OK if **T21** increased 10% or **T22** increased 20% _____ [301.3X][8]

☐ Engineered design req'd when exceeding above limits _____ [301.3]

Live Loads & Allowable Deflection 09 IRC

☐ Min uniformly distributed live load per **T2** _____ [301.5]

☐ Allowable deflection of structural members per **T3** _____ [301.7]

TABLE 2	MIN. UNIFORMLY DISTRIBUTED LIVE LOADS [T301.5]
Use	**Live Load (psf)**
Attic without storage	10
Attic with limited storage (see **p.48**)	20
Habitable attics & attics with fixed stairs	30
Sleeping rooms	30
Balconies & decks	40
Rooms other than sleeping rooms	40
Stairs	40[A]
Concentrated point load on top of handrails or guardrails	200
Guardrail in-fill components	50[B]

A. Individual treads can be designed for 300 lb. load over a 4 sq. in. area.
B. Horizontally applied load over a 1 sq. ft. area

TABLE 3	ALLOWABLE DEFLECTION [T301.7]
Structural Member	**Deflection[A]**
Rafters > 3:12 slope & no finished ceiling attached	L/180
Interior walls & partitions	H/180
Floors & plastered ceilings	L/360
Lintels supporting masonry veneer walls	L/600
Other structural members	L/240

A. L = span length, H = span height

LOCATION ON SITE

Location & Setbacks   09 IRC

☐ 1-hr. rated walls req'd when FSD < 5 ft. EXC **F1** _____ [T302.1]
  • Walls & openings perpendicular to line determining FSD **F1** ____ [302.1X1]
  • Accessory structures that are exempt from permit _____ [302.1X3]
☐ No opening in walls < 3 ft. from PL EXC _____ [T302.1]
  • Foundation vents _____ [302.1X5]
  • Openings perpendicular to line determining FSD **F1** _____ [302.1X1]
☐ Underside of eaves 1-hr. rated if FSD ≥ 2 ft. to 5 ft. _____ [T302.1]
☐ No projections (eaves) < 2 ft. of PL EXC _____ [T302.1]
  • Detached garages within 2 ft. of PL eave projections ≤ 4 in. OK_ [302.1X4]

FIG. 1

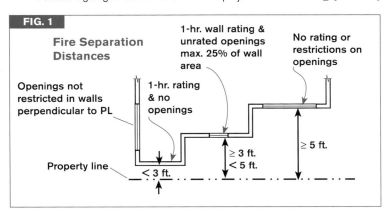

Fire Separation Distances

1-hr. wall rating & unrated openings max. 25% of wall area

No rating or restrictions on openings

Openings not restricted in walls perpendicular to PL

1-hr. rating & no openings

Property line

< 3 ft.

≥ 3 ft. < 5 ft.

≥ 5 ft.

FIRE PROTECTION

Fire-resistive construction materials such as gypsum board provide passive protection against the rapid spread of a fire. Fireblocking slows the spread of fire in small concealed spaces, and draftstopping accomplishes the same function in larger concealed areas.

Separation between Townhouses   09 IRC

☐ Each unit req's its own 1-hr. separation wall to adjacent unit EXC **F2** ____ [302.2]
  • Common 1-hr. wall OK if no plumbing/mechanical in wall cavity _____ [302.2X][9]
  • Electrical boxes meeting penetration rules OK in common wall **F2** ____ [302.2X]
☐ Common walls continue in rated parapet to 30 in. above roof EXC_____ [302.2.2]
  • Noncombustible roof deck or GB wrapback for 4 ft. _____ [302.2.2X]
  • Roofs with > 30 in. elevation difference _____ [302.2.2]

FIG. 2   **Townhouse Separation Wall**

Electrical boxes fire-rated, steel, protected, or separated by mineral wool

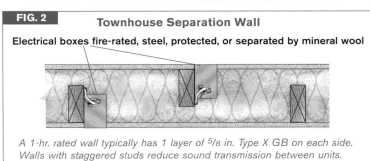

A 1-hr. rated wall typically has 1 layer of 5/8 in. Type X GB on each side. Walls with staggered studs reduce sound transmission between units.

FIRE PROTECTION (CONT.)

Separation in Two-Family Dwellings 09 IRC
☐ 1-hr. common wall req'd from foundation to underside of roof EXC. [302.3]
• ½-hr. OK if building protected by automatic sprinkler system [302.3X1]
• Attic separation can be draft stop if ceilings 5/8 in. Type X [302.3X2]

Penetrations of Fire-Resistive Membranes 09 IRC
☐ Steel electrical boxes allowed in wall membrane if max 16 sq. in. & aggregate area of openings ≤ 100 sq. in. [302.4.2X1]
☐ Steel boxes on opposite sides of wall min 24 in. horizontal separation or protected by insulation, fireblocking, or listed putty pads F2 [302.4.2X1]
☐ L&L fire-rated boxes allowed in walls AMI F2 [302.4.2X2]
☐ Through penetrations req listed firestop penetration system or must be part of approved and tested assembly [302.4.1]

Finish Surfaces & Insulation 09 IRC
☐ Wall & ceiling finishes max smoke-developed index 200, max smoke-developed index 450 in accordance with ASTM E 84 / UL 723 [302.9]
☐ Insulation & facing max flame spread index 25, max smoke-developed index 450 in accordance with ASTM E 84 / UL 723 EXC [302.10.1]
• Facing material exempt when in substantial contact with unexposed surface of wall, floor, or ceiling – i.e., not visible in finished job [302.10.1X1]
☐ Foam plastic max flame spread index 75, max smoke-developed index 450 in accordance with ASTM E 84 / UL 723 [316.3]
☐ Foam plastic not OK to be exposed to building interior [316.4]
☐ Foam req's thermal barrier of min ½ in. GB EXC. [316.4]
• In roof assembly separated by WSPs [316.5.2]
• Alternate lesser covering barriers allowed in crawl spaces & attics entered only for repairs or maintenance [316.5.3&4]

Separation from Garages 09 IRC
☐ Min ½ in. GB or equivalent on garage side of walls & ceilings common to house or shared attic space attic room F3 [T302.6]
• Min 5/8 in. Type X GB ceiling under habitable room F3 [T302.6]
• Min ½ in. GB on walls, beams, or other structures that support ceilings providing separation between house & garage [T302.6]
☐ Garage walls perpendicular to dwelling OK unprotected unless supporting floor/ceiling separations [302.6]
☐ No direct openings between garage & sleeping rooms [302.5.1]
☐ Door to house rated 20-minute, 1⅜ in. solid wood or steel [302.5.1]
☐ Ducts in garage & penetrating common walls min 26-gage steel [302.5.2]
☐ No duct openings in garage [302.5.2]
☐ Seal penetrations of common walls with approved material [302.5.3]
☐ Sealant does not have to comply with ASTM E 136 [302.5.3 & 302.11#4]
☐ Detached garages closer than 3 ft. req ½ in. GB on interior side of garage walls facing house [T302.6]10

GARAGES & CARPORTS

General 09 IRC
☐ Floor surfaces approved noncombustible material EXC [309.1&2]
• Asphalt OK at ground level in carports [309.2X]
• Floor sloped to a drain or to vehicle entry [309.1&2]
☐ Carports not open on 2 sides considered a garage [309.2]

FIG. 3

Fire Separation from Garage

If habitable space over garage, ceiling must be 5/8 in. Type X.

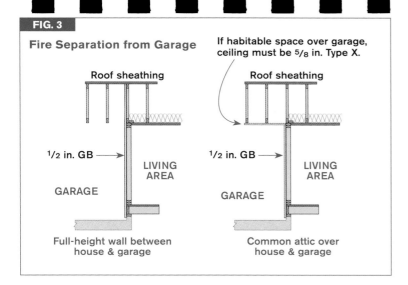

Full-height wall between house & garage

Common attic over house & garage

FIG. 4

Fireblocking at Wall/Ceiling

Pipes, vents, ducts, wires & cables req. fireblocking where the wall intersects the ceiling. If more than 2 NM cables in a single hole are fireblocked, they must be derated (see p.212).

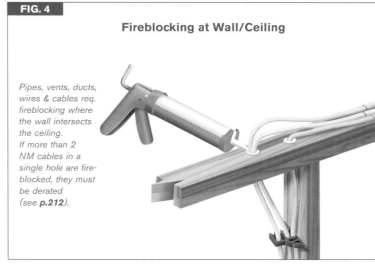

Fireblocking   09 IRC

- ☐ Purpose is to cut off concealed draft openings _____ [302.11]
- ☐ Materials can be 2 in. lumber, 2 thicknesses 1 in. lumber, ¾ in. WSP, ¾ in. particleboard, ½ in. GB, ¼ in. millboard, mineral wool, or glass fiber batts securely retained in place _____ [302.11.1]
- ☐ Unfaced fiberglass must fill entire cavity to height of 16 in. **F6** __ [302.11.1.2]
- ☐ Caulking or other material filling annular space does not have to comply with ASTM E 136 _____ [302.11#4][11]

Fireblocking (cont.)   09 IRC

- ☐ Req'd locations: _____ [302.11]
  - In walls vertically at ceiling & floor levels, horizontally max 10 ft.
  - Intersections of concealed vertical/horizontal spaces (e.g., soffits) **F5,6,7**
  - Concealed spaces between stair stringers at top & bottom of run
  - Openings around vents, ducts, pipes & cables at ceilings & floors **F4**
  - In space between chimneys & combustible framing
  - In 2-family dwelling cornices at line of unit separation

FIRE PROTECTION

FIG. 5

Air Flow through Soffit

Air communicates through soffit to attic or ceiling space above.

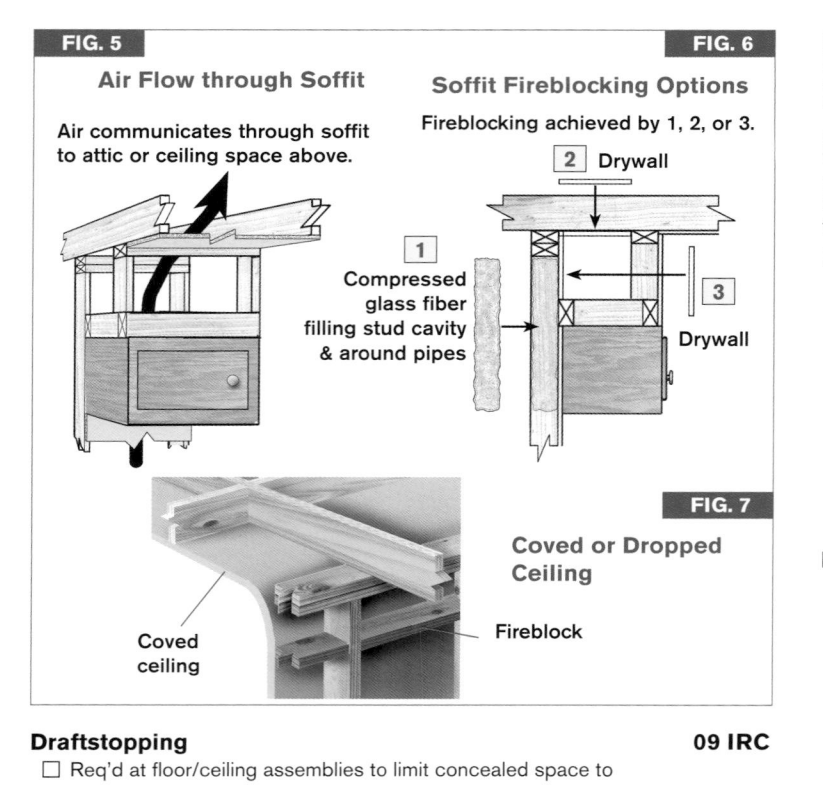

FIG. 6

Soffit Fireblocking Options

Fireblocking achieved by 1, 2, or 3.

2 Drywall

1

Compressed glass fiber filling stud cavity & around pipes

3

Drywall

FIG. 7

Coved or Dropped Ceiling

Fireblock

Coved ceiling

Draftstopping   09 IRC

☐ Req'd at floor/ceiling assemblies to limit concealed space to 1,000 sq. ft. when using suspended ceiling or open-web trusses ___ [302.12]
☐ Materials min ½ in. GB, ⅜ in. WSP, or equivalent _____ [302.12.1]

FIRE SPRINKLER SYSTEMS

The building section of the IRC tells us when we need to install fire sprinklers. The methods for how we do this are in the plumbing section of the code, which offers the alternative of compliance with NFPA 13D. Such fire sprinklers slow a fire sufficiently to allow occupants to escape from the building; their purpose is to protect the occupants, not the building. Local rules vary on the extent of additions or remodels that trigger a need for compliance. Refer to *Code Check 6th edition* for further information.

Required Locations   09 IRC

☐ Required in new townhouses & 1- & 2-family dwellings EXC ____ [313.1&2][12]
 • Additions or alterations of buildings without automatic sprinklers [313.1X&2X]
☐ Must protect all areas of dwelling unit EXC_____ [2904.1.1]
 • Attics, crawl spaces, etc. without fuel-fired appliances _____ [2904.1.1X1]
 • Attics, crawl spaces, etc. with fuel-fired appliances req sprinklers directly over appliance, not elsewhere in that space _____ [2904.1.1X1]
 • GB-surfaced closets ≤ 24 sq. ft. with smallest dimension ≤ 3 ft. [2904.1.1X2]
 • Bathrooms ≤ 55 sq. ft _____ [2904.1.1X3]
 • Garages, carports, exterior porches & unheated entries (mud rooms) that are adjacent to an exterior door _____ [2904.1.1X4]

Methods   09 IRC

☐ May be multipurpose system or stand-alone system _____ [2904.1]
☐ Comply with IRC 2904 or with NFPA 13D_____ [2904.1]
☐ Design flow rate, piping & coverage per either of above standards ___ [2904.1]

Inspections   09 IRC

☐ Preconcealment: req'd areas, clearances, ratings, pipe size & length, listing, manufacturer's instructions & testing _____ [2904.8.1]
☐ Final: Heads not painted or obstructed, pumps (if applicable) automatically start, no impairments to flow such as added filters, owners manual present & warning sign installed **F8** _____ [2904.8.2]

FIG. 8

Required Fire Sprinkler Warning Sign

WARNING!

The water system for this home supplies fire sprinklers that require certain flows and pressures to fight a fire. Devices that restrict the flow, decrease the pressure, or automatically shut off the water to the fire sprinkler system, such as water softeners, filtration systems, and automatic shutoff valves, shall not be added to this system without a review of the fire sprinkler system by a fire protection specialist.

DO NOT REMOVE THIS SIGN.

SMOKE & CARBON MONOXIDE ALARMS

Smoke Alarms   09 IRC

☐ Req'd in each sleeping room, outside each sleeping area & on each additional story, including basements & habitable attics **F9** _____ [314.3]

☐ If split level without intervening door, alarm on upper level sufficient for lower level provided lower level < 1 full story below upper level_____ [314.3]

☐ Interconnect so activation of 1 sets off all other alarms _____ [314.3]

☐ Power must be supplied from building wiring with battery backup ____ [314.4]

☐ NFPA 72 central-station monitored systems allowed_____ [314.2][13]

☐ NFPA 72 system must be permanent fixture of property _____ [314.2][14]

☐ Alterations & additions same rules as new construction EXC_____ [314.3.1]

   • Work only on building exterior does not trigger compliance ____ [314.3.1X1]

   • Interconnection & hardwiring req's waived if no accessible attic or crawl space & no interior finishes removed to allow wiring access [314.4X2]

Carbon Monoxide Alarms   09 IRC

☐ Req'd immediately outside each sleeping room in new dwellings with fuel-fired appliances or with attached garages **F9** _____ [315.1][15]

☐ Req'd in existing homes when work requiring permit is performed__ [315.2][15]

☐ Must comply with UL 2034 & be installed AMI_____ [315.3][15]

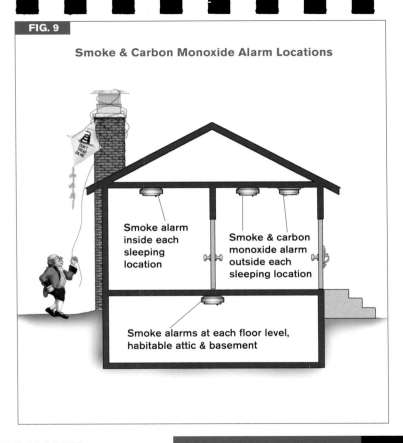

FIG. 9

Smoke & Carbon Monoxide Alarm Locations

Smoke alarm inside each sleeping location

Smoke & carbon monoxide alarm outside each sleeping location

Smoke alarms at each floor level, habitable attic & basement

HABITABILITY

Room Areas & Ceiling Heights 09 IRC
- ☐ Min area of habitable room 70 sq. ft.–except kitchens_____ [304.2]
- ☐ Min horizontal dimension of habitable room 7 ft.–except kitchens ____ [304.3]
- ☐ Min ceiling height 7 ft. EXC_____ [305.1][16]
  - Rooms with sloped ceilings min 50% of room min 7 ft. ceiling___ [305.1X1]
  - Sloped ceiling bathroom min 6 ft. 8 in. at center of req'd clear area in front of fixture_____ [305.1X2][17]
  - Basement areas without habitable space, hallway, laundry, or bathroom 6 ft. 8 in. ceiling OK & beams OK to within 6 ft. 4 in. of floor ____ [305.1.1]
- ☐ Portions of room with sloped ceiling < 5 ft. high, or with horizontal furred ceiling < 7 ft. high, do not count toward req'd room area _____ [304.4]

Heat, Light & Ventilation 09 IRC
- ☐ Habitable rooms req heating facilities capable of maintaining 68°F at 3 ft. above floor & 2 ft. from exterior walls (except Hawaii) _____ [303.8]
- ☐ Portable space heaters not OK as means of compliance _____ [303.8]
- ☐ Habitable rooms req natural ventilation openings to outdoor air ≥ 4% of floor area EXC _____ [303.1]
  - Approved mechanical ventilation system installed _____ [303.1X1]
  - Openings to sunroom additions if 40% open or only screened __ [303.1X3]

ASHRAE 62.2 recommends mechanical ventilation for all kitchens & bathrooms.

Heat, Light & Ventilation (cont.) 09 IRC
- ☐ Habitable rooms req natural light with glazing ≥ 8% of floor area EXC _ [303.1]
  - Mechanically ventilated rooms with artificial light _____ [303.1X2]
  - Borrowed light & ventilation OK from adjoining rooms if common opening min ½ of wall, min 25 sq. ft. & min 10% interior room area _ [303.2]

Bathroom Ventilation *(also see p.160)* 09 IRC
- ☐ Bathroom light from glazed openings min 3 sq. ft. & ½ openable EXC [303.3]
  - Glazed openings not req'd if mechanical ventilation direct to outside min 50 CFM intermittent or 20 CFM continuous _____ [303.3X]
- ☐ Exhaust ducts to outdoors; not to attic, soffit, or crawl space_____ [1501.1]

ESCAPE & RESCUE OPENINGS

Required Locations & Sizes 09 IRC
- ☐ Req'd in habitable attics, basements & sleeping rooms EXC _____ [310.1]
  - Basements ≤ 200 sq. ft. for mechanical equipment only_____ [310.1X]
- ☐ Each basement sleeping room _____ [310.1]
- ☐ Must open directly to public way or yard or court that opens to same_ [310.1]
- ☐ Max height of windowsill 44 in. above floor_____ [310.1]
- ☐ Min net clear area 5.7 sq. ft. EXC_____ [310.1.1]
  - 5.0 sq. ft. OK if grade floor opening (sill ≤ 44 in. above grade) __ [310.1.1X]
- ☐ Min net clear height 24 in., min net clear width 20 in. T4,5 _____ [310.1.2&3]
- ☐ Must be openable without keys, tools, or special knowledge _____ [310.1.4]

TABLE 4	BEDROOM WINDOW EGRESS FOR 5.0 SQ. FT. OPENING: GRADE-FLOOR OPENINGS ONLY (IN.)																				
Width	20	20½	21	21½	22	22½	23	23½	24	24½	25	25½	26	26½	27	27½	28	28½	29	29½	30
Height	36	35	34½	33½	33	32	31½	31	30	29½	29	28½	28	27½	27	26½	26	25½	25	24½	24

TABLE 5	BEDROOM WINDOW EGRESS: MIN. HEIGHT & WIDTH FOR 5.7 SQ. FT. OPENING SIZE (IN.)																											
Width	20	20½	21	21½	22	22½	23	23½	24	24½	25	25½	26	26½	27	27½	28	28½	29	29½	30	30½	31	31½	32	32½	33	33½	34
Height	41	40	39½	38½	37½	36½	35½	35	34½	33½	33	32½	31	31	30½	30	29½	29	28½	28	27½	27	26½	26½	25½	25½	25	24½	24

Window Wells F10   09 IRC

- ☐ Openings with finished sill below adjacent ground req window well __ [310.1]
- ☐ Min horizontal area of window well 9 sq. ft. _____ [310.2]
- ☐ Min horizontal dimension opposite opening 3 ft. EXC _____ [310.2]
  - Ladder may encroach 6 in. into req'd dimensions of window well __ [310.2X]
- ☐ Permanent ladder req'd complying with F10 _____ [310.2.1]

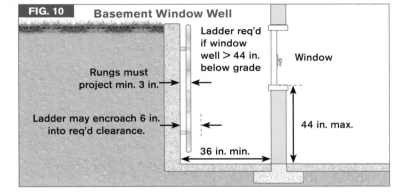

FIG. 10   **Basement Window Well**

Ladder req'd if window well > 44 in. below grade

Window

Rungs must project min. 3 in.

Ladder may encroach 6 in. into req'd clearance.

44 in. max.

36 in. min.

EGRESS

Doors   09 IRC

- ☐ Egress door req'd direct to exterior—not through garage _____ [311.1]
- ☐ Egress door side-hinged & min net clear width 32 in. _____ [311.2][18]
- ☐ Min clear height of egress door 78 in. top of threshold to stop _____ [311.2]
- ☐ OK for interior doors to have other dimensions _____ [311.2]
- ☐ Egress doors req keyless operation from interior side _____ [311.2

Thresholds & Landings at Doors   09 IRC

- ☐ Floor or landing min 36 in. deep on each side of door EXC _____ [311.3]
  - Balconies < 60 sq. ft. OK for landing to be < 36 in. deep _____ [311.3X][19]
  - OK for stair of ≤ 2 risers at exterior door other than req'd egress door provided door does not swing over stair _____ [311.3.2X]
- ☐ Min landing width same as door served by landing _____ [311.3]
- ☐ Max slope of exterior landings 2% _____ [311.3]
- ☐ Max threshold height above landing or floor 1½ in. EXC _____ [311.3.1]
  - 7¾ in. below threshold OK if door not swinging over landing ___ [311.3.1X]
- ☐ Storm & screen doors may swing over lower landing _____ [311.3.3]

Spiral Stairways   09 IRC

- ☐ Min width 26 in., all treads identical, min headroom 6 ft. 6 in. _____ [311.7.9.1]
- ☐ Min tread depth 7½ in. at 12 in. from center post, max riser 9½ in. [311.7.9.1]

Landings at Stairs   09 IRC

- ☐ Min 36 in. deep landing req'd at top & bottom each stair flight EXC _ [311.7.5]
  - Landing not req'd at top if door not swinging over interior stairs _ [311.7.5X]
- ☐ Garage-to-house stair considered interior stair for above rule ____ [311.7.5X]
- ☐ Max 12 ft. vertical between landings or floor levels _____ [311.7.5X]

Stairs: General:
IRC 09
- ☐ Min width above handrail 36 in. except spiral stairways F11 — [311.21]
- ☐ Max handrail projection into stairway 4½ in. F11 — [311.21]
- ☐ Min headroom 6 ft. 8 in. EXC F11 — [311.72]
- • Floor openings above stair OK to project 4¾ in. into req'd headroom at the side of a flight of stairs — [311.72X][20]
- ☐ Riser height max 7¾ in., tread depth min 10 in. EXC F12 — [311.74.1&2]
- • Tread depth min 11 in. if no nosing projection on treads F12 — [311.74.3X]
- ☐ Measure rise & run exclusive of carpets, rugs, or runners — [311.74.2][21]
- ☐ Deepest tread max > 3/8 in. deeper than shortest tread F12 — [311.74.2]
- ☐ Tallest riser max 3/8 in. more than shortest riser F12 — [311.74.1]
- ☐ Max 2% slope on treads & landings — [311.76]
- ☐ Enclosed accessible space below stairs req's min ½ in. GB — [302.7]

Nosings & Risers:
IRC 09
- ☐ Nosing req'd for solid risers with treads < 11 in. deep F12 — [311.74.3]
- ☐ Nosing projection min ¾ in., max 1¼ in. F12 — [311.74.3]
- ☐ Deepest nosing projection max 3/8 in. more than shortest F12 — [311.74.3]
- ☐ Beveling of nosing max ½ in., max nosing radius 9/16 in. — [311.74.3]
- ☐ Risers vertical or sloped from tread above max 30° from vertical — [311.74.3]
- ☐ Open riser treads must prevent passage of 4 in. sphere EXC — [311.74.3]
- • Openings between adjacent treads in stairs with rise ≤ 30 in. — [311.74.3X]

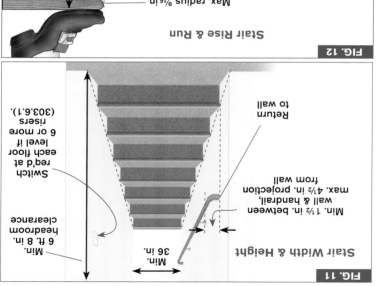

FIG. 12

Stair Rise & Run

Max. radius 9/16 in.

Max. 7¾ in.

Tread min. 10 in.

Nosing projection: min ¾ in., max 1¼ in. Largest projection max. 3/8 in. more than shortest projection

If no nosing & solid risers, tread min. 11 in.

FIG. 11

Stair Width & Height

Min. 6 ft. 8 in. headroom clearance

Min. 36 in.

Min. 1½ in. between wall & handrail, max. 4½ in. projection from wall

Return to wall

Switch req'd at each floor level if 6 or more risers (303.6.1).

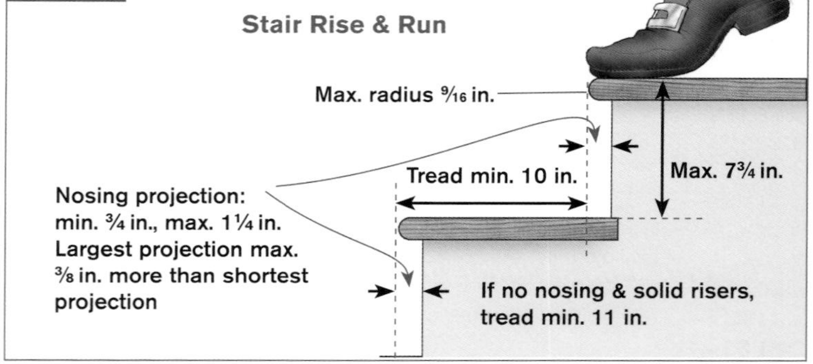

Winding Stairs F13 — 09 IRC

- ☐ Walkline concentric to curvature of stair & measured 12 in. from first clear area on narrow side of winder walking surface _____ [311.7.3][22]
- ☐ Min tread depth 10 in. at walkline_____ [311.7.4.2]
- ☐ Deepest tread max $3/8$ in. more than shortest measured at walkline [311.7.4.2]
- ☐ OK for winder treads to not be within $3/8$ in. of depth of rectangular treads in same flight of stairs _____ [311.7.4.2][23]

FIG. 13

Winding Stairs

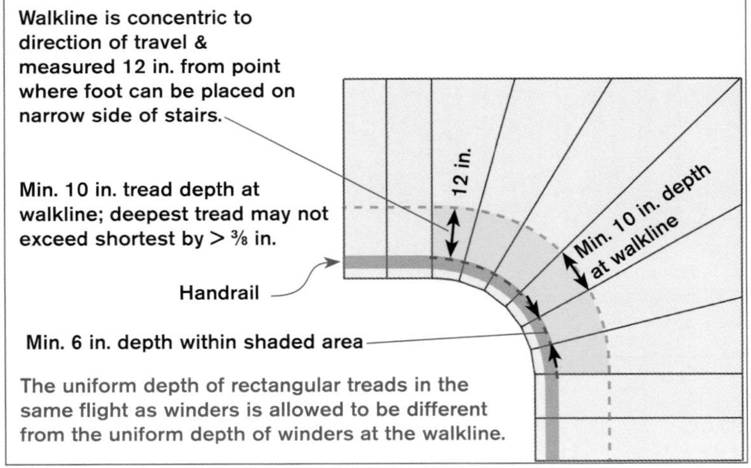

Walkline is concentric to direction of travel & measured 12 in. from point where foot can be placed on narrow side of stairs.

Min. 10 in. tread depth at walkline; deepest tread may not exceed shortest by > $3/8$ in.

Handrail

Min. 6 in. depth within shaded area

The uniform depth of rectangular treads in the same flight as winders is allowed to be different from the uniform depth of winders at the walkline.

12 in.

Min. 10 in. depth at walkline

Handrails — 09 IRC

- ☐ Req'd on at least one side of flights of stairs with ≥ 4 risers **F11,15** _[311.7.7]
- ☐ Top 34–38 in. above line connecting nosings **F15** EXC _____ [311.7.7.1]
  - Volute, turnout, or starting easing OK over lowest tread _____[311.7.7.1X1]
  - Fitting or bending OK to exceed max height at continuous transition between flights, start of flight, or from handrail to guard **F15** _[311.7.7.1X2][24]
- ☐ Ends must return to wall or post or safety terminal **F11,15** _____ [311.7.7.2]
- ☐ Min 1 ½ in. space between wall & handrail **F11** _____ [311.7.7.2]
- ☐ Handrail continuous from line above top & bottom nosings EXC __ [311.7.7.2]
  - May be interrupted by post at landing_____[311.7.7.2X1]
  - Volute, turnout, or starting easing OK over lowest tread **F15** __[311.7.7.2X2]
- ☐ Round handrails min 1 ¼ in.–max 2 in. diameter **F14** _____ [311.7.7.3]
- ☐ Non-round Type I handrails perimeter 4–6 ¼ in. **F14** _____ [311.7.7.3]
- ☐ If perimeter > 6 ¼ in., finger recess req'd both sides **F14** _____ [311.7.7.3]

FIG. 14

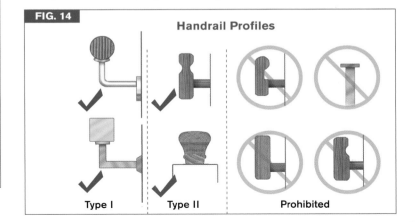

Handrail Profiles

Type I   Type II   Prohibited

Stairway Illumination 09 IRC

☐ Illumination req'd for stairs & landings _____ [303.6&311.7.8]
☐ Interior stair req's switch at each floor level if ≥ 6 risers **F11**_____ [303.6.1]
☐ Exterior stair light at top of landing, control inside dwelling EXC _____ [303.6]
• Lights with automatic controls _____ [303.6.1X]

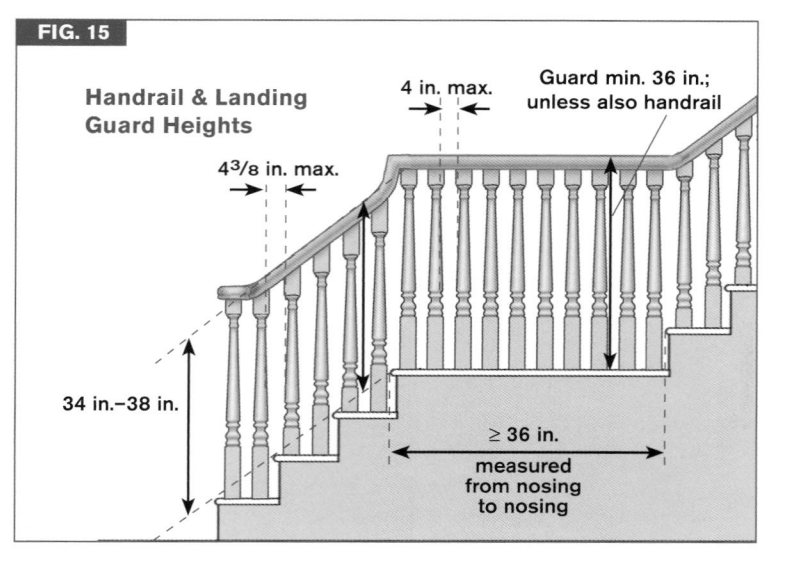

FIG. 15

Handrail & Landing
Guard Heights

4 in. max.

Guard min. 36 in.;
unless also handrail

4³/8 in. max.

34 in.–38 in.

≥ 36 in.
measured
from nosing
to nosing

SAFETY GLASS

Safety glass can be laminated or fully tempered. Proper use of safety glass is critical in areas that are subject to human impact. Typical causes of accidents are failure to see the glass, intentional breakage, or slips and falls.

Identification 09 IRC

☐ Safety glass req's permanent etched label EXC_____ [308.1]
• Non-tempered safety glazing OK to provide certificate to BO ___ [308.1X1]
• Tempered spandrel glass may have removable paper label_____ [308.1X2]
☐ Glass panes ≤ 1 sq. ft. in multipane assemblies OK for
all but 1 to have label that states only "CPSC 16 CFR 1201"
or "ANSI Z97.1 _____ [308.1.1][25]
☐ Glazing not in doors or enclosures of tubs or showers OK to
have ANSI Z97.1 Category A without a CPSC label _____[308.3.1X][26]

Human Impact Loads & Hazardous Locations 09 IRC

☐ Safety glazing req'd in locations per **T6, F16** EXC_____ [308.3&4]
• Louvered windows & jalousies _____ [308.3X1]
• Mirrors mounted on continuous backing support (typically vinyl) exempt
from requirement to meet impact test standards _____ [308.3X2]
• Glass unit masonry (glass block) _____ [308.3X3]

Impact Test Categories 09 IRC

☐ CPSC Category II req'd for doors or enclosures of wet areas (tubs,
showers, spas, hot tubs, saunas, steam rooms, or pools) _____ [308.3.1]
☐ CPSC Category II or ANSI Z97.1 Category A req'd in
non-wet areas EXC _____ [308.3.1]
• CPSC Category I or ANSI Category B also allowed for lites
≤ 9 sq. ft. in doors or sidelites other than patio doors _____ [308.3.1]

FIG. 16

Safety Glass

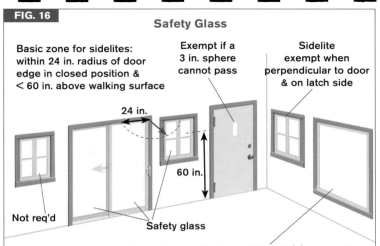

Basic zone for sidelites: within 24 in. radius of door edge in closed position & < 60 in. above walking surface

Exempt if a 3 in. sphere cannot pass

Sidelite exempt when perpendicular to door & on latch side

24 in.

60 in.

Not req'd

Safety glass

Safety glass when walk-through hazard exists: all four of (1) > 9 sq. ft., (2) lower edge < 18 in. above walking surface, (3) upper edge > 36 in. above walking surface & (4) within 36 in. horizontal of walking surface

TABLE 6	REQUIRED SAFETY GLAZING LOCATIONS [308.4]	
	Location	Exceptions
1	Glazing in doors **F16**	1. Openings that prevent passage of a 3 in. sphere 2. Decorative glass
2	Sidelites where any part of glass within 24 in. arc of the door in closed position & < 60 in. above floor or standing surface **F16**	1. Decorative glass 2. When protected by intervening barrier 3. Glass perpendicular to door on latch side 4. When door serves only closet ≤ 3 ft. deep 5. Glass adjacent to fixed side of patio doors

TABLE 6 (cont.)	REQUIRED SAFETY GLAZING LOCATIONS [308.4]	
	Location	Exceptions
3	Walk-through hazard: > 9 sq. ft. & lowest edge < 18 in. from walking surface & upper edge > 36 in. above walking surface & ≤ 3 ft. horizontal from walking surface **F16**	1. Decorative glass 2. When protected by min 1½ in. high horizontal rail, 34–38 in. above walking surface with rail able to resist 50 lb. force without contacting glass 3. Outboard panes ≥ 25 ft. above grade, roof, or other surface below
4	Railings, including infill	None
5	Enclosures or walls facing tubs, showers, hot tubs, whirlpools, saunas & steam rooms where glass < 60 in. above standing or walking surface	1. Walls > 60 in. away from water's edge & facing unenclosed tubs, whirlpools, or hot tubs[27]
6	Glass < 60 in. above walking surface & < 60 in. horizontally from edge of pools, hot tubs & spas	None
7	Glass adjacent to stairways, landings & ramps within 36 in. horizontally of walking surface & < 60 in. above walking surface	1. When protected by min 1½ in. high horizontal rail 34–38 in. above walking surface with rail able to resist 50 lb. force without contacting glass 2. When > 18 in. horizontally from a railing meeting req's of an open-stair guard 3. When solid vertical surfaces installed 34–36 in. above walking surface & top of solid surface resists same loads as a guard
8	Glass within 60 in. horizontally of bottom tread of stair in any direction when glass < 60 in. above nose of tread	Same as exceptions 2 & 3 to item 7

SAFETY GLASS

SKYLIGHTS

09 IRC Identification & Materials
☐ Skylight = glazing at ≥ 15° from vertical [308.6.1]
☐ May be laminated, fully tempered, heat-strengthened or wired glass or approved plastics [308.6.2]

09 IRC Installation
☐ Unit skylights req'd labeling from approved independent laboratory[308.6.9]
☐ Screens req'd under tempered or heat-strengthened glass EXC [308.6.3]
• Glass ≤ 16 sq. ft. & highest point ≤ 12 ft. above walking surface [308.6.5]
• Glass > 16 sq. ft. & max 30° from vertical & highest point 10 ft. [308.6.5]
• Sloped areas ≤ 20 ft. above grade in greenhouses. [308.6.6]
☐ Screen capable of supporting 2x glass weight & max 1 in. mesh. [308.6.7]
☐ Unit skylights min 4 in. curb in roof with < 3:12 slope or AMI [308.6.8]

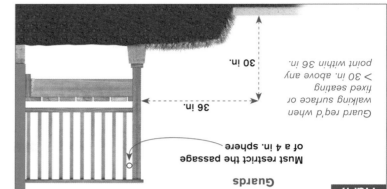

FIG. 17

Guards

☐ Must restrict the passage of a 4 in. sphere

36 in.

30 in.

Guard req'd when walking surface or fixed seating > 30 in. above any point within 36 in.

GUARDS

09 IRC Location & Height
☐ Open-sided walking surfaces including stairs & landings > 30 in. above lower floor or grade within 36 in. horizontally F17 [312.1][28]
☐ Min 36 in. above walking surface or adjacent fixed seating EXC [312.2][29]
• Guard on open side of stair min 34 in. high [312.2X1]
• Handrail as guard 34-38 in. above line connecting nosings F15 [312.2X2]

09 IRC Openings
☐ Openings must prevent passage of 4 in. sphere F15,17 EXC [312.3]
• 6 in. at triangular opening of riser, tread & bottom rail F15 [312.3X1]
• 4-3/8 in. at open sides of stairs F15 [312.3X2]
☐ Open risers in stairs must prevent passage of 4 in. sphere [311.7.4.3]

HILLSIDE CONSTRUCTION

09 IRC Construction Adjacent to Slopes
☐ Setback & clearance to slopes > 1:3 (vertical to horizontal) F18 [403.1.7]
☐ Setback & clearance to slopes > 1:1 (vertical to horizontal) F19 [403.1.71]
☐ Measure height from top of retaining walls at toe of slope F18,19 [403.1.71]
☐ BO may approve alternate setbacks per engineering investigation [403.1.74]

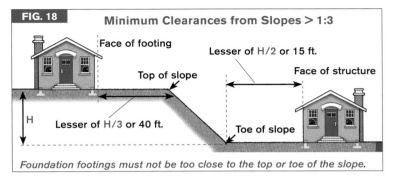

FIG. 18 — **Minimum Clearances from Slopes > 1:3**

Face of footing

Lesser of H/2 or 15 ft.

Top of slope

Face of structure

H

Lesser of H/3 or 40 ft.

Toe of slope

Foundation footings must not be too close to the top or toe of the slope.

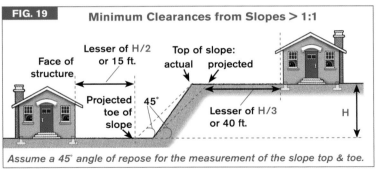

FIG. 19 — **Minimum Clearances from Slopes > 1:1**

Lesser of H/2 or 15 ft.

Top of slope: actual · projected

Face of structure

Projected toe of slope

45°

Lesser of H/3 or 40 ft.

H

Assume a 45° angle of repose for the measurement of the slope top & toe.

SOILS

Soils   09 IRC

☐ BO may req soil tests if expansive, compressible, or questionable ___ [401.4]
☐ BO may allow **T7** in lieu of complete geotechnical evaluation_____ [401.4.1]
☐ Expansive soils per IBC_____ [403.1.8]
☐ Compressible or shifting soils removed to stable level_____ [401.4.2]
☐ Filled soils layered & compacted per accepted engineering practice__ [401.2]

Retaining Walls   09 IRC

☐ Design req'd for retaining walls without lateral support & retaining
    > 24 in. unbalanced backfill_____ [404.4]
☐ Design for safety factor of 1.5 against lateral sliding & overturning ___ [404.4]

TABLE 7	PRESUMPTIVE LOAD-BEARING VALUES OF SOILS
Class of Material	**Load-Bearing Pressure (psf)**
Crystalline bedrock	12,000
Sedimentary & foliated rock	4,000
Sandy gravel &/or gravel	3,000
Sand, silty sand, clayey sand, silty gravel & clayey gravel	2,000
Clay, sandy clay, silty clay, clayey silt, silt & sandy silt	1,500

GRADING & DRAINAGE

Grading F20

IRC 09

- [ ] Grade surface to storm drain or other approved collection point [401.3]
- [ ] Grade away from foundation min 6 in. fall within 1st 10 ft. EXC [401.3]
  - • Use swale if physical barrier or lot line prohibits 6 in. fall in 10 ft. [401.3X]
- [ ] Hardscape within 10 ft. min 2% slope from building [401.3X]

Drainage

IRC 09

- [ ] Top of foundation min elevation above drainage inlet or street gutter 12 in. + 2% slope [403.1.7.3]
- [ ] If water does not readily drain from site, crawl space on same level as outside grade or install approved drainage system [408.6]
- [ ] Roof drain must discharge min 5 ft. from footing or to approved drain system if soils expansive or collapsible [801.3]

TABLE 8	MIN. WIDTH (IN.) OF CONCRETE OR MASONRY FOOTINGS [403.1]			
Construction Type	No. of Stories	Load Bearing Value of Soil^A (psf)			
		1,500	2,000	3,000	≥ 4,000
Conventional light-frame construction	1	12	12	12	12
	2	15	12	12	12
	3	23	17	12	12
4 in. brick veneer over frame or 8 in. hollow-concrete masonry	1	12	12	12	12
	2	21	16	12	12
	3	32	24	16	12
8 in. solid or fully grouted masonry	1	16	12	12	12
	2	29	21	14	12
	3	42	32	21	16

A. See T7 for vertical load-bearing values of different soil types.

FOOTINGS

General

IRC 09

- [ ] Footings supported on undisturbed soil or engineered fill [403.1]
- [ ] Placement in soils min 12 in. below undisturbed ground surface [403.1.4]
- [ ] Extend below frost line or be frost protected [403.1.4.1]
- [ ] Min width for concrete or masonry footings per T8 [403.1.4.1]
- [ ] Min thickness 6 in. F21 [403.1.1]
- [ ] Projection past foundation min 2 in, max = footing thickness F21 [403.1.1]
- [ ] Top surface of all footings level F33 [403.1.5]
- [ ] Bottom surface of footings max 10% slope (step when > 10%) [403.1.5]

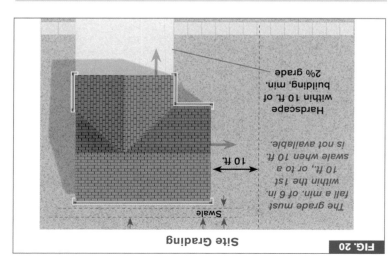

FIG. 20   Site Grading

2% grade

Hardscape within 10 ft. of building, min.

10 ft

Swale

The grade must fall a min. of 6 in. within the 1st 10 ft, or to a swale when 10 ft is not available.

Frost-Protected Shallow Foundations   09 IRC

- ☐ If monthly mean building temp maintained at 64°F, footing not req'd to extend below frost line if protected by insulation _____ [403.3]
- ☐ If unheated slab abuts frost-protected foundation, provide insulation under slab & between slab & protected foundation _____ [403.3.1.1]

SDC D Footings   09 IRC

- ☐ Continuous footings req'd at braced wall lines _____ [403.1.2]
- ☐ Bottom reinforcement min 3 in. clear from bottom of footing EXC __ [403.1.3]
  - Plain concrete footings allowed 1- & 2-family dwellings with studs [403.1.3X]
- ☐ Joint between stem wall & footing min #4 vertical bar at 4 ft. o.c. __ [403.1.3]
- ☐ Footing min 12 in. below top of slab at interior bearing walls _____ [403.1.4.2]

SDC D Footing Reinforcement   09 IRC

- ☐ Foundations with stem walls req #4 bar top & bottom **F21** _____ [403.1.3.1]
- ☐ Slab with footings min #4 bar top & bottom EXC **F25** _____ [403.1.3.2]
  - Monolithic slab 1 #5 bar OK in middle 1/3 of footing depth **F25** [403.1.3.2X]

FIG. 21

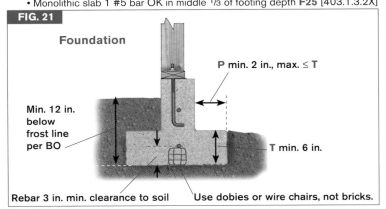

Foundation

P min. 2 in., max. ≤ T

Min. 12 in. below frost line per BO

T min. 6 in.

Rebar 3 in. min. clearance to soil   Use dobies or wire chairs, not bricks.

CONCRETE

Designs in accordance with standards from the American Concrete Institute (ACI 318 or 332) or the Portland Cement Association (PCA 100) are acceptable as alternatives to the methods prescribed by the IRC.

Mixing & Strength   09 IRC

- ☐ Min 2,500 psi in SDC A, B, or C _____ [404.1.2.3.1]
- ☐ Min 3,000 psi in SDC D _____ [404.1.2.3.1]
- ☐ Min compressive strength also to comply with **T9** _____ [402.2]
- ☐ Air-entrained concrete req'd if moderate or severe weathering **T9** ____ [402.2]
- ☐ Max slump 6 in. for concrete in removable forms _____ [404.1.2.3.4]
- ☐ Thoroughly work concrete around rebar & into corners _____ [404.1.2.3.5]
- ☐ Slump of concrete in stay-in-place forms (ICF) > 6 in. _____ [404.1.2.3.4]
- ☐ Vibrate concrete in stay-in-place forms (ICF) _____ [404.1.2.3.5]

TABLE 9	MIN. COMPRESSIVE STRENGTH OF CONCRETE AT 28 DAYS (PSI) [T402.2]		
Type or Location of Concrete Construction	**Weathering Potential**		
	Negligible	Moderate	Severe
Basement walls, foundations & other concrete not exposed to weather	2,500	2,500	2,500[A]
Basement slabs & interior slabs on grade, except garage floor slabs	2,500	2,500	2,500[A]
Basement & foundation walls, exterior walls & other vertical concrete exposed to weather	2,500	3,000[B]	3,000[B]
Porches, carport slabs & steps exposed to weather & garage floor slabs	2,500	3,000[B,C]	3,500[B,C]

A. Must be air-entrained if exposed to freeze−thaw during construction.
B. Air-entrainment req'd. Air content between 5% & 7% by volume of concrete.
C. Garage floor slab air-entrainment may be reduced to 3% if strength increased to 4,000 psi.

Forms   09 IRC

☐ Size per approved plans & tables_____[404.1.1&2]
☐ Pipe penetrations sleeved_____[2603.5]
☐ Excavation free of debris & roots _____[408.5]
☐ Wood beam connections min ½ in. air space 3 sides _____[317.1#4]
☐ Forms to resist deformation during concrete placement _____ [404.1.2.3.6]
☐ Remove all wood forms used for placing concrete_____[408.5]
☐ Foundation wall min 6 in. above finished grade EXC _____ [404.1.6]
   • 4 in. OK if masonry veneer is used _____ [404.1.6]
☐ Concrete wall cold joints req reinforcement at max 24 in. o.c. _[404.1.2.3.7.8]

Reinforcement Methods   09 IRC

☐ Secure with tie wire, dobies, etc., to prevent displacement____[404.1.2.3.7.4]
☐ Min reinforcement cover per **T10** _____[404.1.2.3.7.4]
☐ Splices lapped min 20 in. #4 bar, 25 in. #5 bar, 30 in. #6 bar_[404.1.2.3.7.5]

TABLE 10	REINFORCING STEEL COVER [404.1.2.3.7.4]	
Foundation Surface		**Min. Cover**
Concrete cast against & permanently exposed to earth		3 in.
Concrete exposed to earth or weather after forms removed		1½ in.**A**
Not exposed to weather (e.g., top of slab)		¾ in.
Concrete in stay-in-place forms (ICF)		¾ in.

A. 2 in. cover req'd for #6 or larger bars

ANCHORING TO FOUNDATION

General   09 IRC

☐ All wood sole plates at monolithic slab exterior walls & braced walls,
   & all wood sill plates req bolts to foundation_____ [403.1.6]
☐ Bolts min 7 in. into concrete or grouted CMU cell **F21,25** _____ [403.1.6]
☐ Bolts min ½ in. diameter, nut & washer on each bolt **F21,25** _____ [403.1.6]
☐ Bolt distance from end of plate min 7 diameters, max 12 in. **F33**___ [403.1.6]
☐ Max spacing 6 ft. o.c. & min 2 bolts per plate EXC _____ [403.1.6]
   • Walls ≤ 24 in. connecting offset braced walls1 bolt in center ⅓ [403.1.6X2]
   • Wall section ≤ 12 in. connecting offset braced walls no bolt OK [403.1.6X2]
   (WSP sheathing must be continuous through the offset to allow the above exceptions)
☐ Interior bearing wall sole plates on slab that are not part of BWP
   OK to use other types of approved fasteners _____ [403.1.6]

SDC C Townhouses & SDC C & D 1- & 2-Family   09 IRC

☐ Plate washers or anchor straps needed for full length of BWLs ___[403.1.6.1]
☐ Slotted plate washers permitted if standard washer also used___[602.11.1][30]
☐ Interior braced wall plates on continuous foundation req bolts ____[403.1.6.1]
☐ Interior bearing wall sole plates req anchor bolts _____[403.1.6.1]
☐ Max anchor bolt spacing 4 ft. o.c. if > 2 stories in height_____[403.1.6.1]

BASEMENTS

The IRC contains extensive prescriptive requirements for ICF foundations and walls, and these are not included in this book. When using such a system, we advise consulting with the form manufacturer. The IRC provides design tables for masonry foundation walls based on soils type, height of unbalanced backfill, thickness of CMUs, and amount of reinforcement. A combined version of those tables is available at www.codecheck.com/ccb3/tables.

Foundation Walls   09 IRC

☐ Design with accepted engineering practice if hydrostatic pressure exists or if supporting > 48 in. unbalanced backfill without lateral support at top_____ [404.1.3]
☐ Min thickness ≥ walls supported above EXC _____ [404.1.5]
  • 8 in. masonry walls OK to support veneered walls ≤ 20 ft. _____[404.1.5.1]
  • Concrete foundation with shelf for veneer based on reduced thickness portion of wall_____[404.1.5.2]

SDC C & D   09 IRC

☐ Plain concrete or masonry OK if complying with following: ___ [404.1.4.1&2]
  • Min wall thickness for plain concrete 7½ in., plain masonry 8 in.
  • Max height 8 ft., max unbalanced backfill 4 ft.
  • Min vertical reinforcement 4 ft. o.c. #4 bars concrete, #3 bars masonry
☐ Reinforced masonry per tables _____[404.1.4.1]
☐ Min 2 #4 bars in upper 12 in. of reinforced masonry walls _____[404.1.4.1]
☐ Reinforced concrete per tables _____ [404.1.4.2][31]

Waterproofing & Dampproofing   09 IRC

☐ If high water table exists, waterproofing req'd to finished grade _____[406.2]
☐ Waterproofing membranes lapped & sealed _____[406.2]
☐ Dampproofing req'd for foundations retaining earth & enclosing interior spaces & floors below grade_____[406.1]
☐ Parge CMUs prior to dampproofing EXC _____[406.1]
  • When dampproofing material approved for direct application _____[406.1X]
☐ Drainage systems req'd **F22**_____[405.1]
☐ Drains to extend 1 ft. past footing, 6 in. above & req filter fabric **F22**__ [405.1]

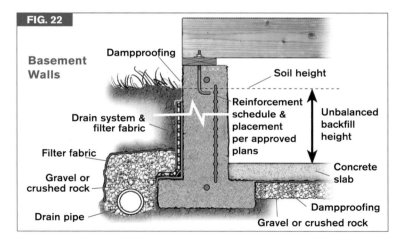

FIG. 22

Basement Walls

Dampproofing

Soil height

Drain system & filter fabric

Reinforcement schedule & placement per approved plans

Unbalanced backfill height

Filter fabric

Concrete slab

Gravel or crushed rock

Dampproofing

Drain pipe

Gravel or crushed rock

Reinforcement

09 IRC

☐ All reinforcing bars fully embedded in mortar or grout [606.13]

☐ ¾ in. min reinforcement cover; 2 in. if exposed to weather or soil [606.13]

☐ If construction stopped for 1 hr. in grouted masonry walls, all tiers set at same elevation & grout within ½ in. of top [609.1.2]

☐ Reinforcement held in position top & bottom & at 200 bar diameters [609.4.1]

☐ Grout poured in lifts max 8 ft. high & if total > 8 ft, special inspections req'd & grout placed in lifts ≤ 5 ft. [609.4.1]

SDC C & D

09 IRC

☐ Min vertical reinforcement #4 bars at 48 in. o.c. [606.12.2.3]

☐ Vertical reinforcement within 16 in. of end of walls [606.12.2.3]

☐ SDC D max spacing vertical & horizontal reinforcement = smaller of ⅓ length, ⅓ height, or 48 in. [606.12.3.2.1]

TABLE 11   MASONRY WALL LATERAL SUPPORT SPACING [606.9]

Construction	Max. Length to Thickness or Height to Thickness Ratio
Bearing walls – solid or solid grouted	20
All other bearing walls	18
Nonbearing exterior walls	18
Nonbearing interior walls	36

CONCRETE MASONRY UNITS (CMUs)

General

09 IRC

☐ 6 in. block OK only for 1 story to 9 ft. + 6 ft. of gable [606.2.1]

☐ Min 8 in. block if > 1 story or > 9 ft. [606.2.1]

☐ Lateral support req'd vertically &/or horizontally per **T11** [606.9]

☐ Horizontal lateral support by cross walls, pilasters, buttresses, or structural frame; vertical lateral support by floors & roofs [606.9]

☐ Anchor walls to floor & roof systems **F23** [606.11]

☐ Beam supports min bearing 3 in. [606.14]

☐ Joist support min bearing 1½ in. on 3 in. nominal ledger with bolt min 4 in. embedment [606.14.1]

FIG. 23

Anchoring to Masonry Walls

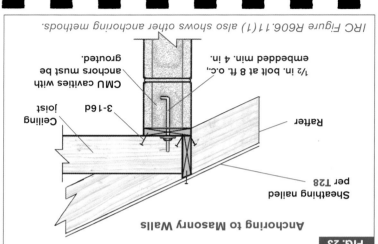

IRC Figure R606.11(1) also shows other anchoring methods.

½ in. bolt at 8 ft. o.c., embedded min. 4 in.

CMU cavities with anchors must be grouted.

3-16d

Ceiling joist

Rafter

Sheathing nailed per T28

CONCRETE SLABS

Concrete Slabs on Ground   09 IRC

- ☐ Min 3½ in. thick **F24** _____ [506.1]
- ☐ Excavation free of debris & roots _____ [506.2]
- ☐ Max fill 24 in. clean sand/gravel or 8 in. earth _____ [506.2.1]
- ☐ Below grade slabs req min 4 in. base course **F24** _____ [506.2.2]
- ☐ Min 6 mil poly or approved vapor retarder req'd EXC **F22,24** _____ [506.2.3]
  - • Detached garages & unheated accessory structures _____ [506.2.3X1][32]
  - • Unheated storage rooms ≤ 70 sq. ft. & carports _____ [506.2.3X2]
  - • Driveways & other unenclosed flatwork _____ [506.2.3X3]
  - • Where approved by BO based on local site conditions _____ [506.2.3X4]
- ☐ Dobies or other support req'd to hold reinforcement in place between center & upper third of slab during concrete placement **F24** _____ [506.2.4]

UNDERFLOOR AREA (CRAWL SPACES)

General   09 IRC

- ☐ Remove all vegetation & organic material _____ [408.5]
- ☐ Wood forms must be completely stripped off foundation _____ [408.5]

Access Openings   09 IRC

- ☐ Through-floor openings min 18 in. × 24 in. _____ [408.4]
- ☐ Perimeter wall openings min 16 in. high × 24 in. wide **F25** _____ [408.4]
- ☐ If access below grade, provide full-depth access well with min 16 in. × 24 in. footprint **F25** _____ [408.4]
- ☐ Through-wall openings not OK under a door _____ [408.4]
- ☐ Opening large enough to remove underfloor mechanical equipment _____ [408.4]

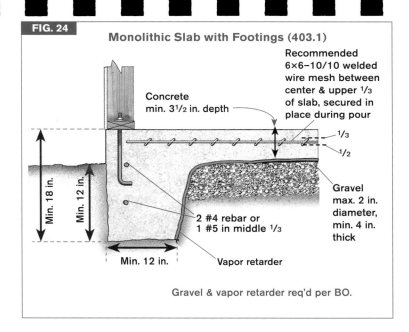

FIG. 24

Monolithic Slab with Footings (403.1)

Recommended 6×6–10/10 welded wire mesh between center & upper ⅓ of slab, secured in place during pour

Concrete min. 3½ in. depth

⅓

½

Min. 18 in.

Min. 12 in.

Gravel max. 2 in. diameter, min. 4 in. thick

2 #4 rebar or 1 #5 in middle ⅓

Vapor retarder

Min. 12 in.

Gravel & vapor retarder req'd per BO.

UNDERFLOOR FRAMING

General   09 IRC

☐ Wood naturally durable against decay = heartwood of redwood, cedar, black locust & black walnut   [202]

☐ Wood in ground contact or embedded in concrete req's specific rating of PT for ground contact   [317.1.2]

☐ Basement furring strips PT, naturally durable, or on vapor retarder   [317.1#7]

☐ Fasteners for preservative-treated wood hot-dipped zinc-coated, stainless steel, silicon bronze, or copper   [317.3.1]

☐ Connector coatings for preservative-treated wood AMI   [317.3.1]

Sills   09 IRC

☐ Sills < 8 in. from exposed ground PT or naturally durable wood   [317.1]

☐ Sills & sleepers on slabs PT or separated by moisture barrier   [317.1]

☐ Min 2x4 nominal size   [404.3]

Posts & Columns   09 IRC

☐ Crawl space & basement wood columns PT or naturally durable EXC   [317.1.4]

• Basement with pedestal ≥ 1 in. above concrete floor or 6 in. above earth that is covered with impervious moisture barrier   [317.1.4X1]

• Crawl space pier ≥ 8 in. above earth & vapor retarder F26   [317.1.4X2]

☐ Steel columns painted or treated to protect against corrosion   [407.2]

☐ Steel columns min 3 in. diameter, wood columns min 4x4 F26   [407.3]

☐ Bottom of columns req restraint to prevent lateral displacement EXC   [407.3]

• SDC A, B & C if on pier & in area enclosed by foundation   [407.3X]

☐ Height of masonry piers max 10x their least dimension   [606.6]

☐ Masonry piers req reinforcement EXC   [F606.11(2&3)]

• Unfilled piers limited to 4x least dimension allowed only in SDC A & B & top 4 in. req's fill   [606.6]

Ventilation   09 IRC

☐ Openings min 1 sq. ft. per 150 sq. ft. of underfloor area EXC   [408.1]

• Reduction to 1 sq. ft. per 1,500 sq. ft. OK with Class I vapor retarder   [408.1]

☐ One vent opening within 3 ft. each corner   [408.1]

☐ Openings may be perforated sheet metal, expanded plates, cast-iron grill, or corrosion-resistant wire with min 1/8 in. mesh   [408.2]

☐ OK to omit ventilation openings if Class I vapor retarder on soils with 6 in. overlaps & 6 in. lapped & sealed to stem wall & [408.3]

• Continuous exhaust 1 cfm/50 sq. ft. & perimeter insulated

• Conditioned air supply 1 cfm/50 sq. ft. & return to common area through duct or air-transfer opening & perimeter insulated

• When underfloor used as plenum in compliance with mechanical code

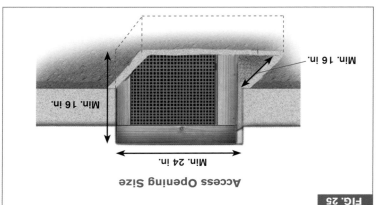

FIG. 25

Access Opening Size

Min. 24 in.   Min. 16 in.   Min. 16 in.

FIG. 26

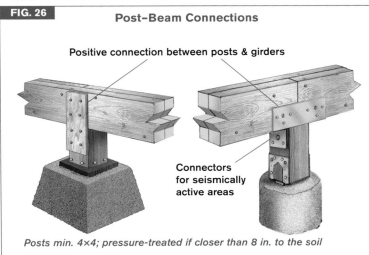

Post-Beam Connections

Positive connection between posts & girders

Connectors for seismically active areas

Posts min. 4×4; pressure-treated if closer than 8 in. to the soil

Girders
09 IRC

- ☐ Built-up girder spans per **T12, 13** _____ [502.5]
- ☐ Built-up girders req 10d nails 32 in. o.c. each layer at top & bottom & staggered & 2 nails at ends at each splice **F26** _____ [T602.3(1)]
- ☐ End bearing min 3 in. on concrete or masonry, 1½ in. on wood or metal _____ [502.6]
- ☐ Positive connection req'd between posts & girders **F26** _____ [502.9]
- ☐ Max offset one joist depth from bearing wall above **F29** _____ [502.4]
- ☐ Notching & boring per **F30, T14** _____ [502.8.1]

TABLE 12	ALLOWABLE GIRDER & HEADER SPANS IN INTERIOR BEARING WALLS [T502.5(2)]					
No. of Floors Supported	Min. Size	Building Width^A				
		20 ft.		28 ft.		36 ft.
		Span^B	NJ^C	Span^B	NJ^C	Span^B	NJ^C
1	2-2×4	3-1	1	2-8	1	2-5	1
	2-2×6	4-6	1	3-11	1	3-6	1
	2-2×8	5-9	1	5-0	2	4-5	2
	2-2×10	7-0	2	6-1	2	5-5	2
	2-2×12	8-1	2	7-0	2	6-3	2
	3-2×8	7-2	1	6-3	1	5-7	2
	3-2×10	8-9	1	7-7	2	6-9	2
	3-2×12	10-2	2	8-10	2	7-10	2
2	2-2×4	2-2	1	1-10	1	1-7	1
	2-2×6	3-2	2	2-9	2	2-5	2
	2-2×8	4-1	2	3-6	2	3-2	2
	2-2×10	4-11	2	4-3	2	3-10	3
	2-2×12	5-9	2	5-0	3	4-5	3
	3-2×8	5-1	2	4-5	2	3-11	2
	3-2×10	6-2	2	5-4	2	4-10	2
	3-2×12	7-2	2	6-3	2	5-7	3

A. Based on built-up #2 grade Douglas fir-larch, hem-fir, southern pine, and spruce-pine fir lumber. Building widths are measured perpendicular to the ridge.
B. Spans are given in feet & inches (ft.–in.).
C. NJ = number of jack studs under each end. If the number is 1, the header is permitted to be supported by framing anchors attached to full-length wall studs & the header.

TABLE 13 — ALLOWABLE GIRDER & HEADER SPANS IN EXTERIOR BEARING WALLS [T502.5(1)]

Support	Min. Size	Building Width[A]					
		20 ft.		28 ft.		36 ft.	
		Span[B]	NJ[c]	Span[B]	NJ[c]	Span[B]	NJ[c]
Roof & Ceiling	2-2×4	3-6	1	3-2	1	2-10	1
	2-2×6	5-5	1	4-8	1	4-2	1
	2-2×8	6-10	1	5-11	2	5-4	2
	2-2×10	8-5	2	7-3	2	6-6	2
	2-2×12	9-9	2	8-5	2	7-6	2
Roof, Ceiling & 1 Center-Bearing Floor	2-2×4	3-1	1	2-9	1	2-5	1
	2-2×6	4-6	1	4-0	1	3-7	2
	2-2×8	5-9	2	5-0	2	4-6	2
	2-2×10	7-0	2	6-2	2	5-6	2
	2-2×12	8-1	2	7-1	2	6-5	2
Roof, Ceiling & 1 Clear-Span Floor	2-2×4	2-8	1	2-4	1	2-1	1
	2-2×6	3-11	1	3-5	2	3-0	2
	2-2×8	5-0	2	4-4	2	3-10	2
	2-2×10	6-1	2	5-3	2	4-8	2
	2-2×12	7-1	3	6-1	3	5-5	3
Roof, Ceiling & 2 Center-Bearing Floors	2-2×4	2-7	1	2-3	1	2-0	1
	2-2×6	3-9	2	3-3	2	2-11	2
	2-2×8	4-9	2	4-2	2	3-9	2
	2-2×10	5-9	2	5-1	2	4-7	3
	2-2×12	6-8	2	5-10	3	5-3	3

A. Based on built-up #2 grade Douglas fir-larch, hem-fir, southern pine, and spruce-pine fir lumber & a 30 lb. ground snow load. Building widths are measured perpendicular to the ridge.

B. Spans are given in feet & inches (ft.-in.).

C. NJ = number of jack studs under each end. If the number is 1, the header is permitted to be supported by framing anchors attached to full-length wall studs & the header.

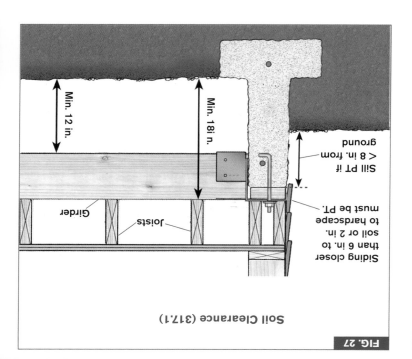

FIG. 27

Soil Clearance (317.1)

Min. 12 in.

Min. 18 in.

Girder

Joists

Siding closer than 6 in. to soil or 2 in. to hardscape must be P.T.

Sill PT if < 8 in. from ground

Joists

- [ ] Joists & subfloor min 18 in. above earth if not PT or naturally durable _ [317.1]
- [ ] Girders min 12 in. above earth if not PT or naturally durable_____ [317.1]
- [ ] Size & span for sleeping areas & attics with stairs per **T15**_____ [502.3.1]
- [ ] Size & span for all other areas per **T16** _____ [502.3.2]
- [ ] Cantilevers see IRC T502.3.3.(1) _____ [502.3.3]
- [ ] Double joists under parallel bearing walls **F31** _____ [502.4]
- [ ] Bearing min 3 in. on concrete or masonry,
  1½ in. on wood or metal EXC_____ [502.6]
  - On 1×4 ribbon strip & nailed to adjacent stud (balloon frame)_____ [502.6]
  - Into side of wood girder on hangers or 2×2 ledger _____ [502.6.2]
- [ ] Min lap across girder 3 in. & min 3-10d face nails **F28** _____ [502.6.1]
- [ ] Notching & boring per **F30, T14**_____ [502.8.1]

Joist Blocking & Bridging

- [ ] Joists blocked or attached to rim joists at all ends _____ [502.7]
- [ ] Blocking min 2× material & full depth of joist _____ [502.7]
- [ ] Blocking also req'd at intermediate supports in SDC D _____ [502.7X2]
- [ ] Joists > 2×12 req bridging at max 8 ft. intervals_____[502.7.1]

Framing at Openings

- [ ] Combustible framing min 2 in. from masonry chimneys _____[1003.18]
- [ ] Double headers & trimmers spanning > 4 ft._____ [502.10]
- [ ] Double trimmers for openings > 3 ft. from trimmer bearing points ___ [502.10]
- [ ] Header joists > 6 ft. must be hung with hardware _____ [502.10]
- [ ] Tail joists > 12 ft. hung with hardware or on 2×2 ledgers _____ [502.10]

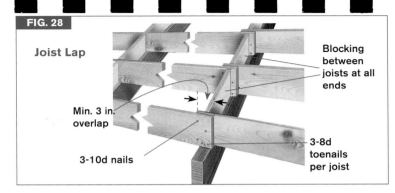

FIG. 28

Joist Lap

Blocking between joists at all ends

Min. 3 in. overlap

3-10d nails

3-8d toenails per joist

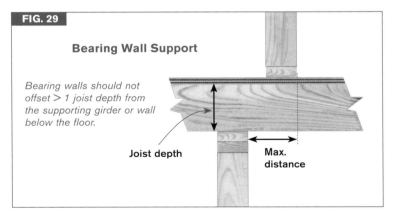

FIG. 29

Bearing Wall Support

Bearing walls should not offset > 1 joist depth from the supporting girder or wall below the floor.

Joist depth

Max. distance

UNDERFLOOR FRAMING

FIG. 31

Double Joists under Parallel Bearing Wall

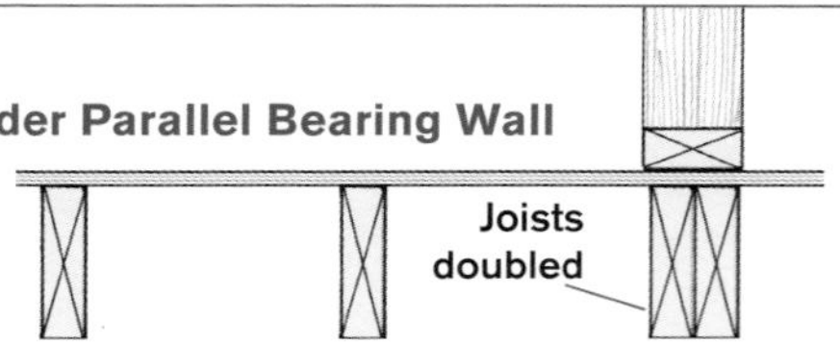

joists doubled

TABLE 16 — JOISTS SPANS FOR 40 LB. LIVE LOAD [T502.3.1(2)]

Size	Douglas Fir-Larch #2 Spacing o.c.			Southern Pine #2 Spacing o.c.		
	12	16	24	12	16	24
2×6	10-9	9-6	8-1	10-9	9-9	8-6
2×8	14-2	12-7	10-3	14-2	12-10	11-0
2×10	17-9	15-5	12-7	18-0	16-1	13-1
2×12	20-7	17-10	14-7	21-9	18-10	15-5

Measurements given in feet & inches (ft.-in.).
Dead load = 10 psf

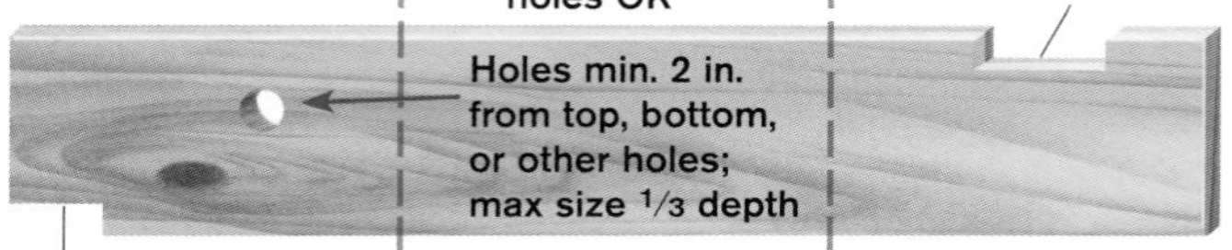

TABLE 15 — JOISTS SPANS FOR 30 LB. LIVE LOAD [T502.3.1(1)]

Size	Douglas Fir-larch #2 Spacing o.c.			Southern Pine #2 Spacing o.c.		
	12	16	24	12	16	24
2×6	11-10	10-9	9-1	11-10	10-9	9-4
2×8	15-7	14-1	11-6	15-7	14-2	12-4
2×10	19-10	17-2	14-1	19-10	18-0	14-8
2×12	23-0	19-11	16-3	24-2	21-1	17-2

Measurements given in feet & inches (ft.-in.).
Dead load = 10 psf

FIG. 30

Notching & Boring Joists & Girders

Notch depth at end max. $^{1}/_{4}$ of joist depth

No notching bottom of < 4x lumber, except at ends

Outer $^{1}/_{3}$

Holes min. 2 in. from top, bottom, or other holes; max size $^{1}/_{3}$ depth

No notching in middle $^{1}/_{3}$; holes OK

Notch depth max. $^{1}/_{3}$ joist depth, length max. $^{1}/_{3}$ joist depth

Notch depth max. $^{1}/_{6}$ joist depth

TABLE 14 — NOTCHING & BORING JOISTS [502.8.1]

Nominal[A] Dimension Joist or Girder	Max. Diameter Bored Hole	Max. Notch Length Outer $^{1}/_{3}$	Max. Notch Depth Outer $^{1}/_{3}$	Max. Notch Depth End Notch
6	1 $^{3}/_{4}$ in.	1 $^{3}/_{4}$ in.	$^{7}/_{8}$ in.	1 $^{3}/_{8}$ in.
8	2 $^{3}/_{8}$ in.	2 $^{3}/_{8}$ in.	1 $^{3}/_{16}$ in.	1 $^{7}/_{8}$ in.
10	3 $^{1}/_{16}$ in.	3 $^{1}/_{16}$ in.	1 $^{1}/_{2}$ in.	2 $^{3}/_{8}$ in.
12	3 $^{3}/_{4}$ in.	3 $^{3}/_{4}$ in.	1 $^{7}/_{8}$ in.	2 $^{7}/_{8}$ in.

A. Table numbers based on actual dimensions: typically 5½, 7¼, 9¼ & 11¼

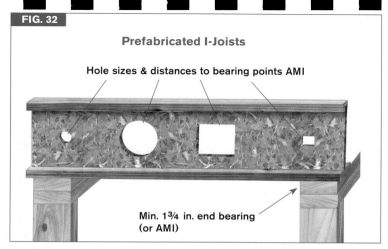

FIG. 32

Prefabricated I-Joists

Hole sizes & distances to bearing points AMI

Min. 1¾ in. end bearing
(or AMI)

Manufactured Lumber & Floor Trusses 09 IRC

☐ Cuts, notches & holes only where specified by manufacturer or
registered design professional **F32,34** _____ [502.8.2]
☐ Point loads & other installation details AMI **F34** _____ [502.7.1X]
☐ Blocking, bridging & other lateral support AMI _____ [502.7.1X]
☐ Truss drawings to include bracing requirements _____ [502.11.2]
☐ No truss alterations without approval of registered design
professional _____ [502.11.3]

CRIPPLE WALLS

Cripple Walls 09 IRC

☐ No smaller than size of studding above cripple wall _____ [602.9]
☐ If < 14 in. high, solid WSP sheathing or solid blocking req'd _____ [602.9]
☐ If > 4 ft. high, size as if additional story _____ [602.9]
☐ SDC A–D₁ bracing length 1.15× req'd length of wall above **T21** __ [602.10.9]
☐ Max spacing between BWPs 18 ft. _____ [602.10.9]
☐ SDC D if interior BWL not over continuous
foundation, parallel exterior BWL lengths must be
increased 1.5× req'd length of **T21,22** _____ [602.10.9.1]
☐ Can be redesignated as 1st story for bracing purposes _____ [602.10.9.2]

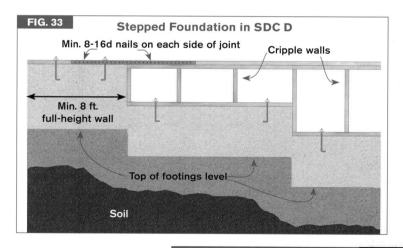

FIG. 33   Stepped Foundation in SDC D

Min. 8-16d nails on each side of joint

Cripple walls

Min. 8 ft.
full-height wall

Top of footings level

Soil

FIG. 34

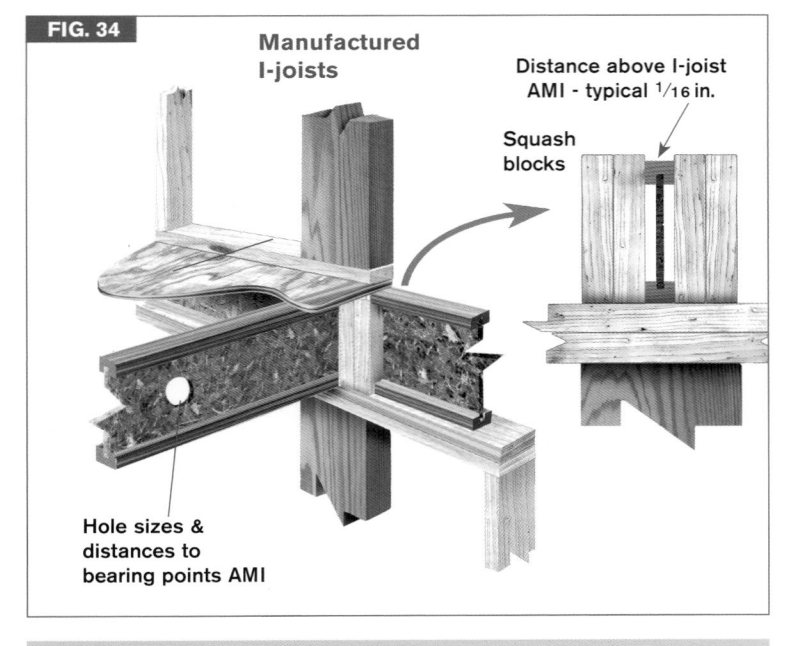

Manufactured I-joists

Distance above I-joist AMI - typical 1/16 in.

Squash blocks

Hole sizes & distances to bearing points AMI

FIG. 35 Wood Structural Panel Grade Mark

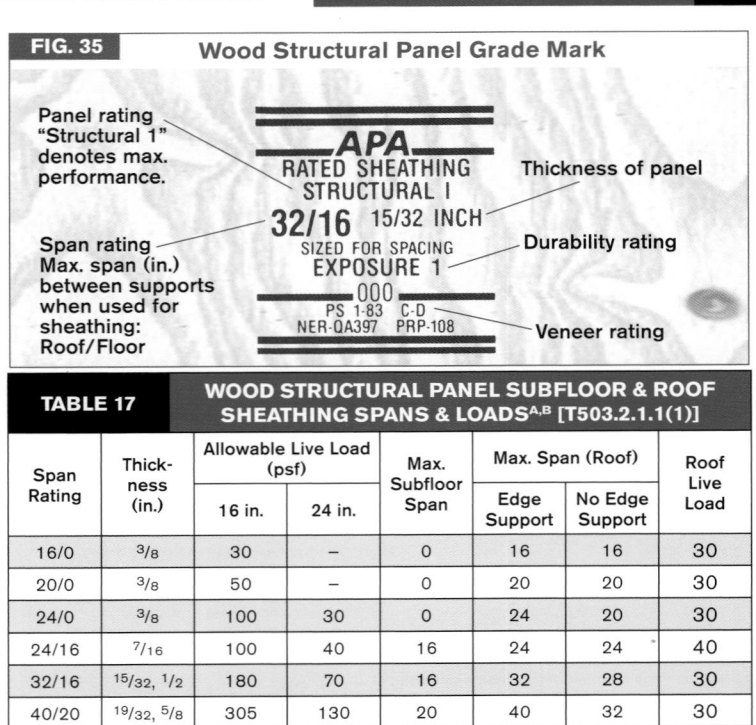

Panel rating "Structural 1" denotes max. performance.

Span rating Max. span (in.) between supports when used for sheathing: Roof/Floor

Thickness of panel

Durability rating

Veneer rating

APA
RATED SHEATHING
STRUCTURAL I
32/16   15/32 INCH
SIZED FOR SPACING
EXPOSURE 1
000
PS 1-83 C-D
NER-QA397 PRP-108

WOOD STRUCTURAL PANELS

WSP Sheathing   09 IRC

☐ WSP sheathing used for structural purposes req's grade stamp from approved agency **F35** _____ [503.2.1 & 803.2.1]
☐ Allowable spans & loads per **T17** _____ [503.2.2 & 803.2.2]
☐ Install in accordance with **T28** _____ [503.2.3 & 803.2.3]

TABLE 17	WOOD STRUCTURAL PANEL SUBFLOOR & ROOF SHEATHING SPANS & LOADS[A,B] [T503.2.1.1(1)]						
Span Rating	Thickness (in.)	Allowable Live Load (psf)		Max. Subfloor Span	Max. Span (Roof)		Roof Live Load
		16 in.	24 in.		Edge Support	No Edge Support	
16/0	3/8	30	–	0	16	16	30
20/0	3/8	50	–	0	20	20	30
24/0	3/8	100	30	0	24	20	30
24/16	7/16	100	40	16	24	24	40
32/16	15/32, 1/2	180	70	16	32	28	30
40/20	19/32, 5/8	305	130	20	40	32	30
48/24	23/32, 3/4	–	175	24	48	36	35
60/32	7/8	–	305	32	60	48	35

A. Based on 10 psf dead load; if more than 10 psf, then the live load should be reduced accordingly.
B. Panels continuous over min. 2 spans with strength axis perpendicular to supports.

WALL FRAMING

The building frame must be capable of supporting the dead loads—the weight of the building and its fixed equipment—and the live loads—the weight of its occupants and furnishings. In addition, the frame must be capable of transmitting lateral loads (from wind or earthquakes) through the vertical support elements to the foundation.

Stud Walls — 09 IRC

- ☐ Stud size & spacing per **T18** EXC _____ [602.3.1]
  - • 12 ft. + floor frame height OK per story if bracing increased ____ [301.3X1]
  - • Height of wall > 10 ft. OK within limits of T602.3.1 _____ [602.3.1X2]
- ☐ End-jointed lumber OK if identified by grade mark _____ [602.1.1]
- ☐ Studs req full bearing on plate at least equal to stud width _____ [602.3.4]
- ☐ Studs continuous from sole plate to top plate EXC _____ [602.3]
  - • Jack studs, trimmer studs & cripple studs _____ [602.3X]

Notching & Boring — 09 IRC

- ☐ Notch depth 25% max in bearing wall, 40% in nonbearing wall **F36** __ [602.6]
- ☐ Bored holes min 5/8 in. from face of stud _____ [602.6]
- ☐ Boring 40% max in bearing wall, 60% nonbearing EXC **F36** _____ [602.6]
  - • 2 successive doubled bearing studs 60% OK **F36** _____ [602.6]

Corners — 09 IRC

- ☐ 3 studs at corners **F37** EXC _____ [F602.3(2)]
  - • 2 studs OK with devices as backing to secure face materials __ [F602.3(2)]
- ☐ Lap plates at corners _____ [602.3.2]

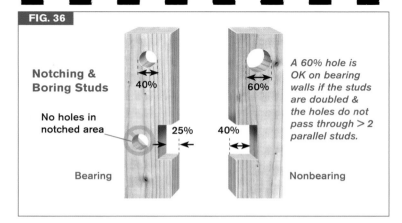

FIG. 36

Notching & Boring Studs

40% 60%

No holes in notched area

25% 40%

Bearing   Nonbearing

A 60% hole is OK on bearing walls if the studs are doubled & the holes do not pass through > 2 parallel studs.

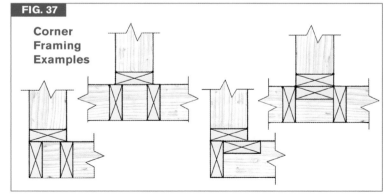

FIG. 37

Corner Framing Examples

FIG. 38

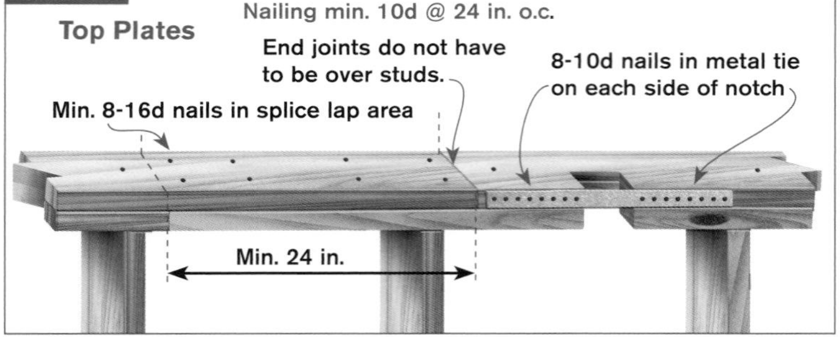

Top Plates

Nailing min. 10d @ 24 in. o.c.

End joints do not have to be over studs.

Min. 8-16d nails in splice lap area

8-10d nails in metal tie on each side of notch

Min. 24 in.

Top Plates & Headers   09 IRC

- ☐ Double top plates req'd EXC _____ [602.3.2]
  - Single plate OK with metal tie at joints & joists or rafters centered over studs with max. tolerance of 1 in. _____ [602.3.2X]
- ☐ Plates at least same width as studs _____ [602.3.2]
- ☐ End joints offset min 24 in., need not occur over studs **F38** _____ [602.3.2]
- ☐ Nailing 10d max 24 in. o.c. **F38, T28** _____ [602.3.2]
- ☐ Min. 8 16d nails in lapped area of end joints **F38, T28** _____ [602.3.2]
- ☐ Min 1½ in. strap 6 in. past cuts or holes > 50% of top plate width **F38** EXC _____ [602.6.1]
  - Not req'd if entire side of wall with notch/hole covered with WSP [602.6.1X]
- ☐ Strap secured with min 8-10d nails each side of notch/hole **F38** _ [602.6.1][33]
- ☐ Nonbearing walls do not req headers at openings _____ [602.7.2]
- ☐ Built-up headers at bearing walls see **T12,13** _____ [602.7]

TABLE 18	**STUD SIZE, HEIGHT, NOTCHING & BORING [602.3(5)]**		
Bearing Wall Studs ≤ 10 ft. High Between Supports Perpendicular to Wall[A]			
	2×4	**3×4**	**2×6**
Supporting roof + ceiling	24 in. o.c.	24 in. o.c.	24 in. o.c.
Supporting 1 floor + roof + ceiling	16 in. o.c.	24 in. o.c.	24 in. o.c.
Supporting 2 floors + roof + ceiling	n/a	16 in. o.c.	16 in. o.c.
Max. notch depth	$7/8$ in.	$7/8$ in.	$1 3/8$ in.
Max. diameter bored hole	$1 3/8$ in.	$1 3/8$ in.	$2 3/16$ in.
Max. diameter bored hole doubled studs	2 in.	2 in.	$3 1/4$ in.
Nonbearing Walls			

	2x3[B]	**2x4**	**3x4**	**2x6**
Max. laterally unsupported stud height	10 ft.	14 ft.	14 ft.	20 ft.
Max. notch depth	1 in.	$1 3/8$ in.	$1 3/8$ in.	$2 3/16$ in.
Max. diameter bored hole	$1 1/2$ in.	2 in.	2 in.	$3 1/4$ in.

A. See Table 602.3.1 for stud walls > 10 ft. in height.
B. 2×3s max. 16 in. o.c. & not in exterior walls, other sizes max. 24 in. o.c.

WALLS OTHER THAN WOOD FRAME

The IRC contains extensive prescriptive requirements and tables for cold-formed steel framing. These are based on *AISI S230, Standard for Cold-Formed Steel Framing–Prescriptive Method for One- and Two-Family Dwellings*, 2007 edition. In the latest edition and the IRC, the scope was expanded to include 3-story dwellings.

The IRC also has extensive sections on ICF (Insulating Concrete Form) foundation and wall structures. These methods should not be considered as an alternative to conventional foundations; they should be integrated into an overall "green" building design. Another new method–Structural Insulated Panel Wall Construction–was added to the 2009 IRC. Through the prescriptive inclusion of this method the project drawings are not required to bear the stamp of a design professional unless required by state law (as in California) or by the local jurisdiction. The U.S. Department of Housing & Urban Development has a free downloadable design guide for this method.

WALL BRACING

Wall bracing resists the forces imposed by winds and earthquakes. The type and amount of bracing must be adequate to resist whatever is the stronger of those two forces at the building site. The horizontal elements such as floors, ceilings, and roofs collect lateral forces and must be properly connected to the walls to transmit those forces to the braced elements. When the prescriptive limits are not adequate, designs per the IBC or the reference documents at the beginning of this book should be used.

Bracing: General   09 IRC

☐ Comply with prescriptive IRC bracing or IBC or documents
 referenced in 301 (listed in introduction of this book) _____ [602.10]
☐ Bracing length greater of req'd amount for seismic or wind EXC [602.10.1.2]
 • 1- & 2-family in SDC C req's bracing only for wind _____ [602.10X]³⁴

Braced Wall Lines (BWLs)   09 IRC

☐ BWL distance measured to perpendicular BWL or exterior walls
 or projection of same EXC _____ [602.10.1]
 • Angled corners with diagonal ≤ 8 ft. included in BWL per **F39** [602.10.1.3]³⁵

Braced Wall Panel (BWP) Locations   09 IRC

☐ BWPs may offset ≤ 4 ft. from braced wall line **F39** _____ [602.10.1.4]³⁶
☐ BWP max 25 ft. o.c., max 12.5 ft. total from ends of BWL **F40** [602.10.1.4]³⁷
☐ SDC D BWP must start at corners EXC_____ [602.10.1.4.1]³⁸
 • Start 8 ft. from corner OK if 2 ft. wide WSPs attached
 each side of corner **F40** _____ [602.10.1.4.1X1]
 • WSP at 8 ft. begins with min 1,800 lb. hold-down **F40** __ [602.10.1.4.1X2]
☐ Min. total length of bracing 48 in. in any BWL _____ [602.10.1.2]
☐ Length where wind is determining factor based upon spacing _ [602.10.1.2]³⁹
☐ SDC D spacing between BWLs max 25 ft. EXC _____ [602.10.1.5]⁴⁰
 • 35 ft. allowed to accommodate 1 room ≤ 900 sq. ft._____ [602.10.1.5X]
 • 35 ft. allowed with increased bracing & fastening _____ [602.10.1.5X]

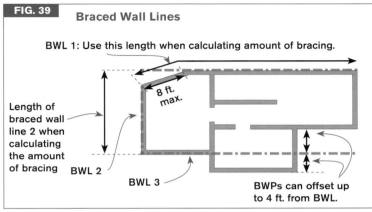

FIG. 39 Braced Wall Lines

BWL 1: Use this length when calculating amount of bracing.

8 ft. max.

Length of braced wall line 2 when calculating the amount of bracing

BWL 2

BWL 3

BWPs can offset up to 4 ft. from BWL.

Bracing Methods   09 IRC

☐ Acceptable methods include intermittent bracing methods **T19**
 & continuous sheathing (CS) methods _____ [602.10.1.1]⁴¹
☐ Mixed bracing methods allowed story to story _____ [602.10.1.1]⁴²
☐ Mixed bracing methods OK in different BWL of same story ___ [602.10.1.1]⁴²
☐ Mixed bracing methods in same BWL only in SDC A, B & C __ [602.10.1.1]⁴²
☐ Methods DWB, WSP, SFB, PBS & PCP req GB on interior side of wall
 unless bracing length multiplied by factor of 1.5 EXC **T19** __[602.10.2.1X3]⁴³
 • Approved interior finish with shear capacity = to GB _____ [602.10.2.1X2]
☐ SDC C & D adhesive not OK to fasten GB _____ [602.10.2.2]

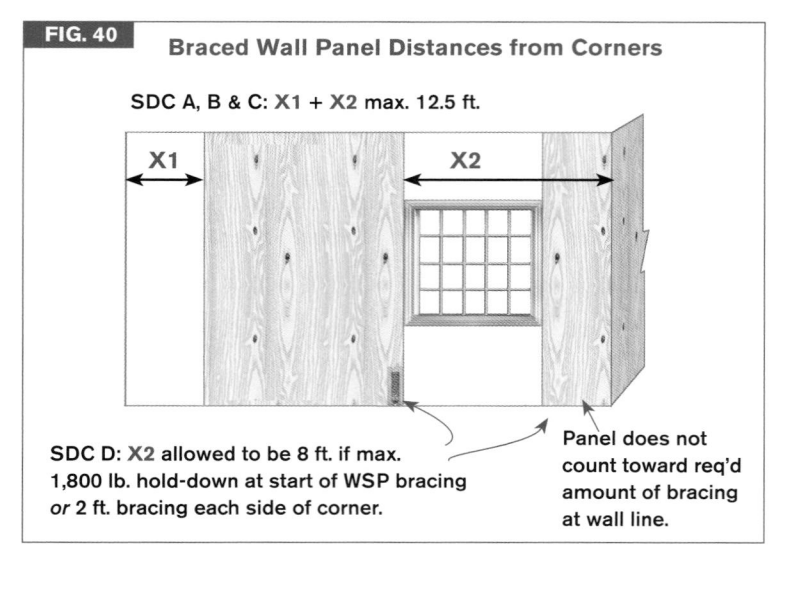

FIG. 40 Braced Wall Panel Distances from Corners

SDC A, B & C: X1 + X2 max. 12.5 ft.

X1

X2

SDC D: X2 allowed to be 8 ft. if max. 1,800 lb. hold-down at start of WSP bracing *or* 2 ft. bracing each side of corner.

Panel does not count toward req'd amount of bracing at wall line.

TABLE 19	INTERMITTENT BRACING METHODS [T602.10.2]
Abbreviation	**Bracing Method**
ABW	Alternate braced wall
DWB	Diagonal wood boards
GB	Gypsum board
HPS	Hardboard panel siding
LIB	Let-in bracing[A]
PBS	Particleboard sheathing
PCP	Portland cement plaster
PFG	Portal frame at garage
PFH	Portal frame with hold-downs
SFB	Structural fiberboard sheathing
WSP	Wood structural panels

A. Let-in bracing has few allowable applications & is not recommended.

FIG. 41

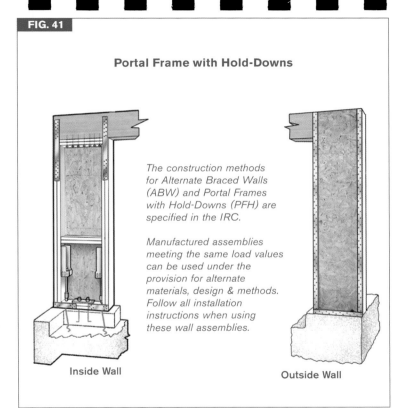

Portal Frame with Hold-Downs

The construction methods for Alternate Braced Walls (ABW) and Portal Frames with Hold-Downs (PFH) are specified in the IRC.

Manufactured assemblies meeting the same load values can be used under the provision for alternate materials, design & methods. Follow all installation instructions when using these wall assemblies.

Inside Wall

Outside Wall

Portal Frame with Hold-Downs (PFHs) F41   09 IRC
- ☐ May substitute for 4 ft. sections of bracing req'd by **T21,22** _____ [602.10.3.3]
- ☐ Use only below full-length headers_____ [602.10.3.3]
- ☐ $5/8$ in. anchor bolt & 2 embedded-type hold-down straps req'd _ [602.10.3.3]
- ☐ Min 16 in. wide for single story, 24 in. 1st story of 2-story_____ [602.10.3.3]

Braced Wall Panel (BWP) Lengths   09 IRC
- ☐ Min length each BWP 48 in. (96 in. for GB) EXC _____ [602.10.3]
  - ABW can replace 4 ft. bracing lengths of other methods _____[602.10.3X2]
  - Each PFH replaces 4 ft. lengths next to openings with headers[602.10.3X3]
  - PFG either side of garage door opening in SDC A, B & C __ [602.10.3X3][44]
  - SDC A, B & C BWPs between 36 in. & 48 in. methods DWB, WSP, SFB, PBS, PCP & HPS can count if adjusted per **T20** ____ [602.10.3X4][45]
- ☐ None of above reductions allowed if wall has masonry veneer in SDC D **T25**_____ [T602.10.1.2(3) & 602.12][46]
- ☐ Min bracing length per wind tables **T21** _____[T602.10.1.2(1)][47]
- ☐ Min bracing length per seismic tables **T22** _____ [T602.10.1.2(2)][48]

TABLE 20	BWPs < 48 IN. WIDE IN SDC A, B & C [T602.10.3]		
Actual Length (in.)	**Assigned Value[A] (in.)**		
	8 ft. wall height	**9 ft. wall height**	**10 ft. wall height**
48	48	48	48
42	36	36	n/a
36	27	n/a	n/a

A. Interpolation of table values is allowed.

WALL BRACING

Alternate Braced Wall Panels (ABWs)　　　09 IRC

- ☐ May substitute for 4 ft. sections of bracing req'd by **T21,22** ____ [602.10.3.2]
- ☐ Max height, min length & hold-down force per T602.10.3.2 ____ [602.10.3.2]

Portal Frame at Garage Door Openings in SDC A, B & C　　09 IRC

- ☐ Allowed only in single story or supporting one story + roof ____ [602.10.3.4]
- ☐ Counts as 1.5× its width for bracing for **T21** ____ [602.10.3.4]
- ☐ Constructed with following provisions: ____ [602.10.3.4]
  - Min width 24 in., max height 10 ft.
  - Min 7/16 in. sheathing extends over solid or glulam garage header
  - Sheathing nailed with 8d @ 3 in. o.c.
  - Min 1,000 lb. strap header to inner studs on opposite side from sheathing
  - BWP directly on foundation with min of two ½ in. bolts with plate washers
  - If panel on only one side of header, jack stud on other side req's 1,000 lb. straps to header & foundation

TABLE 21	BRACING BASED ON WIND SPEED[A] [T602.10.1.2(1)]			
Story Location	Min. Total Length of BWP at each BWL (ft.)			
	BWL Spacing (ft.)	Method GB (double-sided)	Methods DWB, WSP, SFB, PCP & HPS	Continuous Sheathing
	10	3.5	2.0	1.5
	20	6.0	3.5	3.0
	30	8.5	5.0	4.5
	40	11.5	6.5	5.5
	50	14.0	8.0	7.0
	60	16.5	9.5	8.0
	10	6.5	3.5	3.0
	20	11.5	6.5	5.5
	30	16.5	9.5	8.0
	40	21.5	12.5	10.5
	50	26.5	15.0	13.0
	60	31.5	18.0	15.5
	10	9.0	5.5	4.5
	20	17.0	10.0	8.5
	30	24.5	14.0	12.0
	40	32.0	18.0	15.5
	50	39.0	22.5	19.0
	60	46.5	26.5	22.5

A. Table based on wind speed ≤ 85 mph, exposure category B, 30 ft. mean roof height, 10 ft. eave-to-ridge height, 10 ft. wall height & 2 BWLs; see full code tables for situations beyond these limits.

TABLE 22 — BRACING BASED ON SEISMIC DESIGN CATEGORY[A] [T602.10.1.2(2)]

SDC & Story Location	BWL length (ft.)	Method GB (double-sided)	Methods DWB, WSP, SFB, PCP & HPS	Continuous Sheathing
D₀ or D₁	10	3.0	2.0	1.7
	20	6.0	4.0	3.4
	30	9.0	6.0	5.1
	40	12.0	8.0	6.8
	50	15.0	10.0	8.5
D₀ or D₁	10	6.0	4.5	3.8
	20	12.0	9.0	7.7
	30	18.0	13.5	11.5
	40	24.0	18.0	15.3
	50	30.0	22.5	19.1
D₀ or D₁	10	8.5	6.0	5.1
	20	17.0	12.0	10.2
	30	25.5	18.0	15.3
	40	34.0	24.0	20.4
	50	42.5	30.0	25.5

(SDC & Story Location — D_0 or D_1; Min. Total Length of BWP at each BWL (ft.))

TABLE 22 (Cont.) — BRACING BASED ON SEISMIC DESIGN CATEGORY[A] [T602.10.1.2(2)]

SDC & Story Location	BWL length (ft.)	Method GB (double-sided)	Methods DWB, WSP, SFB, PCP & HPS	Continuous Sheathing
D₂	10	4.0	2.5	2.1
	20	8.0	5.0	4.3
	30	12.0	7.5	6.4
	40	16.0	10.0	8.5
	50	20.0	12.5	10.6
D₂	10	7.5	5.5	4.7
	20	15.0	11.0	9.4
	30	22.5	16.5	14.0
	40	30.0	22.0	18.7
	50	37.5	27.5	23.4

A. Table based on 10 ft. wall height, 10 psf floor dead load, 15 psf roof/ceiling dead load & BWL spacing ≤ 25 ft.

Continuous Sheathing (CS)

☐ All BWLs on exterior walls on same story CS EXC   [602.10.4][49]
- SDC A, B & C with basic wind speed < 100mph OK to use other methods on other BWLs on same story

☐ CS req's WSP on all sheathable surfaces of BWL, including those above & below openings & gable end walls   [602.10.4.1][49]

☐ CS methods: [T602.10.4.1][50]
- CS-WSP: min 3/8 in. WSP, nailing 6d, 6 in. o.c. edges, 12 in. o.c. field
- CS-G: same as CS-WSP & adjacent to garage & supporting roof only
- CS-PF: max 4 panels in BWL, header lengths max 22 ft.

☐ CS length based on adjacent clear opening height F42, T23   [602.10.4.2]

09 IRC

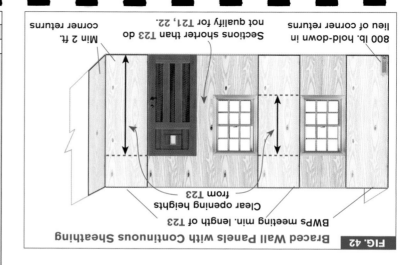

FIG. 42   **Braced Wall Panels with Continuous Sheathing**

- 800 lb. hold-down in lieu of corner returns
- Sections shorter than T23 do not qualify for T21, 22.
- Min 2 ft. corner returns
- Clear opening heights from T23
- BWPs meeting min. length of T23

Continuous Sheathing (CS) (cont.)

☐ CS panels max 25 ft. o.c. & BWL ends min 2 ft. return EXC   [602.10.4.4][51]
- 800 lb. hold-down at end OK in lieu of 2 ft. return F42   [602.10.4.4]
- 1st BWP 12.5 ft. from end in SDC A, B & C or 8 ft. in SDC D & full-height 2 ft. BWP at both sides of corners or 800 lb. hold-down at BWP closest to each side of corner   [602.10.4.4X]

09 IRC

TABLE 23 — LENGTH REQUIREMENTS FOR CONTINUOUS SHEATHING BWPs^A [T602.10.4.2]

Method	Adjacent Clear Opening Height (in.)	8 ft. Wall Height	9 ft. Wall Height	10 ft. Wall Height
CS-WSP	64	24	27	30
	68	26	27	30
	72	28	27	30
	76	29	30	30
	80	31	33	30
	84	35	36	33
	88	39	39	36
	92	44	42	39
	96	48	45	42
	100	–	48	45
	104	–	51	48
	108	–	54	51
	112	–	–	54
	116	–	–	57
	120	–	–	60
CS-G	≤120	24	27	30
CS-PF	≤120	16	18	20

A. Interpolation of table values allowed.

Panel Joints in Braced Wall Panels (BWPs) 09 IRC

☐ Vertical joints fastened on common studs, horizontal joints fastened on common blocking min 1½ in. thick EXC_____ [602.10.8]
- Blocking on horizontal joints not req'd if BWP length doubled__[602.10.8X2]
- Method GB installed horizontally does not req blocking _____[602.10.8X3]

BALCONIES & DECKS

Balconies 09 IRC

☐ Landings, balconies & decks positively anchored or self-supporting [311.5.1]
☐ Attachment not with toenails or means subject to withdrawal_____ [311.5.1]
☐ Where positive connection not verifiable during construction, decks must be self-supporting _____ [502.2.2]
☐ Deck ledger attachment with min ½ in. lag screws or bolts **T24**_ [502.2.2.1][52]
☐ Lag screws & bolts hot-dipped galvanized or stainless steel_____[502.2.2.1]
☐ Lag screws & bolts req washers **F43**_____[502.2.2.1]
☐ Lag screws & bolts 2 in. from top & bottom, staggered_____ [502.2.2.1.1]
☐ Alternative connections to accepted engineering practice: girders not on ledgers & ledgers not supported on masonry veneer _____[502.2.2.2]
☐ Lateral connection can be done with horizontal hold-downs ____ [502.2.2.3][53]

TABLE 24	DECK LEDGER ATTACHMENT & DETAILS [T502.2.2.1]						
Joist Span (ft.)	≤ 6	≤ 8	≤ 10	≤ 12	≤ 14	≤ 16	≤ 18
Connection Details	On-center Fastener Spacing (in.)						
½ in. lag screw **F43** – 1	30	23	18	15	13	11	10
½ in. bolt **F43** – 2	36	36	34	29	24	21	19
½ in. bolt with spacer washers **F43** – 3	36	36	29	24	21	18	16

Note: Designs based on max. 15/32 in. sheathing, southern pine or hemlock fir ledger, and 2 in. nominal band joist. Min. 1×9½ laminated veneer lumber can substitute for band joist.

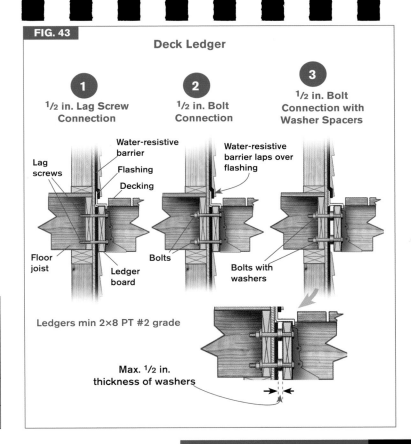

FIG. 43

Deck Ledger

1 ½ in. Lag Screw Connection

2 ½ in. Bolt Connection

3 ½ in. Bolt Connection with Washer Spacers

Water-resistive barrier
Lag screws
Flashing
Decking
Floor joist
Ledger board
Water-resistive barrier laps over flashing
Bolts
Bolts with washers

Ledgers min 2×8 PT #2 grade

Max. ½ in. thickness of washers

MASONRY VENEER

Bracing in SDC C
09 IRC

☐ 1st of 2 stories & 1st or 2nd of 3 stories, multiply T21 BWP by 1.5 [T602.12.1][54]

Bracing in SDC D
09 IRC

☐ Cripple walls not allowed [602.12]
☐ Bracing amount per T25 for both interior & exterior walls [602.12.1.1][55]
☐ BWP must begin within 8 ft. of end of BWL [602.12.1.2]
☐ Max spacing of BWPs 25 ft. o.c. [602.12.1.2]
☐ BWP min 7/16 in. with nailing max 4 in. o.c. edges, 12 in. o.c. field [602.12.1.3]
☐ Hold-downs req'd per T25 [602.12.1.3]

Masonry Veneer Height & Support
09 IRC

☐ SDC A, B & C max thickness 5 in., max height 30 ft. + 8 ft. gable [T703.7(1)]
☐ SDC D_0 & D_1 max thickness 4 in., max height 20 ft. + 8 ft. gable [T703.7(2)]
☐ SDC D_2 max thickness 3 in., max height 20 ft. + 8 ft. gable [T703.7(2)]
☐ Not OK to support additional loads on masonry veneer. [703.73]
☐ Steel angle support min 4 in. deep x 6 in. vertical x 5/16 in. thick [703.72.1]
☐ Framing behind steel support angle min doubled 2x4s 16 in. o.c. [703.72.1]
☐ Steel lintels min 4 in. bearing & protected with shop coat primer [703.73][56]
☐ Steel angles or lintels max deflection L/600 [703.72]

TABLE 25	SDC D MASONRY VENEER WALL BRACING [T602.12(2)]			
SDC	No. of Stories	Story	Min. Length of BWPs[A]	Hold-Down Force[B]
D_0	1	1	35	n/a
	2	Top	35	1900
	2	Bottom	45	3200
	3	Top	40	1900
	3	Middle	45	3500
	3	Bottom	60	3500
D_1	1	1	45	2100
	2	Top	45	2100
	2	Bottom	45	3700
	3	Top	45	2100
	3	Middle	45	3700
	3	Bottom	60	3700
D_2	1	1	55	2300
	2	Top	55	2300
	2	Bottom	55	3900

A. Length = percentage of total BWL.
B. Hold-down force in middle & bottom story is cumulative when BWPs align with those in story above.

Masonry Veneer Attachment & Flashing F44   09 IRC

- ☐ Ties min 1½ in. embedded in veneer mortar, min ⅝ in. from face ___ [703.7.4]
- ☐ Min 1 in. air space req'd behind veneer _____ [703.7.4.2]
- ☐ Max 1 in. air space for ties, max 4½ in. for metal strand tie wires ____ [703.7.4]
- ☐ Ties max 24 in. o.c. vertical & horizontal_____ [703.7.4.1]
- ☐ Max supported wall area of each tie 2.67 sq. ft. EXC_____ [703.7.4.1]
  - • SDC D each tie max 2 sq. ft. supported wall area _____ [703.7.4.1X]
- ☐ Flashing req'd beneath first course above ground _____ [703.7.5]
- ☐ Flashing req'd at all other points of support such as lintels _____ [703.7.5]
- ☐ Weepholes req'd immediately above flashing at max 33 in. o.c. ____ [703.7.6]
- ☐ Weepholes min ³/₁₆ in. diameter _____ [703.7.6]
- ☐ Adhered masonry AMI_____ [703.12]

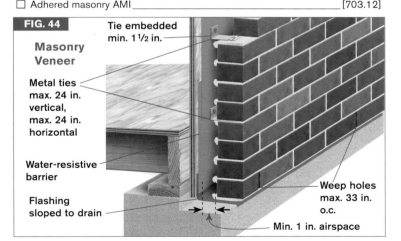

FIG. 44

Masonry Veneer

Tie embedded min. 1½ in.

Metal ties max. 24 in. vertical, max. 24 in. horizontal

Water-resistive barrier

Flashing sloped to drain

Weep holes max. 33 in. o.c.

Min. 1 in. airspace

EXTERIOR WALL COVERS

General   09 IRC

- ☐ Exterior sheathing must be dry before installing exterior cover _____ [701.2]
- ☐ Wall coverings must resist wind loads_____ [703.1.2][57]
- ☐ All fasteners corrosion-resistant _____ [703.4]
- ☐ Exterior wall construction must prevent water accumulation in wall _ [703.1.1]
- ☐ WRB req'd behind all exterior veneers EXC _____ [703.1.1]
  - • Concrete or masonry walls with proper flashing_____ [703.1.1X]
- ☐ WRB min No. 15 asphalt felt complying with ASTM D 226 _____ [703.2]
- ☐ WRB req'd over studs or sheathing at all exterior walls_____ [703.2]
- ☐ Install WRB shingle fashion min 2 in. horizontal lap, 6 in. at vertical joints _____ [703.2]

Flashings   09 IRC

- ☐ Install shingle fashion to prevent moisture entry into wall cavities or to structural framing components _____ [703.8]
- ☐ Flashings must extend to exterior _____ [703.8]
- ☐ Self-adhered flashing must comply with AAMA 711 (Note: caulking used with assembly must be compatible with self-adhered flashing)_____ [703.8]
- ☐ Req'd locations of flashing:_____ [703.8]
  - • Exterior door & window openings **F47**
  - • Intersections of chimneys with frame or stucco walls
  - • Under & at ends of masonry, wood, or metal copings & sills
  - • Continuously above all projecting wood trim
  - • Where porches, decks, or stairs attach to a wood-framed wall or floor **F43**
  - • At wall & roof intersections **F56**
  - • At built-in gutters

Exterior Insulation Finish Systems (EIFS)   09 IRC

- ☐ Barrier EIFS systems to comply with ASTM E 2568 _____ [703.9.1][58]
- ☐ EIFS with drainage also to comply with ASTM E 2568 _____ [703.9.2][58]
- ☐ Drainage-type EIFS req's WRB between EIFS & sheathing _____ [703.9.2.2]
- ☐ Drainage-type EIFS end min 6 in. above finished ground level ____ [703.9.4.1]
- ☐ Decorative trim not OK to face-nail through EIFS _____ [703.9.4.2]

Stucco   09 IRC

- ☐ Must comply with ASTM C 926 & ASTM C 1063 **T27** _____ [703.6]
- ☐ Lath fastener spacing max 6 in. _____ [703.6.1]
- ☐ Min 3-coat system over metal or wire lath, 2-coat over masonry ____ [703.6.2]
- ☐ Proportions per **T26** _____ [T702.1(3)][59]
- ☐ Intervals between coats **T26** _____ [703.6.5][60]
- ☐ Maintain moist min 48 hr. before subsequent coats _____ [703.6.4]

*The IRC typically references only specific sections of other standards, such that the entire standard does not apply. That is not the case with ASTM C 926 & C 1063—those standards apply in their entirety. All of the items in **T27** are mandatory.*

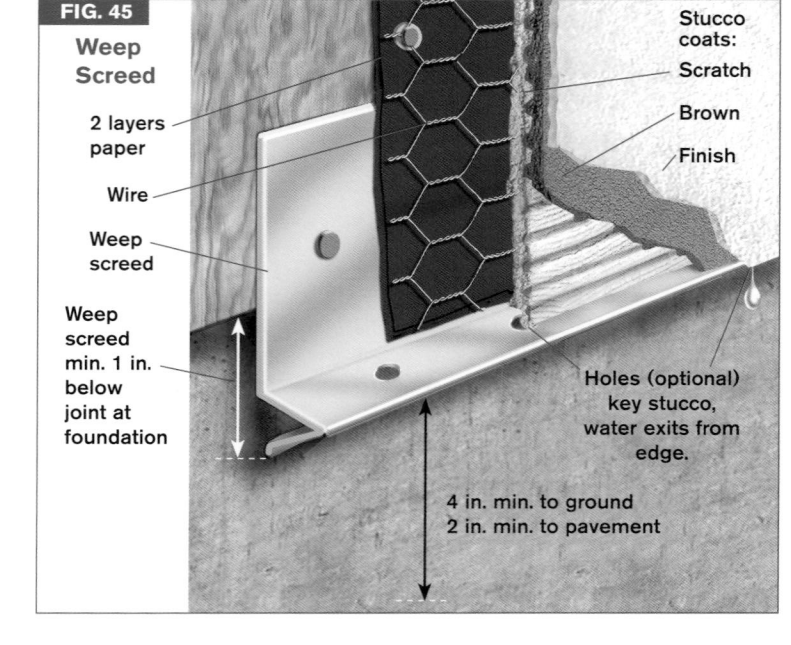

FIG. 45

Weep Screed

2 layers paper

Wire

Weep screed

Weep screed min. 1 in. below joint at foundation

Stucco coats:
Scratch
Brown
Finish

Holes (optional) key stucco, water exits from edge.

4 in. min. to ground
2 in. min. to pavement

TABLE 26	3-COAT STUCCO (VERTICAL SURFACES) [703.6]			
Coat	Thickness	Lime-to-Cement Volume Ratio	Sand-to-Cement Volume Ratio[A]	Interval before Next Coat[B]
Scratch	3/8 in.	3/4 to 1 1/2[C]	2 1/2 to 4	Min. 7 days
Brown	3/8 in.	3/4 to 1 1/2[C]	3 to 5[D]	Min. 48 hrs.
Finish	1/8 in.	3/4 to 2	1 1/2 to 3	Min. 7 days

A. Ratio of sand to combined volume of cement & lime.
B. ASTM C 926 allows lesser curing times depending on climate.
C. Max. 3/4 unless over absorbent surface (scratch of 2-coat system over CMU).
D. Same or greater proportion of sand in 2nd coat as used in 1st coat.

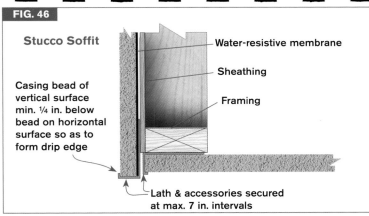

FIG. 46

Stucco Soffit

Casing bead of vertical surface min. ¼ in. below bead on horizontal surface so as to form drip edge

Water-resistive membrane

Sheathing

Framing

Lath & accessories secured at max. 7 in. intervals

TABLE 27	SELECTED ASTM C 926 & ASTM C 1063 REQUIREMENTS
C 926	**Summary of Requirement**
7.1.5	Install each coat without interruption or cold joints.
8.1	Continuously hydrate between coats.
8.1	Time between coats depends on climatic & job conditions.
12.3.2	Apply when ambient temperature > 40°F.
A2.2.3	Vertical-to-horizontal intersections req casing beads both surfaces, with vertical ¼ in. below horizontal to provide drip edge. Horizontal casing bead held back min. ¼ in. **F46**.
A2.3.1.2	Control joints to be included in plans & specifications.

(table continues in next column)

TABLE 27	SELECTED ASTM C 926 & ASTM C 1063 REQUIREMENTS (CONT.)
C 1063	**Summary of Requirement**
7.10.1.4	Lath stopped & tied each side of control joints.
7.10.1.5	Ceilings (soffits) req casing bead at intersections to walls or columns (no cornerite).
7.10.1.6	Load-bearing walls req casing bead or similar (no cornerite) at wall ends abutting structural walls or columns.
7.10.2.1	Lath must attach to framing members, not just to sheathing.
7.10.2.2	Diamond-mesh lath to horizontal framing with min 1½ in. nails.
7.11.1.1	Flanges of accessories secured at max. 7 in. intervals.
7.11.2	Install corner beads, corner reinforcement, or wrap lath around corners for min 1 support.
7.11.3	Casing beads to isolate nonload-bearing members from load-bearing.
7.11.4	Control joint separation spacing min ⅛ in.
7.11.4.1	Control joints to delineate areas not > 144 sq. ft.
7.11.4.2	Max 18 ft. distance between control joints.
7.11.4.2	Max 2½ to 1 ratio of length to width between control joints.
7.11.4.2	Control joint req'd where ceiling framing or furring changes direction.
7.11.4.4	Wall or partition height door frames considered control joints.
7.11.5	Weep screed req'd at bottom of all steel or wood framed walls **F45**.
7.11.5	Bottom edge of weep screed min. 1 in. below joint between foundation & framing **F46**.
7.11.5	Nose of screed min. 4 in. above earth or 2 in. above paving **F45**.

EXTERIOR WALL COVERS

Wood Panel Siding   09 IRC

- [ ] Vertical joints must be over framing members or WSP _____ [703.3.1]
- [ ] Vertical joints shiplapped or covered with batten _____ [703.3.1]
- [ ] Horizontal joints over solid blocking or over wood or WSP_____ [703.3.1]
- [ ] Horizontal joints lapped 1 in., shiplapped, or Z-bar flashing _____ [703.3.1]

Horizontal Lap Siding   09 IRC

- [ ] Install AMI _____ [703.3.2][61]
- [ ] In lieu of manufacturer instructions, lap 1 in. or ½ in. if rabbeted, vertical joints at ends caulked or flashed _____ [703.3.2]

Fiber Cement Lap Siding   09 IRC

- [ ] Must conform to ASTM C 1186 _____ [703.10.2][62]
- [ ] Min lap 1¼ in., vertical joints sealed, flashed with H-section, located over flashing, or installed over other barrier _____ [703.10.2]
- [ ] Nail heads can be concealed or exposed AMI _____ [703.10.2]

Wood Shakes & Shingles   09 IRC

- [ ] WRB req'd over sheathing behind shingles or shakes _____ [703.5.1]
- [ ] Weather exposure < ½ of shingle length for single course _____ [T703.5.2]
- [ ] 2 fasteners per shake or shingle_____ [703.5.3]
- [ ] Bottom course doubled, other courses single or double _____ [703.5.4]

Vinyl Siding   09 IRC

- [ ] Siding & accessories labeled for conformity to ASTM D 3679 _____ [703.11]
- [ ] Install AMI _____ [703.11.1]
- [ ] Soffit panels individually fastened to supports or AMI _____ [703.11.1.1][63]
- [ ] When installed over foam plastic sheathing, min 1¼ in. fastener penetration into wood framing EXC _____ [703.11.2.1][64]
  - AMI if instructions specific for basic wind speed _____ [703.11.2X&3]
- [ ] Adjust design wind pressure by 0.39 if basic wind speed > 90mph; 0.27 if no GB on interior side _____ [703.11.2.2][65]

TABLE 28	FASTENER SCHEDULE [602.3(1)]	
Connection	**Fastener[A]**	**Method**
FLOORS		
Built-up girders & beams, 2 in. lumber layers	10d	32 in. o.c. top & bottom staggered + 2 @ ends & splices
Joist to sill or girder	3-8d	toe nail
Ledger strip supporting joists or rafters	3-16d	at each joist or rafter
Rim joist to top plate	8d at 6 in. o.c.	toe nail
WALLS		
Built-up corner studs	10d	24 in. o.c.
Built-up header, two pieces with ½ in. spacer	16d	16 in. o.c. along each edge
Continuous header to stud	4-8d	toe nail
Double studs or double top plates	10d at 24 in. o.c.	face nail
Double top plate splice lap area (min 48 in.)	8-16d	face nail each side
Sole plate to joist or blocking	16d at 16 in. o.c.	typical face nail
Sole plate to joist or blocking	3-16d at 16 in. o.c.	BWPs
Stud to sole plate	3-8d or 2-16d	toe nail
Top or sole plate to stud	2-16d	end nail
Top plate laps & intersections	2-10d	face nail

(table continues on next page)

TABLE 28	FASTENER SCHEDULE (CONT.) [602.3(1)]	
Connection	FastenerA	Method
ROOF & CEILING CONSTRUCTION		
Blocking between joists or rafters to top plates	3-8d	toe nail
Ceiling joists to plate	3-8d	toe nail
Ceiling joists, laps over partitions	3-10d	face nail
Ceiling joists to parallel rafters	depends upon rafter slope	See IRC T802.5.1(9)
Collar tie to rafter	3-10d	face nail
Rafter tie to rafter	depends upon rafter slope	See IRC T802.5.1(9)
Rafter to plate	2-16d	toe nail
Rafter to ridge, valley, or hip rafters	4-16d 3-16d	toe nail face nail
WOOD STRUCTURAL PANELS		
Roof sheathing up to 1 in.	8d common	6 in. o.c. edges, 12 in. o.c. field
Wood structural panels, subfloor & wall sheathing up to 1/2 in.	6d common	6 in. o.c. edges, 12 in. o.c. field
Wood structural panels, subfloor & wall sheathing 19/32–1 in.	8d common	6 in. o.c. edges, 12 in. o.c. field
1 1/8 in. subfloor	10d common	6 in. o.c. edges, 12 in. o.c. field
A. Common or box nails unless otherwise noted		

INTERIOR WALL SURFACES

Gypsum wallboard installed for fire-resistance or shear values should be inspected prior to taping. In areas subject to direct moisture, cement boards are typically used. Water-resistant gypsum board is not rated for such areas.

Gypsum Board   09 IRC

☐ Protect from adverse weather during construction_____ [701.2]
☐ Do not install interior GB where exposed to weather or water _____ [702.3.5]
☐ Install only after all rough inspections complete _____ [109.1.2]
☐ Edges & ends over framing unless perpendicular to framing_____ [702.3.5]
☐ Fastening per **T29** _____ [702.3.5]
☐ GB ceiling diaphragms perpendicular to framing members & ends of adjacent courses of GB not on same joist _____ [702.3.7]

TABLE 29	FASTENING SCHEDULE FOR GBA [T702.3.5]			
Location	Orientation to framing	Max. frame spacing	Nails	Screws
CeilingsB	Perpendicular	24 in.	7	12
CeilingsB	Either	16 in.	7	12
Walls	Either	24 in.	8	12
Walls	Either	16 in.	8	16

A. For either 1/2 in. or 5/8 in. GB
B. Type X GB on garage ceiling beneath habitable space must be perpendicular to framing & fastened max 6 in. o.c. with 1 7/8 in. nails or equivalent screws.

Water-Resistant Gypsum Backing Board (Greenboard)   09 IRC

☐ No 1/2 in. greenboard on ceilings > 12 in. o.c. framing_____ [702.3.8]
☐ No 5/8 in. greenboard on ceilings > 16 in. o.c. framing _____ [702.3.8]
☐ Not allowed over vapor retarder in tub or shower_____ [702.3.8]
☐ May be used as backer for adhesive applications of tile in areas where there is no direct exposure to water or high humidity _____ [702.3.8.1]

INTERIOR WALL SURFACES (CONT.)

Cement Board
IRC 09

☐ Fiber-cement, fiber-mat reinforced cement, glass mat gypsum backers, or fiber-reinforced gypsum backers in compliance with ASTM C 1288, C 1325, C 1178, or C 1278 OK as backer for shower & tub tile if installed AMI [702.4.2]

WINDOWS & EXTERIOR DOORS

Performance & Labeling
IRC 09

☐ Window & door openings installed & flashed AMI F47 [612.1]⁶⁶
☐ Written installation instructions req'd for each window & door [612.1]⁶⁶
☐ Must be designed for wind loads [612.5]
☐ Windows & sliding doors req labeling from approved agency to indicate compliance with AAMA standards [612.6]
☐ Window & glass door assemblies must be anchored AMI [612.10.1]
☐ Anchor direct or with frame clip to wood frame structure [612.10.2]

Child Fall Prevention
IRC 09

☐ Openable windows > 72 in. above finished grade or surface below req lowest clear part of opening min 24 in. above floor EXC [612.2]
  • Windows that will not allow 4 in. sphere in fully open position [612.2X1]
  • Openings with window fall prevention devices per ASTM F 2090 [612.2X3]
  • Approved window-opening limiting devices [612.2X4]
☐ Window-opening limiting devices must prevent passage of 4 in. sphere when installed AMI [612.4.1]
☐ Window-opening limiting devices req release mechanisms max 15 lb. force & operable without tools or special knowledge [612.4.2]
☐ Window-opening limiting devices may not reduce net clear area req'd for escape & rescue T4.5 [612.4.2]

ROOF & CEILING FRAMING

An attic with limited storage is one that is provided with an entrance opening, no insulation above the joists or bottom truss chord, and where a 24 in. wide by 42 in. high rectangle parallel to the rafters or trusses would fit within the openings of the attic framing.

General
IRC 09

☐ Lumber req'd grade mark or agency approval certificate [802.1]
☐ Conventional framing provisions apply to roof slopes ≥ 1:3 [802.2]
☐ Each rafter or joist min 1½ in. end bearing on wood, 3 in. on masonry [802.6]
☐ Cutting, boring & notching dimensional lumber per T14, F30 [802.7.1]
☐ Cutting, boring & notching of engineered wood products AMI F32 [802.7.2]
☐ Fastening per T28 [802.2]
☐ > 5:1 dimension ratio rafters or joists req blocking at bearing points [802.8]
☐ > 6:1 dimension ratio for joists req solid blocking, diagonal bridging, or 1x3 backer at max 8 ft. intervals [802.8.1]

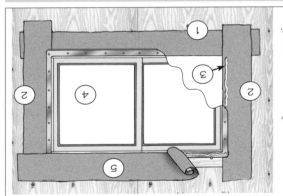

FIG. 47

Window Flashing

The window flashing essentially extends the window flange. WRB is integrated shingle fashion to the flange. Numbers indicate flashing & caulking sequence.

Ceiling Joists   09 IRC

☐ Spans per **T30,31** EXC_____ [802.4]
 • Use **T15** for attics with fixed stairs_____ [502.3.1]
☐ 3 in. lap over partitions or butted & toenailed to bearing member___ [802.3.2]
☐ Butted joists acting as rafter restraint req ties_____ [802.3.2]

TABLE 30	CEILING JOIST SPANS: ATTICS WITHOUT STORAGE 10 LB. LIVE LOAD [T802.4(1)]							
Size	Douglas Fir-Larch #2 Spacing o.c.				Southern Pine #2 Spacing o.c.			
	12 in.	16 in.	19.2 in.	24 in.	12 in.	16 in.	19.2 in.	24 in.
2×4	12–5	11–3	10–7	9–10	12–5	11–3	10–7	9–10
2×6	19–6	17–8	16–7	14–10	19–6	17–8	16–8	15–6
2×8	25–8	23–0	21–0	18–9	25–8	23–4	21–11	20–1
2×10	>26	>26	25–8	22–11	>26	>26	>26	23–11

Measurements given in feet & inches (ft.–in.).
Dead load = 5 psf

TABLE 31	CEILING JOIST SPANS: UNINHABITABLE ATTICS WITH LIMITED STORAGE, 20 LB. LIVE LOAD [T802.4(2)]							
Size	Douglas Fir-Larch #2 Spacing o.c.				Southern Pine #2 Spacing o.c.			
	12 in.	16 in.	19.2 in.	24 in.	12 in.	16 in.	19.2 in.	24 in.
2×4	9–10	8–9	8–0	7–2	9–10	8–11	8–5	7–8
2×6	14–10	12–10	11–9	10–6	15–6	13–6	12–3	11–0
2×8	18–9	16–3	14–10	13–3	20–1	17–5	15–10	14–2
2×10	22–11	19–10	18–2	16–3	23–11	20–9	18–11	16–11

Measurements given in feet & inches (ft.–in.).
Dead load = 10 psf

Rafters   09 IRC

☐ Roofs < 3:12 slope design ridges, valleys & hips as beams _____ [802.3]
☐ Rafter horizontal spans per **T33,34** EXC_____ [802.5]
 • Rafter span can be measured from purlin support _____ [802.5.1]
☐ Purlins ≥ dimension of rafters they support _____ [802.5.1]
☐ Purlin supports (kickers) min 2×4, max spacing 4 ft. o.c._____ [802.5.1]
☐ Purlin supports min 45° from horizontal_____ [802.5.1]
☐ Ridge min 1× material & full depth of cut rafter ends _____ [802.3]
☐ Valleys & hip rafters min 2× material & full depth of cut rafter ends ___ [802.3]
☐ Hip & valley rafters adequate to support load: max deflection L/180 __ [802.3]

Openings   09 IRC

☐ Single trimmer OK for single header within 3 ft. of trimmer bearing ___ [802.9]
☐ Doubled header & trimmer joists req'd if header > 4 ft. _____ [802.9]
☐ Hangers req'd for header-trimmer connections if header > 6 ft. _____ [802.9]
☐ Hangers or ledger strips req'd at header for tail joists > 12 ft. _____ [802.9]

Roof Sheathing   09 IRC

☐ Lumber sheathing min 5/8 in. net thickness at 24 in. o.c. supports ___ [T803.1]
☐ Min 1½ in. net thickness at > 24 in. o.c. supports _____ [T803.1]
☐ Spaced lumber sheathing not allowed in SDC D_2 _____ [803.1]
☐ WSP req's grade mark from approved agency **F35**_____ [803.2.1]
☐ WSP spans per **T17** _____ [803.2.2]
☐ WSP sheathing OK to be permanently exposed on underside
 (such as eaves) if identified as Exposure 1 _____ [803.2.1.1]
☐ Fire-retardant-treated plywood req's grade mark from approved
 agency & derived values & fastening req's approved method _____ [803.2.1.2]

Rafter Ties & Collar Ties

09 IRC

☐ Rafter ties req'd if joists not tied to parallel rafters [802.3.1]
☐ Rafter ties min 2×4 & located in lower 1/3 of attic F48 [802.3.1]
☐ Span tables must be adjusted per T32 if $H_C > H_R \div 0.133$ F48 [802.3.1]
☐ Rafter tie & ceiling joist nailing to rafters min 3-16d (more nails req'd per T802.5.1 (9) as spacing increased & slope decreased) [802.3.1]
☐ Collar ties to resist wind uplift req'd in upper 1/3 of attic F48 [802.3.1]
☐ Collar ties min 1×4 max spacing 4 ft. o.c. [802.3.1]

TABLE 32 — RAFTER SPAN ADJUSTMENT FACTORS [802.5.1] (SEE F48)

H_C / H_R	Adjustment	H_C / H_R	Adjustment
1/3	0.67	1/5	0.83
1/4	0.76	1/6	0.90

FIG. 48

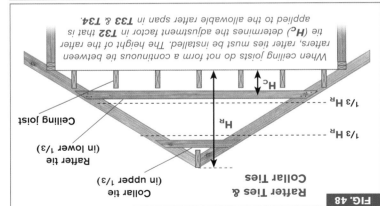

Rafter Ties & Collar Ties

Collar tie (in upper 1/3)
Rafter tie (in lower 1/3)
$^1/_3\ H_R$
H_R
H_C
Ceiling joist

When ceiling joists do not form a continuous tie between rafters, rafter ties must be installed. The height of the rafter tie (H_C) determines the adjustment factor in T32 that is applied to the allowable rafter span in T33 & T34.

TABLE 33 — RAFTER HORIZONTAL SPANS A,B [802.5.1(1)]
10 LB. DEAD LOAD, 20 LB. LIVE LOAD C

Size	Douglas Fir-Larch #2 Spacing o.c. 12 in.	16 in.	19.2 in.	24 in.	Southern Pine #2 Spacing o.c. 12 in.	16 in.	19.2 in.	24 in.
2×4	10-10	9-10	8-11	8-0	10-10	9-10	9-3	8-7
2×6	16-7	14-4	13-1	11-9	17-0	15-1	13-9	12-3
2×8	21-0	18-2	16-7	14-10	22-5	19-5	17-9	15-10
2×10	>26	>26	20-3	18-2	25-8	23-2	21-2	18-11
2×12	>26	25-9	23-6	21-0	>26	>26	24-10	22-2

A. Measurements given in feet & inches (ft.-in.) before T32 adjustment factor.
B. Ceiling not attached to rafters, deflection max. L/180.
C. Loads here typical for asphalt shingles with no snow load; check with local BO for snow load & applicable table for your area.

TABLE 34 — RAFTER HORIZONTAL SPANS A,B [802.5.1(3)]
20 LB. DEAD LOAD, 30 PSF SNOW LOAD C

Size	Douglas Fir-Larch #2 Spacing o.c. 12 in.	16 in.	19.2 in.	24 in.	Southern Pine #2 Spacing o.c. 12 in.	16 in.	19.2 in.	24 in.
2×4	8-5	7-5	6-8	5-11	9-0	7-10	7-1	6-4
2×6	12-4	10-8	9-9	8-8	12-11	11-2	10-2	9-2
2×8	15-7	13-6	12-4	11-0	16-8	14-5	13-2	11-9
2×10	19-1	16-6	15-1	13-1	19-11	17-3	15-9	14-1
2×12	22-1	19-2	17-6	15-7	23-4	20-2	18-5	16-9

A. Measurements given in feet & inches (ft.-in.) before T32 adjustment factor.
B. Ceiling not attached to rafters, deflection max. L/180.
C. Loads here would be typical for slate with 30 psf snow load; check with local BO for snow load & applicable table for your area.

TRUSSES

Most roof trusses have two bearing points and do not bear weight on interior walls. The outside members are chords, and the interior members are the web (**F49**). Trusses must not be cut or altered from their original design and must be installed in accordance with instructions included in the truss shipment. Because of possible seasonal truss movement, connections to interior walls are made with hardware that allows vertical movement (**F50**). Web bracing must be installed in accordance with the plans or *BCSI 1-06, Guide to Good Practice for Handling, Installing and Bracing Metal Plate Connected Wood Trusses.* It is published by the Structural Building Components Association and the Truss Plate Institute. Loads should not be added to trusses other than those for which they are designed. Alterations cannot be made without the written concurrence of a registered design professional.

Trusses   09 IRC

☐ Design drawings must be approved by BO prior to installation ___ [802.10.1]
☐ Design drawings must be included with truss shipment at jobsite _ [802.10.1]
☐ Design drawing must include following information: _____ [802.10.1]
  • Slope or depth, span & spacing
  • Location of all joints
  • Req'd bearing widths
  • Design loads, including top chord live & dead loads, bottom chord live
    & dead loads, concentrated loads & wind and earthquake loads
  • Adjustments to connector design values for conditions of use
  • Each reaction force & direction
  • Joint connector type & description
  • Lumber size, species & grade for each member
  • Connection requirements
  • Calculated deflection ratio
  • Max axial compression forces (for design of lateral bracing)
  • Req'd permanent bracing locations
☐ Where not otherwise specified, bracing per BCSI 1-06 _____ [802.10.3]

Trusses (cont.)   09 IRC

☐ Consult BCSI 1-06 for handling procedures & temporary bracing_ [802.10.3]
☐ No alterations without approval of registered design professional _ [802.10.4]
☐ No added loads (such as HVAC) without verification of capacity __ [802.10.4]
☐ Bearing point connectors min uplift rating 175 lb. **F51** _____ [802.10.5]

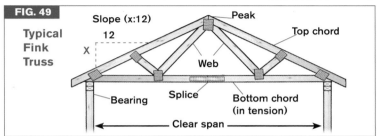

FIG. 49

Typical Fink Truss

Slope (x:12)
12
X
Peak
Top chord
Web
Bearing
Splice
Bottom chord (in tension)
Clear span

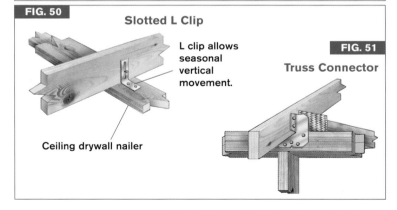

FIG. 50

Slotted L Clip

L clip allows seasonal vertical movement.

Ceiling drywall nailer

FIG. 51

Truss Connector

09 IRC 60

Unvented Attic Assemblies

☐ OK between top-story ceiling & roof if all 5 of following are met: [806.4][71]

1. Unvented space completely within building thermal envelope
2. No vapor retarders on ceiling side of unvented attic assembly
3. If wood shingles or shakes, $1/4$ in. air space beneath them
4. In climate zones 5, 6, 7 & 8 air-impermeable insulation must act as vapor retarder or have vapor retarder installed on underside
5.1 Air-impermeable insulation contacting underside of roof sheathing
5.2 Air-permeable insulation under roof sheathing with air-impermeable above meeting R-value of T35
5.3 Air-impermeable insulation meeting T35 under roof sheathing with air-permeable insulation under air-impermeable insulation

TABLE 35	**MIN. INSULATION FOR CONDENSATE CONTROL [T806.4][A]**
Climate Zone	**Rigid Board on Air-impermeable Insulation R-value**
2B & 3B tile roof	None required
1, 2A, 2B, 3A, 3B, 3C	R-5
4C	R-10
4A, 4B	R-15
5	R-20
6	R-25
7	R-30
8 (Alaska)	R-35

A. These values only for condensation control in unvented attic assemblies & do not supersede energy code requirements.

ATTICS

09 IRC 60

Access

☐ Access req'd if attic area > 30 sq. ft. & > 30 in. high measured from top of ceiling framing to underside of roof framing [807.1][67]
☐ Rough-framed opening min 22 in. × 30 in. & readily accessible [807.1]
☐ Attic opening in wall min 22 in. wide × 30 in. high [807.1][68]
☐ Attic opening in ceiling min 30 in. headroom at some point above opening measured from bottom of ceiling framing [807.1]
☐ Opening must be large enough to remove mechanical equipment. [1305.1.3]

Ventilation

☐ Vent each enclosed attic & rafter bay [806.1]
☐ Openings least dimension $1/16$ in., max dimension $1/4$ in. [806.1][69]
☐ Openings < $1/4$ in. protected with screening $1/16$–$1/4$ in. [806.1][69]
☐ Total area of ventilation $1/150$ of vented space EXC [806.2]
• Reduction to $1/300$ OK if 50–80% of venting provided by openings in upper portion of space min 3 ft. above eave or cornice vents & balance from eave or cornice F52 [806.2]
• Reduction to $1/300$ OK if Class I or II vapor retarder on warm-in-winter side of ceiling [806.2][70]
☐ Min 1 in. space between insulation & sheathing at eave/cornice vents [806.3]

FIG. 52

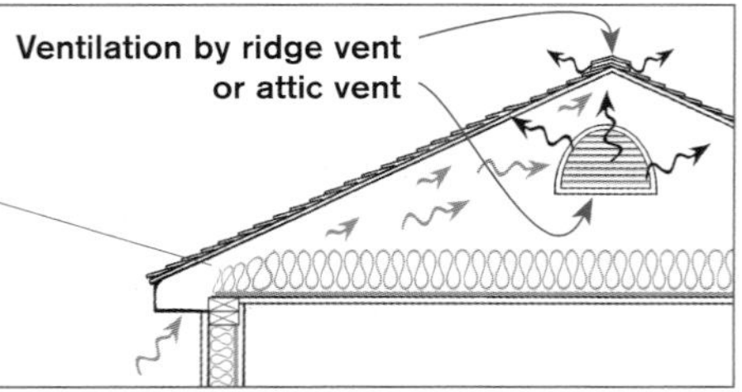

Attic Ventilation

Ventilation by ridge vent or attic vent

Insulation held back to allow 1 in. air space. Use baffles with loose insulation.

ROOFS

General   09 IRC
- ☐ Roof materials must be installed AMI_____ [903.1 & 904.1]
- ☐ Materials req conformity to recognized standards _____ [904.3]
- ☐ Materials req identification & test agency labels _____ [904.4]
- ☐ Materials must resist design wind loads per T301.2(2) & (3) _____ [905.1]
- ☐ Consider hail exposure in material selection _____ [903.5]

Fire Ratings   09 IRC
- ☐ Class A, B or C req'd per local laws or if < 3 ft. of PL_____ [902.1]
- ☐ Roof decks with masonry, brick, or concrete considered Class A _ [902.1X1]
- ☐ Copper or ferrous metal sheets or shingles & concrete or clay roof tile & slate over noncombustible deck considered Class A _____ [902.1X2]
- ☐ Fire-retardant-treated wood roofs req test agency label each bundle _ [902.2]

Flashing & Drainage   09 IRC
- ☐ Flashing req'd to prevent moisture entry to roof & walls_____ [903.2]
- ☐ Flashing req'd at wall & roof intersections, changes of roof slope or direction & around roof openings_____ [903.2.1]
- ☐ Metal flashing corrosion-resistant min 26-gage galvanized steel ___ [903.2.1]
- ☐ Crickets req'd on ridge side of penetrations > 30 in. wide _____ [903.2.2]
- ☐ Parapet walls req noncombustible coping ≥ thickness of parapet ____ [903.3]
- ☐ Drains at each low point of roof unless designed to run over edges __ [903.4]
- ☐ Overflow drains req'd with inlets 2 in. above low points of roof ____ [903.4.1]
- ☐ Overflow can be scupper in parapet wall _____ [903.4.1]
- ☐ Size of roof drains & leaders to comply with plumbing code_____ [903.4.1]
- ☐ Overflow drains must discharge separately from main roof drains __ [903.4.1]

Ice Barriers   09 IRC
- ☐ In areas with history of ice forming along eaves causing backup of water, ice barrier req'd for asphalt shingles, metal roof shingles, mineral-surfaced roll roofing, slate and slate shingles & wood shingles and shakes EXC_____ [905.2.7.1]
  - • Detached accessory structures with no conditioned area_____ [905.2.7.1X]
- ☐ Ice barrier = 2 layers underlayment cemented together or self-adhering polymer sheet to 24 in. inside exterior wall line_____ [905.2.7.1]

Code citations above are for asphalt shingles; citations for ice barriers with other roofing types are within subsections for those roofs.

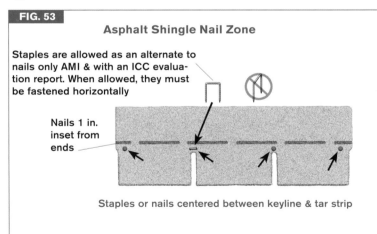

FIG. 53

Asphalt Shingle Nail Zone

Staples are allowed as an alternate to nails only AMI & with an ICC evaluation report. When allowed, they must be fastened horizontally

Nails 1 in. inset from ends

Staples or nails centered between keyline & tar strip

Asphalt Shingles 09 IRC

- ☐ Roof deck req's solid sheathing **T17** _____ [905.2.1]
- ☐ Min slope 2:12, double underlayment if < 4:12 _____ [905.2.2]
- ☐ Underlayment for slopes < 4:12 installed shingle fashion, 19 in. starter strip & successive 36 in. wide sheets lapped 19 in._____ [905.2.7]
- ☐ Underlayment for slopes ≥ 4:12 lapped min 2 in., end laps 6 in. & end laps offset by 6 ft. _____ [905.2.7]
- ☐ Underlayment min ASTM D 226 Type I felt _____ [905.2.3]
- ☐ Fastener penetration min ¾ in. or through if sheathing < ¾ in. **F54**_ [905.2.5]
- ☐ Fasteners AMI & min 4 per strip or 2 per individual shingle **F53**____ [905.2.6]
- ☐ Drip flashing below underlayment, rake flashing above **F55** _____ [905.2.8.1]
- ☐ Wall & pipe jack flashings AMI _____ [905.2.8.4]
- ☐ Sidewall step flashing min 4 in. high & 4 in. wide, length AMI **F56**_ [905.2.8.3]
- ☐ Sidewall flashings must terminate in kickout flashing **F56** _____ [905.2.8.3][72]
- ☐ Valley linings AMI; typical open valley req's metal 24 in. wide or 2 plies rolled mineral roofing, bottom layer 18 in., top layer 36 in. _[905.2.8.2]
- ☐ Closed valleys req min 1 layer 36 in. wide roll mineral roofing or self-adhering modified bitumen as underlayment _____ [905.2.8.2]

FIG. 54

Asphalt Shingle Nailing Method

Staples are allowed only AMI & with an ICC ES report; nails are preferred.

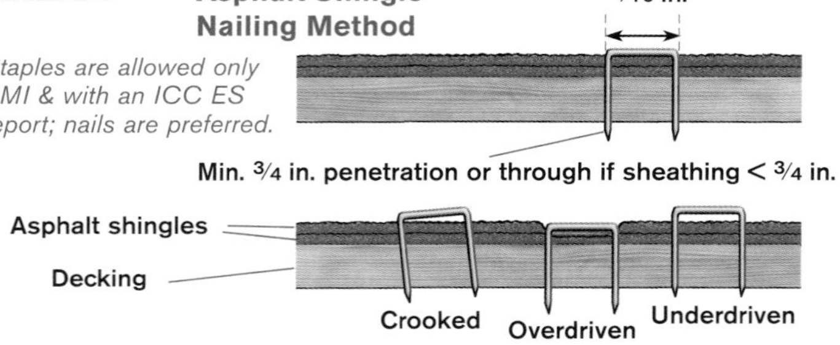

15/16 in.

Min. ¾ in. penetration or through if sheathing < ¾ in.

Asphalt shingles

Decking

Crooked   Overdriven   Underdriven

FIG. 55

Roof Edge Flashing

Underlayment laps over the drip flashing; rake edge flashing laps over the underlayment.

Drip flashing

Underlayment

Rake edge flashing

FIG. 56

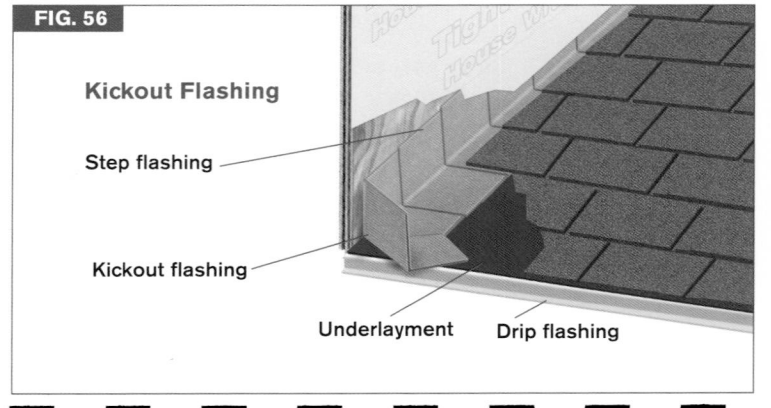

Kickout Flashing

Step flashing

Kickout flashing

Underlayment   Drip flashing

Clay & Concrete Tile .... 09 IRC

Note: The Tile Roofing Institute publishes installation manuals for cold & snow regions, for moderate climate regions & for Florida. Manufacturer's requirements are typically more stringent than the IRC for membranes & underlayment.

☐ Application AMI based on climate, slope, underlayment & tile type _ [905.3.7]

☐ Min slope 2½:12, double underlayment if < 4:12 _____ [905.3.2]

☐ Roof-to-wall flashings min 26-gage corrosion-resistant metal_____ [905.3.8]

☐ Valley flashing min 11 in. each way from centerline **F57** _____ [905.3.8]

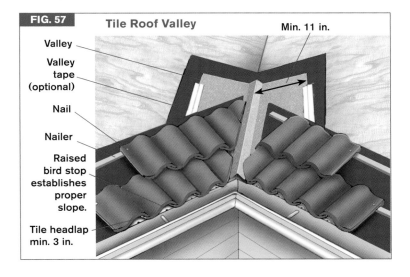

FIG. 57

Tile Roof Valley

Min. 11 in.

Valley

Valley tape (optional)

Nail

Nailer

Raised bird stop establishes proper slope.

Tile headlap min 3 in.

Metal Roof Panels .... 09 IRC

☐ Deck solid sheathing unless roofing AMI for spaced supports _ [905.10.1]

☐ Min slope 3:12 for lapped nonsoldered seam without lap
   sealant _____ [905.10.2]

☐ Min slope ½:12 (4%) with lap sealant installed AMI _____ [905.10.2]

☐ Min slope for standing seam ¼:12 (2%) _____ [905.10.2]

☐ Materials req'd to be corrosion resistant _____ [905.10.3]

☐ Fasteners AMI; if not specified, use the following: _____ [905.10.4]
   • Galvanized or stainless steel fasteners for steel roofs
   • Copper, brass, bronze, or 300 series stainless for copper roofs

☐ Underlayment AMI _____ [905.10.5]

Slate Roofing .... 09 IRC

☐ Min slope 4:12, sheathing solid _____ [905.6.1&2]

☐ Underlayment AMI (ASTM D 226 or D 4869 Type I or II) _____ [905.6.3]

☐ Min headlap 4 in. if < 8:12 slope, 3 in. if ≥ 8:12 & < 20:12 _____ [905.6.5]

Roll Roofing .... 09 IRC

☐ Mineral-surface roll roofing only on solid sheathing _____ [905.5.1]

☐ Mineral-surface roll roofing AMI & min 1:12 slope (8%) _____ [905.5.2&5]

Low-Slope Roofs .... 09 IRC

Note: In three of the code citations below, the letter X substitutes for the specific code number of the roofing material as follows: BUR = 9, modified bitumen = 11, EPDM = 12, PVC = 13, sprayed polyurethane foam = 14.

☐ Min slope ¼ in. per ft. (2% slope) EXC _____ [905.X.1]

☐ Coal tar BUR OK at ⅛ in. per ft. (1% slope) _____ [905.9.1]

☐ Install in compliance with applicable ASTM standards _____ [905.X.2]

☐ Install AMI _____ [905.X.3]

Wood Shingles

09 IRC

- ☐ Bundles to include label of approved grading bureau _____ [905.7.7]
- ☐ Min slope 3:12, sheathing solid or spaced _____ [905.7.1&2]
- ☐ Underlayment, if installed, min ASTM D 226 Type I _____ [905.7.3]
- ☐ Sidelap min 1½ in., no aligned keyways any 3 adjacent courses **F58** _____ [905.7.5]
- ☐ Keyways min ¼ in. max ⅜ in. **F58** _____ [905.7.5]
- ☐ Exposure per shingle grade & length **T36** _____ [905.7.5]
- ☐ 2 fasteners per shingle, min fastener penetration ½ in. **F58** ____ [905.7.5]
- ☐ Fasteners max ¾ in. from edge & 1 in. above exposure line **F58** [905.7.5]

Wood Shakes

09 IRC

- ☐ Bundles to include label of approved grading bureau _____ [905.8.9]
- ☐ Min slope 3:12, sheathing solid or spaced _____ [905.8.1&2]
- ☐ Interlayment min No. 30 felt (ASTM D 226, Type II) _____ [905.8.4&7]
- ☐ Interlayment strips 18 in. wide & no felt exposed to sun _____ [905.8.7]
- ☐ Sidelap min 1½ in., keyways min ⅜ in., max ⅝ in. **F59** _____ [905.8.6][73]
- ☐ Exposure per shake grade & length **T37** _____ [905.8.6]
- ☐ Max exposure standard #1 grade shake 10 in. at 4:12 slope _____ [905.8.6]
- ☐ 2 fasteners per shake, ≤ 2 in. above exposure line, 1 in. from edge **F59** _____ [905.8.6]

For further information on cedar shakes & shingles, see www.cedarbureau.org.

FIG. 58

Wood Shingle Application

Shingle keyways: min. ¼ in., max. ⅜ in.

Adjacent keyways min. sidelap 1½ in.

2 fasteners per shingle. No aligned keyways any 3 successive courses.

Exposure line

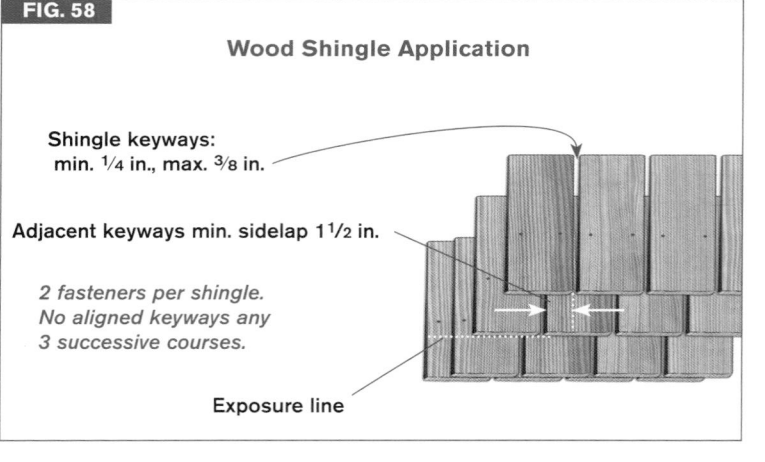

FIG. 59

Wood Shakes Application

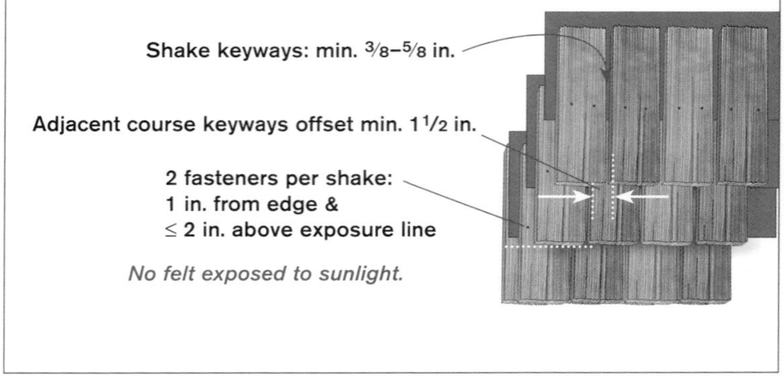

Shake keyways: min. ⅜–⅝ in.

Adjacent course keyways offset min. 1½ in.

2 fasteners per shake: 1 in. from edge & ≤ 2 in. above exposure line

No felt exposed to sunlight.

TABLE 36	WOOD SHINGLE WEATHER EXPOSURE [T905.7.5]		
Shingle Length (in.)	Grade & Label Color	Exposure (in.)	
		3:12 slope to < 4:12	4:12 slope or steeper
16	No. 1 – blue	$3^3/_4$	5
	No. 2 – red	$3^1/_2$	4
18	No. 1 – blue	$4^1/_4$	$5^1/_2$
	No. 2 – red	4	$4^1/_2$
24	No. 1 – blue	$5^3/_4$	$7^1/_2$
	No. 2 – red	$5^1/_2$	$6^1/_2$

TABLE 37	WOOD SHAKE WEATHER EXPOSURE [T905.8.6]		
Material	Length (in.)	Grade	Exposure[A] (in.)
Naturally durable wood (cedar)	18	No. 1	$7^1/_2$
	24	No. 1	10
PT taper-sawn Southern Pine	18	No. 1	$7^1/_2$
	24	No. 1	10
	18	No. 2	$5^1/_2$
	24	No. 2	$7^1/_2$
Taper-sawn naturally durable wood (cedar)	18	No. 1	$7^1/_2$
	24	No. 1	10
	18	No. 2	$5^1/_2$
	24	No. 2	$7^1/_2$

A. Assumes a 4:12 or greater slope.

ROOFS

Reroofing   09 IRC

☐ Same requirements as for new roofs EXC   [907.1]
• Low slope OK < 2% if providing positive drainage   [907.1X]
☐ Remove all existing layers of old roofing if 2 or more layers present of any type of roofing   [907.3]
☐ Metal or tile OK over existing shake roof   [907.3X2]
☐ Sprayed polyurethane foam new protective coat OK over old   [907.3X3]
☐ Prime flashings prior to application of bituminous materials   [907.6]

MASONRY FIREPLACES & CHIMNEYS

Masonry fireplaces are declining in popularity due to energy and environmental issues, and because of the rising popularity of gas-burning appliances that provide an aesthetically pleasing alternative. The heat generated by a masonry fireplace typically does not offset the amount of heated indoor air that is required to maintain the fire, and as a result the codes require outdoor combustion air to be brought into the fireplace. Many jurisdictions restrict the construction of new fireplaces. More information on gas-burning fireplaces can be found in other books in the Code Check series.

Masonry Fireplace & Chimney Construction   09 IRC

☐ Footing min 12 in. thick & 6 in. beyond sides   [1001.2 & 1003.2]
☐ Footing min 12 in. below finished grade   [1001.2 & 1003.2]
☐ Framing min 2 in. clearance from chimney & front & sides of fireplace & 4 in. from back of fireplace EXC   [1001.11 & 1003.18]
• Combustible trim, siding, flooring & sheathing can touch if 12 in. from firebox or flue liner   [1001.11X3 & 1003.18X3]
☐ Combustible trim cannot overlap chimney corners > 1 in.   [1003.18X3]
☐ Clear air space not filled except for fireblocking   [1001.11 & 1003.18]
☐ Chimneys not to support loads other than their own weight   [1003.8]

FIG. 60

3-Ply Built-Up Roof (BUR)

Inter-ply bitumen must be installed in a continuous firmly bonded film with no voids between the plies of material. Approximately 25 lb. of asphalt per square is req'd. The temperature must be maintained at the proper range for the specific type of asphalt.

NRCA Recommendations & ASTM Requirements for BUR F60

• Store rolls on ends not sides to prevent deformation
• Protect water-based materials from freezing prior to installation
• Protect insulation from moisture
• Do not install roofing while ice, rain, or snow are present
• Use cant strips to limit bends to 45° at horizontal-to-vertical intersections
• Sample temperature (typical 350°F to 425°F for Type I asphalt)
• Aggregate must be clean & dry to adhere to hot bitumen

SDC D Reinforcement

- ☐ Min 4 #4 vertical bars **F61** _____ [1001.3.1 & 1003.3.1]
- ☐ If > 40 in. wide, 2 additional bars req'd _____ [1001.3.1 & 1003.3.1]
- ☐ Min ¼ in. horizontal ties 18 in. o.c. around vertical bars [1001.3.2 & 1003.3.2]
- ☐ Horizontal ties req'd at each bend in vertical bars ____ [1001.3.2 & 1003.3.2]
- ☐ Grout must enclose rebar & not bond with flue liner __ [1001.3.1 & 1003.3.1]
- ☐ Anchor at each floor, ceiling, or roof > 6 ft. above
  grade EXC _____ [1001.4 & 1003.4]
  - • Chimneys completely inside exterior walls _____ [1001.4 & 1003.4]
- ☐ Anchor straps hooked around outer bars **F61** _____ [1001.4.1 & 1003.4.1]
- ☐ Fasten each strap to min 4 joists with 2
  ½ in. bolts **F61** _____ [1001.4.1 & 1003.4.1]

FIG. 61

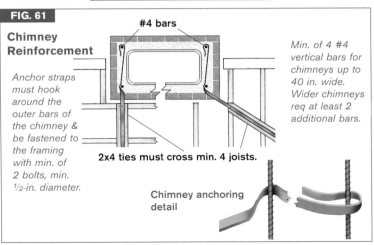

Chimney Reinforcement

#4 bars

Anchor straps must hook around the outer bars of the chimney & be fastened to the framing with min. of 2 bolts, min. ½-in. diameter.

Min. of 4 #4 vertical bars for chimneys up to 40 in. wide. Wider chimneys req at least 2 additional bars.

2x4 ties must cross min. 4 joists.

Chimney anchoring detail

Flues

- ☐ Design for proper draft _____ [1003.15]
- ☐ Terminate min 3 ft. above roof & 2 ft. above building within 10 ft. **F62** [1003.9]
- ☐ Spark arresters net free area min 4× flue opening size **F62** _____ [1003.9.1]
- ☐ Spark arrester screening mesh > ⅜ in. & < ½ in. _____ [1003.9.1]
- ☐ Spark arrester removable for cleaning **F62** _____ [1003.9.1]

FIG. 62

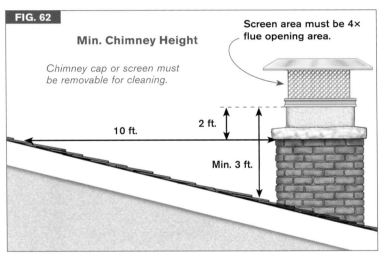

Min. Chimney Height

Chimney cap or screen must be removable for cleaning.

Screen area must be 4× flue opening area.

10 ft.

2 ft.

Min. 3 ft.

Masonry Fireplaces   09 IRC

- ☐ Firebrick-lined firebox walls min 8 in. thick including liner _____ [1001.5]
- ☐ Min depth 20 in. EXC _____ [1001.6]
  - • Rumford fireplaces 12 in. deep for 1/3 width of opening _____ [1001.6X]
- ☐ Throat min 8 in. above fireplace opening & min 4 in. deep _____ [1001.6]
- ☐ Masonry over opening req's lintel with min 4 in. bearing each side __ [1001.7]
- ☐ Operable damper req'd min 8 in. above fireplace opening _____ [1001.7.1]
- ☐ Smoke chamber parged smooth with refractory mortar _____ [1001.8][74]
- ☐ Hearth & extension reinforced to carry their own weight _____ [1001.9]
- ☐ Remove all combustible material from under hearth & extension ____ [1001.9]
- ☐ Min hearth thickness of 4 in., min thickness of extension
  2 in. EXC_____ [1001.9.1&2]
  - • 3/8 in. thick noncombustible extension OK if bottom of fireplace
    opening ≥ 8 in. above extension_____ [1001.9.2X]
- ☐ If opening < 6 sq. ft., extension depth min 16 in. front,
  8 in. side **F63** _____ [1001.10]
- ☐ If opening ≥ 6 sq. ft., extension depth min 20 in. front,
  12 in. side **F63** _____ [1001.10]
- ☐ No combustible material within 6 in. of opening _____ [1001.11X4]
- ☐ Combustible material < 12 in. from opening limited to projection
  of 1/8 in. for each inch distance from opening _____ [1001.11X4]
- ☐ All combustion air from exterior air supply _____ [1006.1]
- ☐ Exterior air intake no higher than elevation of firebox _____ [1006.2]
- ☐ Intake not from garage or basement, screen over opening _____ [1006.2]

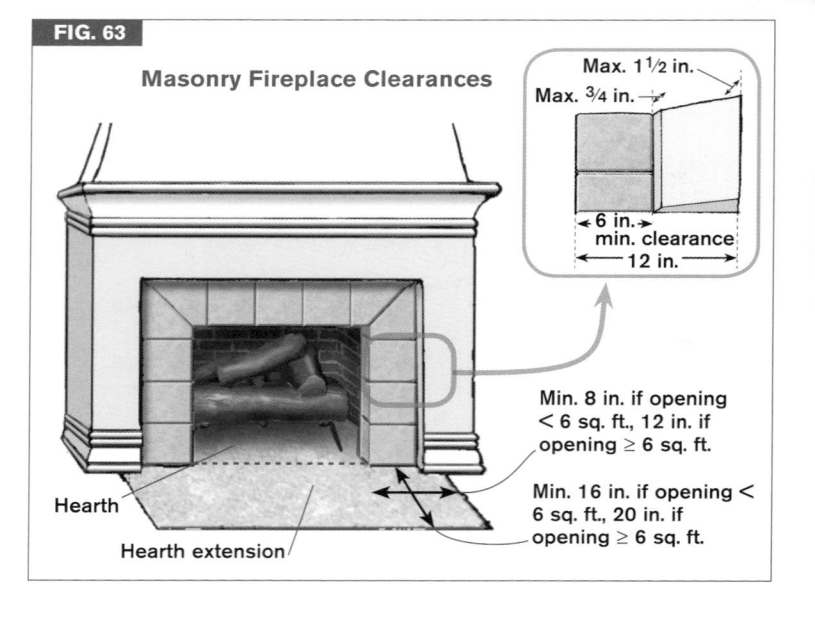

FIG. 63

Masonry Fireplace Clearances

Max. 1 1/2 in.
Max. 3/4 in.
6 in. min. clearance
12 in.

Hearth

Hearth extension

Min. 8 in. if opening < 6 sq. ft., 12 in. if opening ≥ 6 sq. ft.

Min. 16 in. if opening < 6 sq. ft., 20 in. if opening ≥ 6 sq. ft.

FACTORY-BUILT FIREPLACES & CHIMNEYS

Factory-Built Fireplaces & Chimneys   09 IRC
- ☐ Must be L&L & installed AMI_____ [1004.1 & 1005.1]
- ☐ Must be listed to UL 127_____ [1004.1 & 1005.4]
- ☐ Hearth extensions AMI & distinguishable from surrounding floor ____[1004.2]
- ☐ Decorative shrouds only if specifically L&L & AMI _____ [1004.3 & 1005.2]
- ☐ Combustion air ducts must be L&L for specific fireplace & AMI ___ [1006.1.1]
- ☐ Firestop spacer AMI per UL 127 where passing through ceilings ___ [1005.4]
- ☐ Insulation shield per UL 127 if passing through loose-fill in attic ____[1005.4]

FINAL INSPECTION REMINDERS

Final Inspection   09 IRC
- ☐ Address numbers visible from street _____ [319.1]
- ☐ Min 4 in. high Arabic numerals, min ½ in. stroke width _____ [319.1]
- ☐ Fire department signoff on automatic sprinklers prior to issuance of Certificate of Occupancy _____ [local]
- ☐ Smoke & CO alarms provided with battery backups & functional __ [314&315]
- ☐ Finish surfaces affecting stairs or landing completed_____ [109.1.6]
- ☐ Grading to provide req'd slopes & clearances; 6 in. siding to soil, 2 in. siding to hardscape **F27** _____ [317.1#5][75]

Benjamin Franklin was chosen as the main character for our Code Check illustrations for a number of reasons. The "First American's" insatiable curiosity, scientific genius, and civic mindfulness drove him to study fire safety, safe exiting, public sanitation, improved heating methods, and, of course, electricity.

Franklin made major contributions to each of the four main disciplines of building inspection: building, plumbing, mechanical, and electrical. Franklin's first attempt to safeguard the public through building codes came in 1735 with his call for minimum standards in the design of fireplace hearths, hearth extensions, and combustible material clearance. The principles Franklin proposed are codified in all the modern building codes, which prescribe these clearances in detail.

In 1736, Franklin organized the first volunteer fire department in Philadelphia, which still remains the model for our modern fire departments. He also understood the importance of building design in slowing the spread of a fire and was proud that his final home—built after his return from France in 1785—did not have concealed spaces where fire could spread. Thus, by judicious use of plaster, Franklin anticipated the fireblocking rules in today's codes. He also took an interest in designing stairways that were the proper pitch. Building and plumbing codes exist to safeguard persons and property. At Code Check, we feel that purpose to be a continuation of the work of Benjamin Franklin. His ideas are still alive in today's building codes and are carried on by code-making organizations and the people who practice those codes.

In 1736, after an extensive fire in Philadelphia, Benjamin Franklin created the first fire department—a fire brigade named The Union Fire Company.

TABLE 38   SIGNIFICANT CHANGES IN THE 2009 IRC CODE CYCLE

#	Page	Code	Description
1	4	202	New definition of habitable attic not to be considered as a story.
2	4	202	Braced wall line redefined as a straight line drawn through building plan, rather than as series of braced wall panels.
3	4	202	A townhouse in the 2006 IRC req'd open space on at least 2 sides. In the 2009 IRC, townhouses req'd a yard or public way on at least 2 sides.
4	5	106.1.1	BO may req plans to include braced wall lines & methods.
5	5	105.2	Size increased from 120 to 200 sq. ft. for accessory structures not requiring permits.
6	5	105.2	No decks were exempt in 2006 IRC.
7	6	301.2.1.5	New consideration of topographic wind speed-up.
8	6	301.3X	Increase in bracing for 12 ft. story height reduced to 10% when wind load determines bracing amount.
9	7	302.2X	2006 IRC req'd 2-hr. wall between townhouse units.
10	8	T302.6	Detached garages < 3 ft. from dwelling req same protection as attached garage (introduced in 2006 IRC).
11	9	302.11#4	Clarification that fireblocking does not need to comply with ASTM E136 firecaulking standard.
12	10	313	Automatic fire sprinklers now req'd.
13	11	314.2	Alarm must be monitored at central station to qualify.
14	11	314.2	Central-station alarm must be the owner's property or must supplement other installed alarms.
15	11	315	Carbon monoxide alarms now req'd.
16	12	305.1	2006 IRC allowed 4 ft. o.c. beams to project 6 in. into req'd ceiling height.

#	Page	Code	Description
17	12	305.1X1	2006 IRC req'd 6 ft. 8 in. clearance above the fixture; 2009 IRC allows fixtures under sloped ceilings such as below stairs.
18	13	311.2	2009 IRC specifies clear width rather than 3 ft. 0 in. door.
19	13	311.3X	Personal viewing balconies OK to have < 36 in. deep landing.
20	14	311.72X	New allowance for floors to project over edge of stairwell opening.
21	14	311.74	Stair dimensions exclusive of carpeting or rugs.
22	15	311.73	Walkline in 2006 IRC was measured from narrow side of tread; new measurement is from beginning of first walkable surface on narrow side.
23	15	311.74.2	Clarification that winder treads do not req dimensional uniformity with rectangular treads in same flight.
24	15	311.77.1X2	Handrail can exceed 38 in. height at transition fittings.
25	16	308.1.1	New recognition of ANSI Z97.1 designation as alternate to CPSC designation.
26	16	308.3.1X	Glazing with only an ANSI designation now allowed except in wet areas such as tubs & showers.
27	17	308.4#5X	Clarification that glass facing a tub exempt at 60 in.
28	18	312.1	Height measurement above grade now extends 36 in. horizontally.
29	18	312.2	Height of guard to be measured from top of fixed seating.
30	22	602.11.1	Slotted washers permitted for anchor bolts when supplemented with standard washers (introduced in 2006).
31	23	404.1.4.2	Reinforcement tables for concrete walls now combined with ICF tables.
32	25	506.2.3X1	Attached garages now req vapor retarder under slab.

| TABLE 38 | SIGNIFICANT CHANGES IN THE 2009 IRC CODE CYCLE (CONT.) |

#	Page	Code	Description
33	34	602.6.1	Nailing reduced from 8-16d to 8-10d (to avoid splits).
34	35	602.10X	No longer req'd to calculate seismic bracing for SDC A, B, or C except townhouses in SDC C.
35	35	602.10.1.3	New methodology for measurements of angled corners.
36	35	602.10.1.4	Though each exterior wall needs bracing, the braced wall line is not necessarily at the exterior wall.
37	35	602.10.1.4	2006 IRC allowed the braced wall lines to begin 12.5 ft. from *each* end in SDC A–C. In the 2009 IRC, 12.5 ft. is the max *combined* distance from the 2 ends for intermittent bracing.
38	35	602.10.1.4.1	The hold-down in this section was referred to as a tie-down in the 2006 IRC.
39	35	602.10.1.2	Spacing of braced wall lines included in wind tables & could be as much as 60 ft.
40	35	602.10.1.5	2006 IRC allowed 25 ft. spacing with exceptions allowing ≤ 35 ft.
41	35	602.10.1.1	The term *"intermittent bracing"* is new & distinguishes all of its bracing methods as one type in contrast to the continuous sheathing method.
42	35	602.10.1.1	2006 IRC did not address mixed bracing methods.
43	35	602.10.2.1X3	New rule explicit that values assigned to these bracing methods also rely on the value of the interior gypsum board.
44	37	602.10.3X3	Method PFG is new in 2009 IRC.
45	37	602.10.3X4	2006 IRC did not assign any value to sections of bracing < 48 in.

#	Page	Code	Description
46	37	602.12	2006 IRC did not specifically prohibit substituted lengths for bracing walls with masonry veneer.
47	37	602.10.1.2	New separate table for wind bracing based upon spacing of braced walls not on their length.
48	37	602.10.1.2	2006 IRC had a multiplier for continuous sheathing method & 2009 IRC gives it a separate column in wind & seismic tables.
49	40	602.10.4	The section on CS has been completely rewritten. In the 2006 IRC, CS req'd application on all exterior & interior braced walls.
50	40	602.10.4.1	The 3 methods for CS are reformatted & the table of CS length requirements has been modified.
51	40	602.10.4.4	New alternatives to 2 ft. returns at corners for CS.
52	41	502.2.2.1	New table & specifications for deck ledger connection.
53	41	502.2.2.3	New alternative lateral load connection with 2 horizontal hold-downs, min. 1500 lb. each.
54	42	602.12.1	Relocated from 2006 IRC T703.7(1) to chapter 6 in 2009 IRC.
55	42	602.12.1.1	The 2009 IRC does not allow any braced wall panel to be < 4 ft. wide; substitutions with CS or ABW are not OK.
56	42	703.7.3	Steel lintels req shop coat of paint or other corrosion protection.
57	43	703.1.2	Wind resistance must be considered in selection & application of wall coverings.
58	44	703.9	Recognition of both drainable & nondrainable EIFS.
59	44	T702.1(3)	Table on stucco proportions revised & includes blended cements & requirements from ASTM C 926.

TABLE 38	SIGNIFICANT CHANGES IN THE 2009 IRC CODE CYCLE (CONT.)		
#	Page	Code	Description
60	44	703.6.5	Curing times were relocated from table to code text to clarify their application to exterior plaster.
61	46	703.3.2	Lap siding manufacturer instructions are allowed to differ from specifications in code section.
62	46	703.10.2	Fiber-cement siding now req'd to meet ASTM C 1186.
63	46	703.11.1.1	Backing or nailing strips req'd for soffits with vinyl siding or as specified by manufacturer.
64	46	703.11.2.1	Min. fastener penetration to improve performance in high wind conditions.
65	46	703.11.2	New wind performance requirements for vinyl siding.
66	48	612.1	Doors as well as windows must be installed & flashed AMI.
67	52	807.1	Clarification on how attic height is measured.
68	52	807.1	Clarification on size of opening when it is in a wall.
69	52	806.1	$1/16$ in. openings now allowed for attic ventilation.
70	52	806.2	Class I & II vapor retarder designations replace 2006 IRC perm ratings for vapor retarders.
71	52	806.4	*Conditioned attic assemblies* now *unvented attic assemblies*; new table clarifies insulation methods for condensation control.
72	54	905.2.8.3	Kickout flashings req'd at low end of roof/sidewall intersections.
73	56	905.8.6	Min. keyway space for all types of shakes increased to $3/8$ in.
74	60	1001.8	Smooth parging of smoke chamber req'd to minimize turbulence.
75	61	317.1#5	Clearance of non-PT wood to hardscape now specified.

For updates, corrections, and additional tables for this book, visit
www.codecheck.com/CCComplete2nd.html

Code ✓Check® Plumbing Fourth Edition

By DOUGLAS HANSEN & REDWOOD KARDON
Illustrations & Layout by Paddy Morrissey

For more information on the building, electrical, and mechanical codes, valuable resources, and why Benjamin Franklin is featured in the Code Check series, visit www.codecheck.com.

Code Check Plumbing 4th edition is an illustrated guide to common code questions in residential plumbing, heating, ventilation, and air conditioning systems. The book emphasizes the safety principles that are at the heart of the codes for these systems.

The primary code used in this book is the 2009 edition of the *International Residential Code for One- and Two-Family Dwellings*®, published by the International Code Council (ICC). It is the most widely used residential code in the United States. The other major code referenced in this book is the 2009 *Uniform Plumbing Code*®, published by the International Association of Plumbing & Mechanical Officials (IAPMO). For most topics, these codes are in agreement. Each of these codes references standards, many of which are maintained by the organizations in Table 2 (**T2**).

Additional codes for specialized items are listed in **T1**. The National Fire Protection Association (NFPA) publishes *NFPA 54–The National Fuel Gas Code*, which forms the basis of the fuel gas provisions in the IRC, UPC, and UMC.

The 2009 cycle of codes is likely to remain in effect in most areas for at least 3 or 4 years after the cover date. Energy codes vary greatly from one area to another, and may modify or overrule the code requirements shown in this book. Before beginning any project, check with your local building department to determine the codes that apply in your area.

Thanks to Hamid Naderi of ICC for his editorial input.

TABLE 1	CODES & STANDARDS USED IN THIS BOOK	
Organization	Edition	Code
ICC	2009	International Residential Code
ICC	2009	ISPDC–International Private Sewage Disposal Code
IAPMO	2009	Uniform Plumbing Code
IAPMO	2009	Uniform Mechanical Code
NFPA	2009	NFPA 54 National Fuel Gas Code
NFPA	2011	NFPA 58 Liquefied Petroleum Gas Code

TABLE 2	ORGANIZATIONS
Acronym	Name
ASSE	American Society of Sanitary Engineering
IAPMO	International Association of Plumbing & Mechanical Officials
ICC	International Code Council
NFPA	National Fire Protection Association
NSF	National Sanitation Foundation

KEY TO USING CODE CHECK

Code Check Plumbing condenses large amounts of code information by using several "shorthand" conventions that are explained here. Each rule described in Code Check begins with a checkbox and ends with the code citations. When only one code is shown, the code citation is inside of brackets, and when two codes are shown, the second code is shown inside of braces, as in the following example from **p.84**:

☐ All fixture traps req venting _____ [3101.2.1] {901.0}

This line is stating that all fixture traps require venting, and the rule is found in section 3101.2.1 of the IRC and section 901.0 of the UPC.

References to figures and tables are preceded by an F or a T as in the following example from **p.74**:

☐ Changes in direction req appropriate fittings **F9–12, T9** [3005.1] {706.1}

This line is stating that changes of directions must use appropriate fittings, as illustrated in Figures 9–12 and in Table 9.

A change from the previous code edition is shown by a code citation in a different color. The superscript after the code citation refers to the table on **p.121**, where more information about the change is found. The following example is from **p.78**:

Backwater Valves F17   09 IRC   09 UPC

☐ COs for drains through backwater valve req label _____ [n/a] {710.1}²

This line says that when a cleanout would allow a snake to pass through a backwater valve, the UPC requires a label at the cleanout. This is explained further as code change No. 2 on page121, where the exact wording of the label is stated.

A line ending in EXC means that an exception to the rule is contained in the line that follows, and that begins with a bullet rather than a checkbox. The following example is from **p.103**:

☐ Ignition source 18 in. above garage floor EXC **F53** [2801.6] {508.14}
• Flammable vapor ignition-resistant (FVIR) WHs **F54** [2408.2X] {508.14}

These lines are stating that water heaters in garages, as shown in Figure 53, must be elevated so the ignition source is at least 18 inches above the floor, unless the water heater is an FVIR type, as shown in Figure 54.

If a rule does not apply to a particular code, that will be indicated by "n/a" in the code citation column, as in this example from **p.103**:

☐ WH also used for space heating must be L&L for both [2448.2] {n/a}

This line is stating that a water heater used for space heating must be listed and labeled for both purposes. The rule is in section 2448.2 of the IRC and it does not apply when using the UPC.

Rules that are not explicitly stated in a model code are sometimes based on other local ordinances, as indicated in this example from **p.70**:

☐ Building sewer depth per local ordinance _____ [2603.6.1] {local}

This line is saying that IRC section 2603.6.1 directs us to consult local ordinances for required sewer depth. The UPC does not have this rule, and the local building department should be consulted for its requirements.

ABBREVIATIONS

AAV	=	air admittance valve
ABS	=	acrylonitrile-butadiene-styrene drain (black plastic pipe)
ACH	=	air changes per hour
AHJ	=	Authority Having Jurisdiction
AMI	=	in accordance with manufacturer's instructions
AWG	=	American Wire Gage
BO	=	building official
Btu	=	British thermal unit(s)
cfm	=	cubic feet per minute
CO	=	cleanout
CPVC	=	chlorinated PVC pipe
CSST	=	corrugated stainless steel tubing (for gas)
cu.	=	cubic, as in cu. ft.
Cu	=	copper
CW	=	clothes washer
CW&V	=	combination waste & vent
DFU	=	drainage fixture unit
DW	=	dishwasher
DWV	=	drain, waste & vent
EXC	=	exception to rule will follow in the next line
FLR	=	flood level rim
ft.	=	foot/feet
FVIR	=	flammable-vapor ignition-resistant
gal	=	gallon(s)
GPF	=	gallons per flush
gpm	=	gallons per minute
hr.	=	hour(s)
in.	=	inch(es)

KS	=	kitchen sink
lav	=	lavatory sink
L&L	=	listed & labeled
LP	=	liquefied petroleum (LP gas)
LT	=	laundry tub
manu	=	manufacturer, manufacturer's
max	=	maximum
min	=	minimum
MP	=	medium pressure
o.c.	=	on center
PE	=	polyethylene tubing
PEX	=	crossed-link polyethylene tubing
PP	=	polypropylene plastic tubing
PRV	=	pressure relief valve
psf	=	pounds per square foot
psi	=	pounds per square inch
psig	=	pounds per square inch gage
PVC	=	polyvinyl chloride pipe
req	=	require
req'd	=	required
req's	=	requires
SDC	=	Seismic Design Category
sq.	=	square, as in sq. ft.
TPRV	=	temperature & pressure relief valve
WC	=	water closet (toilet)
WH	=	water heater
WSFU	=	water supply fixture unit
Zi	=	zinc, galvanized

TABLE OF CONTENTS

GENERAL RULES FOR ALL PIPING

Materials

		09 IRC	09 UPC
☐ Materials must be 3rd party tested or certified		[2608.4]	{301.1.1}
☐ All pipes & fittings marked by manu		[2608.1]	{301.1.2}

Pipe Support

		09 IRC	09 UPC
☐ Hangers must prevent distortion & maintain alignment (no wires, no metal straps contacting plastic pipe) **F1**		[2605.1]	{314.2&4}
☐ Insulate Zi hangers from contact with Cu pipes		[2605.1]	{314.4}
☐ Max support intervals for water pipe **T3,4**		[2605.1]	{314.1}

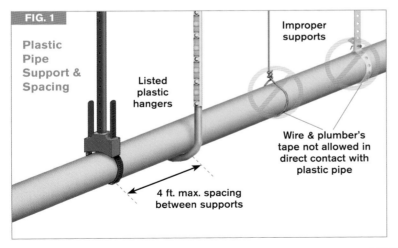

FIG. 1

Plastic Pipe Support & Spacing

Listed plastic hangers

Improper supports

Wire & plumber's tape not allowed in direct contact with plastic pipe

4 ft. max. spacing between supports

TABLE 3	IRC MAX. SUPPORT SPACING OF WATER PIPE [T2605.1]	
Pipe Material	**Horizontal**	**Vertical**
ABS/PVC DWV	4 ft.	10 ft.[A]
Threaded steel	12 ft.	15 ft.
Cast-iron hubless	5 ft. (10 ft. OK for 10 ft. lengths of pipe)	15 ft.
Cu water tubing	6 ft. for ≤ 1¼ in. pipe 10 ft. for ≥ 1½ in. pipe	10 ft.
CPVC	3 ft. for ≤ 1 in. pipe 4 ft. for ≥ 1¼ in. pipe	10 ft.[A]
PEX	32 in.	10 ft.[A]
PEX-AL-PEX	32 in.	4 ft.[A]

A. Provide mid-story guides for pipes ≤ 2 in.

TABLE 4	UPC MAX. SUPPORT SPACING OF WATER PIPE [T3-2]	
Pipe Material	**Horizontal**	**Vertical**
ABS/PVC DWV	4 ft.	Base & each floor[A]
Threaded steel	per AHJ	per AHJ
Cast-iron hubless	Within 18 in. of joints[B] (every other joint if 4 ft.)	Base & each floor 15 ft.
Cu water tubing	6 ft. for ≤ 1½ in. pipe 10 ft. for ≥ 2 in. pipe	Each floor 10 ft.
CPVC	3 ft. for ≤ 1 in. pipe 4 ft. for ≥ 1¼ in. pipe	Base & each floor[A]
PEX	32 in.	Base & each floor[A]
PEX-AL-PEX	98 in.[C]	Base & each floor[A]

A. Provide mid-story guides.
B. Includes horizontal branch connections. Hangers not OK directly on couplings.
C. Manu may req closer support spacing.

TRENCHES & PIPE PROTECTION

Pipes in soil must be supported for their entire length. Smooth, self-compacting backfill such as sand or pea gravel helps to eliminate sags that could cause water to be trapped and lead to blockage. Pipes must be protected from sharp rocks or other debris when backfill is placed. Piping encased in concrete requires protection. Pipes in walls and floors must be protected against damage from fasteners.

Piping in Concrete or Masonry   09 IRC   09 UPC

- ☐ Wrap embedded piping to prevent corrosion — [2603.3] {313.2}
- ☐ Provide for movement (expansion & contraction) — [2606.3] {313.2}
- ☐ Sleeve req'd to prevent structural load on pipes through foundation walls or under footings EXC. — [2603.5] {313.10}
  - • Not req'd for bored or drilled openings — [n/a] {313.10.1X}
- ☐ Seal spaces between pipes & sleeves — [2603.4] {313.10.3}
- ☐ Sleeve min 2 sizes larger than pipe through foundation — [2603.5] {n/a}

Piping in Trenches   09 IRC   09 UPC

- ☐ Pipe supported on firm bed for entire length — [2604.1] {314.3}
- ☐ No rocks supporting or touching pipes — [2604.1] {315.4}
- ☐ No rocks or debris in first 12 in. of backfill over pipe — [2604.3] {315.4}
- ☐ Trenches not to undermine footings (within 45°) F4 — [2604.4] {315.1}
- ☐ Water pipe min 12 in. cover below finished grade — [2603.6] {609.1}
- ☐ Water pipe min 6 in. (12 in. UPC) below frost line — [2603.6] {609.1}
- ☐ Building sewer depth per local ordinance — [2603.6.1] {local}

Piping in Common Trench   09 IRC   09 UPC

- ☐ Water & sewer OK in same trench if sewer materials approved for use within building F5 — [2905.4.2] {609.2}
- ☐ Water & sewer min. 5 ft. apart if sewer materials not approved for use within building F2 EXC — [2905.4.2] {609.2}
  - • Water min 12 in. above & to side of sewer pipe F3 — [2904.4.2] {609.2}
- ☐ Water pipe crossing sewer min 12 in. above EXC — [2905.4.2] {609.2}
  - • Water service sleeved 5 ft. each way from sewer pipe — [2905.4.2X] {n/a}

FIG. 2

Separate Trenches

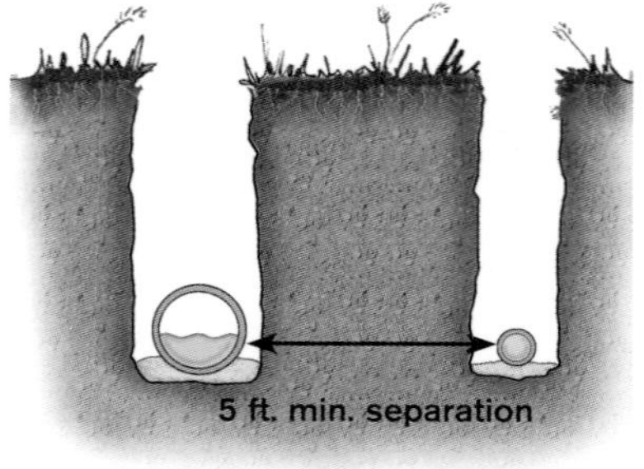

Sewer material type not approved within building.

5 ft. min. separation

FIG. 3

Pipes in Trench

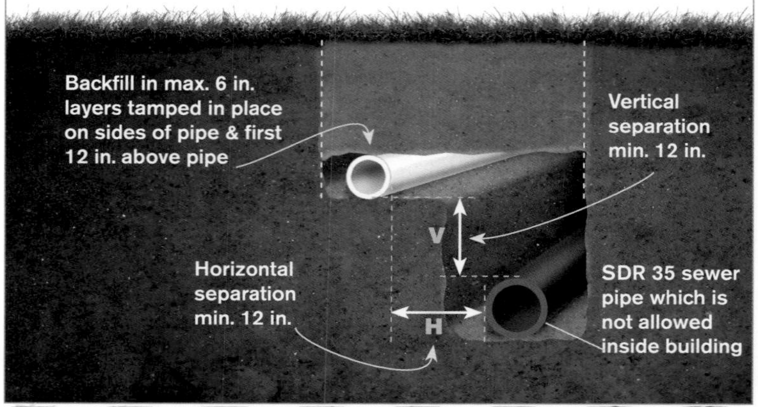

- Backfill in max. 6 in. layers tamped in place on sides of pipe & first 12 in. above pipe
- Vertical separation min. 12 in.
- SDR 35 sewer pipe which is not allowed inside building
- Horizontal separation min. 12 in.
- V
- H

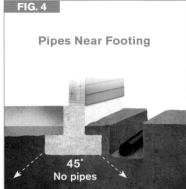

FIG. 4

Pipes Near Footing

45°
No pipes

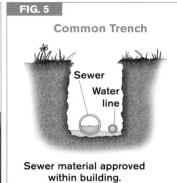

FIG. 5

Common Trench

Sewer
Water line

Sewer material approved within building.

DRAINAGE

Drain and waste pipes must have adequate slope and size, based on the number of drainage fixture units (DFUs) each pipe must serve. Changes of direction in pipes must be done with fittings that will not cause an obstruction in flow. The change from vertical to horizontal requires the greatest sweep, as the liquid and solids in the vertical pipe have greater velocity than in the horizontal pipe.

General	09 IRC	09 UPC
☐ Materials per **T5** _____	[3002.1&2]	{T7-1}
☐ Size per DFU loads **T6–8** _____	[3005.4]	{703.1}
☐ Min slope ¼ in./ft. EXC _____	[3005.3]	{708.0}
• ⅛ in./ft. OK for 3 in. or larger pipe _____	[3005.3]	{Ø}
• ⅛ in./ft. OK for 4 in. pipe if structurally necessary per BO _[n/a]		{708.0}
☐ Building drain & building drain branches max load based on slope per **T8** _____	[3005.4.2]	{n/a}

General (cont.)

	09 IRC	09 UPC
☐ Building drain & building drain branches max load based on slope per **T8** _____	[3005.4.2]	{n/a}
☐ No reductions in direction of flow **F6,8** EXC [3002.3.1, 3005.1.7]		{316.4.1}
• 3 in. × 4 in. WC bend OK **F7** _____	[3005.1.7]	{316.4.2}
☐ No drilled or tapped connections (e.g., saddle fitting) __ [3003.2]		{311.2}
☐ Different types of plastic not glued together EXC _____[3003.2(5)]		{316.1.6}
• ABS & PVC drain to sewer OK with listed transition solvent_[Ø]		{316.1.6}

TABLE 5	DRAINAGE MATERIALS [T3002.1] {T7-1}					
Material	**Above Ground**		**Underground**		**Building Sewer**	
	IRC	UPC	IRC	UPC	IRC	UPC
ABS schedule 40	✔	✔	✔	✔	✔	✔
PVC schedule 40	✔	✔	✔	✔	✔	✔
PVC 3¼ in. O.D.	✔	Ø	✔	Ø	✔	Ø
Cu tubing K or L	✔	Ø	✔	Ø	✔	Ø
Cu tubing M	✔	Ø	✔	Ø	Ø	Ø
Cu tubing DWV	✔	✔	✔	✔	Ø	✔
Cast-iron hubless	✔	✔	✔	✔	✔	✔
Galvanized steel[A]	✔	✔	Ø	Ø	Ø	Ø
Vitrified clay	Ø	Ø	Ø	Ø	✔	✔
ABS/PVC SDR 35[B]	Ø	Ø	Ø	Ø	✔	Ø

A. Maintain at least 6 in. above soil.
B. The IRC also accepts other plastics < schedule 40.

DRAIN SIZING

Drains must be sized to carry the maximum anticipated simultaneous load. Begin by drawing an isometric diagram of all the fixtures, and assign each the number of drainage fixture units from **T6**. Starting at the highest point of the system and working down to the building drain, size each pipe for the number of DFUs). Note that the IRC allows the use of bathroom and kitchen groups that take into account some fixtures not being used simultaneously. Though kitchen sink trap arms may be 1½ in., the vertical drain must be at least a 2 in. diameter pipe.

TABLE 6 — DFUs & TRAP SIZE [T3004.1 & T3201.7] {T7-3}

Fixture	DFUs		Min. Trap Size (in.)	
	IRC	UPC	IRC	UPC
Bar sink	1	1	1½	1½
Bathtub	2	2	1½	1½
Bidet (1½ in. outlet)	1	1	1¼	1¼
Bidet	1	2	1½	1½
CW standpipe	2	3	2	2
DW (not on KS trap)	2	2	1½	2
Floor drain	0	0	2	2
KS[A]	2	2	1½	1½
Lavatory	1	1	1¼	1¼
LT	2	2	1½	1½
Single head shower stall	2	2	1½	2
Additional shower heads	2	1	Note B	2
WC 1.6 GPF	3	3	n/a	n/a

(table continues on next page)

FIG. 8 — Durham Fittings

Short sweep cast iron Durham fitting

Water supply elbow

Durham systems include sections of threaded galvanized steel piping that must be kept at least 6 in. above soil to prevent rust. Durham systems have a smooth interior waterway, whereas threaded water supply pipes have an internal ledge. Durham & cast-iron systems have the advantage of being quieter than plastics, & are often used for upper floors of a dwelling, with plastics on the 1st floor.

FIG. 6 — Closet Bend Reductions

4 in. closet flange

3 in. reduction

FIG. 7

3 in. arm

Smooth 4 × 3 reducer OK

No fittings with internal ledge

TABLE 6 — DFUs & TRAP SIZE (CONT.) [T3004.1 & T3201.7] {T7-3}

Fixture	DFUs IRC	DFUs UPC	Min. Trap Size (in.) IRC	Min. Trap Size (in.) UPC
Full-bath group	5	5[c]	n/a	n/a
Half-bath group	4	3[c]	n/a	n/a
Mutliple-bath groups (1 ½)	7	n/a	n/a	n/a
Multiple-bath groups	Note D	n/a	n/a	n/a
Kitchen group	2	n/a	n/a	n/a
Laundry group	3	n/a	n/a	n/a

A. With or without DW or food waste grinder.
B. The IRC bases the trap size on the flow rate. > 5.7 gpm & ≤ 12.3 gpm = 2 in., ≤ 25.8 gpm = 3 in.
C. The UPC does not have bath groups in the main code text. They are included in appendix L, which reqs local adoption to be in effect.
D. For each additional bath beyond 1 ½ baths, add 1 DFU per half bath, 2 DFUs per full bath.

TABLE 7 — MAX. DFUs ON BRANCHES & STACKS [T3005.4.1] {T7-5}

Pipe Size (in.)	IRC DFUs Horizontal	IRC DFUs Vertical	UPC DFUs Horizontal	UPC DFUs Vertical
1 ¼[A,B]	1	1	1	1
1 ½[B]	3	4	1	2[c]
2[B]	6	10	8	16
2 ½[B]	12	20	14	32
3	20	48	35[D]	48
4	160	240	216	256

A. 1 ¼ in. pipe is limited to a single fixture drain or trap arm.
B. Drains < 3 in. may not receive discharge from WCs.
C. No sinks, urinals, or DW > 1 DFU.
D. Max. 3 WCs on any horizontal branch or drain.

FIG. 9

Drains Entering at Same Level

Back-to-back fitting

Double sanitary tee

A back-to-back fixture fitting should be used for fixtures or trap arms entering at the same level. The IRC also allows a double sanitary tee to be used for this purpose, while the UPC allows it only for branch drains entering at the same level & into a barrel that is a minimum of 2 pipe sizes larger than the inlets.

TABLE 8 — IRC DFUs ON BUILDING DRAIN & BRANCHES [T3005.4.2]

Pipe size (in.)	Slope (in./ft.) ⅛	Slope (in./ft.) ¼	Slope (in./ft.) ½
1 ½[A]	n/a	Note A	Note A
2[B]	n/a	21	27
2 ½[B]	n/a	24	31
3	36	42	50
4	180	216	250

A. 1 ½ in. horizontal branches to building drains limited to 1 pumped fixture (included food waste grinder) or 2 non-pumped fixtures.
B. Drains < 3 in. may not receive discharge from WCs.

DRAIN SIZING

PLUMBING

Fittings & Connections

	09 IRC	09 UPC
☐ Changes in direction req appropriate fittings **F9–12, T9** [3005.1]		{706.1}
☐ Use double sanitary tees or equivalent for 2 fixture inlets at same level **F9** _____ [3005.1.1]		{706.2}
☐ Double sanitary tee barrel min 2 sizes larger than inlets **F9** __[n/a]		{706.2}
☐ No horizontal-horizontal fittings within 10 pipe diameters downstream of stack base or horizontal offset_____ [3005.5]		{n/a}

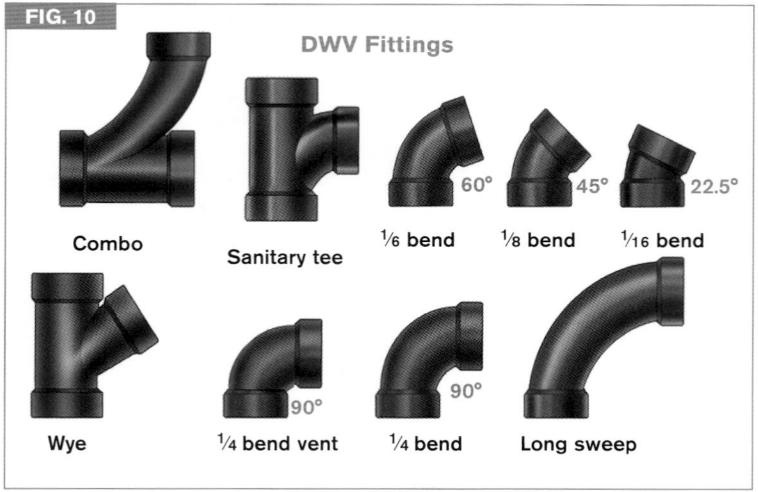

TABLE 9	FITTINGS FOR CHANGE OF DIRECTION [T3005.1] {706}		
Fitting	Horizontal to Vertical	Vertical to Horizontal	Horizontal to Horizontal
1/16 bend	✔	✔	✔
1/8 bend	✔	✔	✔
1/6 bend	✔	✔	IRC ✔ • UPC Ø
1/4 bend	✔	IRCᴬ • UPC Ø	IRCᴬ • UPC Ø
Short sweep (cast iron)	✔	✔ ᴮ	✔ ᴬ
Long sweep	✔	✔	✔
Sanitary tee	✔ ᶜ,ᴰ	Ø	Ø
Wye	✔	✔	✔
Combo wye & 1/8 bend	✔	✔	✔

A. IRC max. 2 in. diameter.
B. IRC fixture drain max 2 in. diameter, fitting min. 3 in. diameter.
C. Double sanitary tees not to receive discharge from pumped waste or from WCs unless min. 18 in. between WC & fitting.
D. Double sanitary tees in UPC must have barrel 2 pipes sizes larger than inlets.

FIG. 10

DWV Fittings

Combo · Sanitary tee · 1/6 bend 60° · 1/8 bend 45° · 1/16 bend 22.5° · Wye · 1/4 bend vent 90° · 1/4 bend 90° · Long sweep

FIG. 11

Sanitary Tees

Not OK on back

OK only for horizontal to vertical

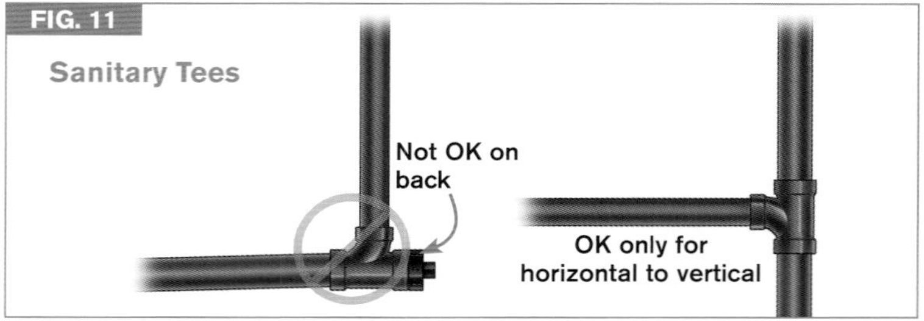

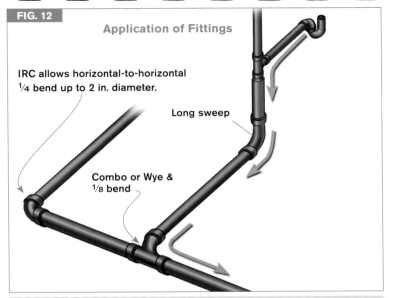

FIG. 12

Application of Fittings

IRC allows horizontal-to-horizontal ¼ bend up to 2 in. diameter.

Long sweep

Combo or Wye & ⅛ bend

CLEANOUTS

Cleanouts are necessary for clearing drain obstructions and for inspecting the building sewer with a sewer camera. While each code allows cleanouts to be in crawl spaces, a preferred method is to extend them to an area that is more readily accessible. The UPC requires that underfloor cleanouts be no farther than 20 ft. from the access opening, with a 30 in. wide, 18 in. high path from the access to the cleanout. When those conditions cannot be met, the cleanout must be extended to the exterior. The IRC allows drain cleaning through a removable fixture trap or by removing the toilet.

IRC Cleanout (CO) Requirements
09 IRC

- ☐ COs liquid & gas tight; plugs brass or plastic _____ [3005.2.1]
- ☐ Size same as drain pipes up to 4 in. diameter EXC_____ [3005.2.9]
  - Traps as CO OK 1 size smaller than drain (e.g., kitchen) _____ [3005.2.9X1]
  - CO in stacks OK 1 size smaller than stack _____ [3005.2.9X1]
- ☐ Removable trap OK as CO _____ [3005.2.10]
- ☐ COs req'd not > 100 ft. apart in each horizontal drain line_____ [3005.2.2]
- ☐ COs for underground drains req extensions above grade **F17** _____ [3005.2.3]
- ☐ Req'd in horizontal drains, building drains & building sewer for each change of direction > 45° **F14** EXC _____ [3005.2.4]
  - Only 1 cleanout req'd per 40 ft. of run **F14** _____ [3005.2.4]
- ☐ Req'd at base of all stacks _____ [3005.2.6][1]
- ☐ Junction of building drain & building sewer req's CO brought to finished grade or lowest floor level EXC_____ [3005.2.7]
  - CO can be on 3 in. soil stack within 10 ft. _____ [3005.2.7]
- ☐ 2-way CO at junction of building drain & building sewer can be req'd CO for both _____ [3005.2.7]
- ☐ Install COs to allow cleaning in direction of flow_____ [3005.2.8]
- ☐ Pipes < 3 in. req 12 in. clearance; 3 in. req's 18 in. _____ [3005.2.5]
- ☐ CO openings not OK for new fixtures without new CO **F15** _____ [3005.2.11]

UPC Cleanout Requirements
09 UPC

- ☐ COs liquid & gas tight _____ {707.3}
- ☐ Plugs brass or plastic with raised head or countersunk slot_____ {707.1}
- ☐ Size: 2 in. pipe = 1½ in.; 2½ or 3 in. pipe = 2½ in.; 4 in. pipe = 3½ in. {707.10}
- ☐ Req'd at upper terminal of all horizontal runs **F14** EXC _____ {707.4}
  - Horizontal runs < 5 ft. (unless serving sinks or urinals)_____ {707.4X1}
  - Horizontal pipes 72° from vertical (⅕ bend) _____ {707.4X2}
  - Pipes above lowest floor of building _____ [707.4X3]
  - No upper terminal CO req'd if 2-way CO at junction of building drain & building sewer **F13** _____ {707.4X4}

UPC Cleanout Requirements (cont.)   09 UPC

- ☐ Req'd for runs with aggregate change of direction > 135° **F14** _____ {707.4}
- ☐ Trap arm bends < 90° do not req CO _____ {707.14}
- ☐ Takeoff above flow line unless wye branch or end of line **F30** _____ {707.5}
- ☐ Pipes 2 in. req 12 in. clearance; > 2 in. req's 18 in. clearance _____ {707.10}
- ☐ Underfloor CO must extend above finished floor or outside building if > 20 ft. from access door or if < 18 in. vertical clearance or if passageway to CO < 30 in. wide _____ {707.9}

FIG. 14

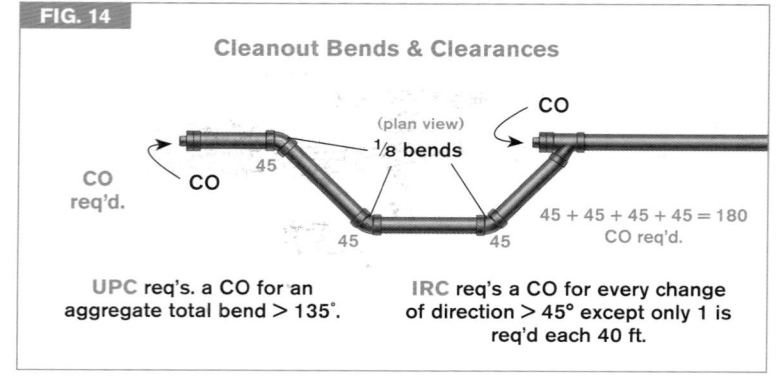

Cleanout Bends & Clearances

(plan view)

CO req'd.

CO   CO   ⅛ bends   CO

45   45   45   45 + 45 + 45 + 45 = 180 CO req'd.

UPC req's. a CO for an aggregate total bend > 135°.

IRC req's a CO for every change of direction > 45° except only 1 is req'd each 40 ft.

FIG. 13

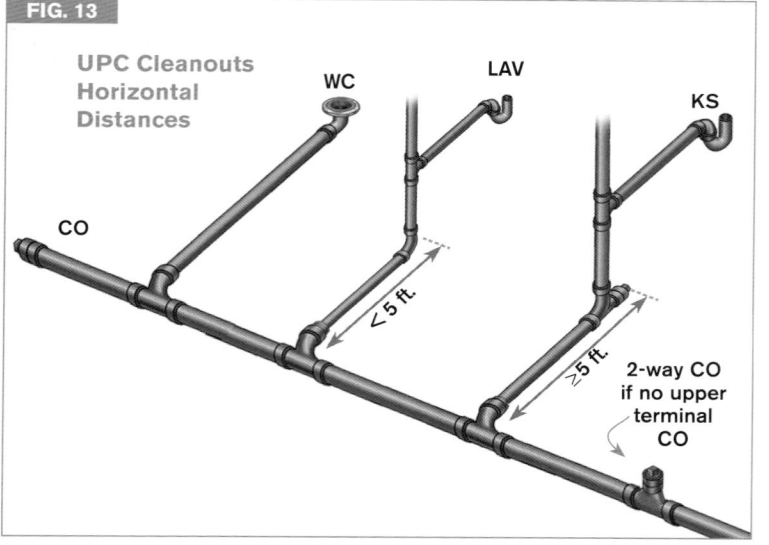

UPC Cleanouts Horizontal Distances

WC   LAV   KS

CO

< 5 ft.

≥ 5 ft.

2-way CO if no upper terminal CO

FIG. 15

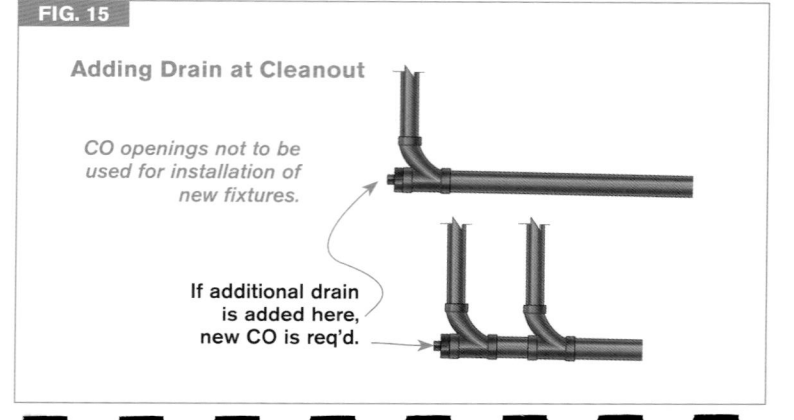

Adding Drain at Cleanout

CO openings not to be used for installation of new fixtures.

If additional drain is added here, new CO is req'd.

WASTE STACKS & VENTS

A waste stack provides a convenient way of discharging clustered fixtures on different floor levels. Toilets and urinals may not discharge into the waste stack. The stack must be undiminished in size to its vent, & offsets are not allowed unless at least 6 in. above the highest fixture draining to the stack. The principle of a waste stack is that water will travel on the walls of the pipe, leaving the center with air to function as a vent. The UPC does not allow waste stacks and vents in single-family residential structures; there is a potential for suds in the lower fixture, or for air pressures from a falling slug of water to create pressure across the lower trap seal. The IRC addresses this by limiting the DFUs discharging into one branch interval.

General

	09 IRC	09 UPC
☐ Waste stack must be vertical with no offsets _____	[3109.2]	{Ø}
☐ No WCs or urinals allowed on waste stack _____	[3109.2]	{Ø}
☐ Stack vent above waste stack may have offsets _____	[3109.3]	{Ø}
☐ Stack vent offsets min 6 in. above FLR off highest fixture	[3109.3]	{Ø}
☐ Waste & vent stack same size for entire length **F16** __	[3109.3&4]	{Ø}
☐ Size waste stack per total DFUs discharging into it **T10**	[3109.4]	{Ø}

TABLE 10	IRC WASTE STACK VENT SIZE [T3109.4]	
Stack Size (in.)	Total Discharge into 1 Branch Interval (DFUs)	Total Discharge for Stack (DFUs)
1½	1	2
2	2	4
2½	No limit	8
3	No limit	24
4	No limit	50

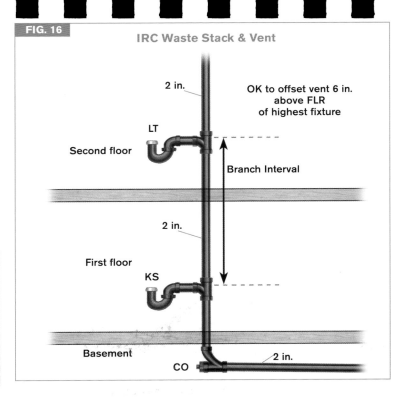

FIG. 16

IRC Waste Stack & Vent

2 in.

OK to offset vent 6 in. above FLR of highest fixture

LT

Second floor

Branch Interval

2 in.

First floor

KS

Basement

CO

2 in.

FIXTURES BELOW MANHOLE COVER OR SEWER

Fixtures above the sewer will drain by gravity. If the fixtures or their drains are also below the manhole cover, sewage could back up to the fixtures when the sewer is blocked. Backup can be prevented by installing backwater valves on the drain lines below the manhole cover. Backwater valves must remain accessible for maintenance. Cleanouts that could lead to a backwater valve should be labeled to avoid damage to the backwater valve.

Backwater Valves F17   09 IRC   09 UPC

☐ Fixtures below next upstream manhole req backwater valve (measured from FLR in IRC, from floor level in UPC) ___ [3008.1] {710.1}

☐ Fixtures above elevation of manhole cover not allowed to discharge through backwater valve ___ [3008.1] {710.1}

☐ Backwater valves req'd to be accessible for service ___ [3008.5] {710.6}

☐ COs for drains through backwater valve req label ___ [n/a] {710.1}[2]

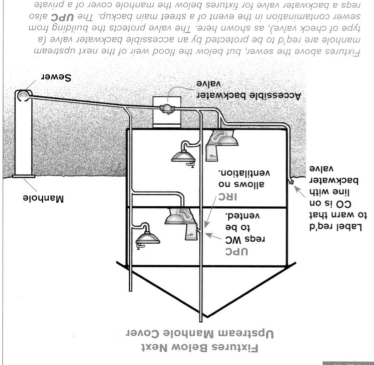

FIG. 17

Fixtures Below Next Upstream Manhole Cover

Fixtures above the sewer, but below the flood weir of the next upstream manhole are req'd to be protected by an accessible backwater valve (a type of check valve), as shown here. The valve protects the building from sewer contamination in the event of a street main backup. The UPC also reqs a backwater valve for fixtures below the manhole cover of a private sewer system.

Fixtures Below Sewer

	09 IRC	09 UPC
☐ Fixtures to drain by gravity where practical _____	[3007.1]	{709.0}
☐ Sump discharge must be lifted above gravity drain _____	[3007.1]	{710.2}
☐ Connect to wye in top of horizontal gravity drain **F19** ___	[3007.3.5][3]	{710.4}
☐ Backwater valve req'd on ejector discharge pipe **F18,19**	[3007.2]	{710.3.2}
☐ Gate valve req'd on discharge side of check valve **F18,19**	[3007.2]	{710.3.2}
☐ Ball valves OK for ejectors not serving WCs _____	[n/a]	{710.4}
☐ Valve bodies cast iron or brass (no plastic ball valves) _____	[n/a]	{710.4}
☐ Sumps req's water & gas tight removable cover **F18** __	[3007.3.2]	{710.10}
☐ Min pump capacity 21 gpm {20 gpm in UPC} _____	[3007.6][4]	{710.3.1}
☐ Min 2 in. discharge piping **T11** EXC _____	[T3007.6][4]	{710.3.2}
• Grinder pumps min 1¼ in. discharge _____	[3007.6]	{710.12}
☐ Gravity drains receiving discharge from ejector sized at 1.5 DFU for each gpm of pump {2 DFU in UPC} **F19** _____	[T3004.1]	{710.5}

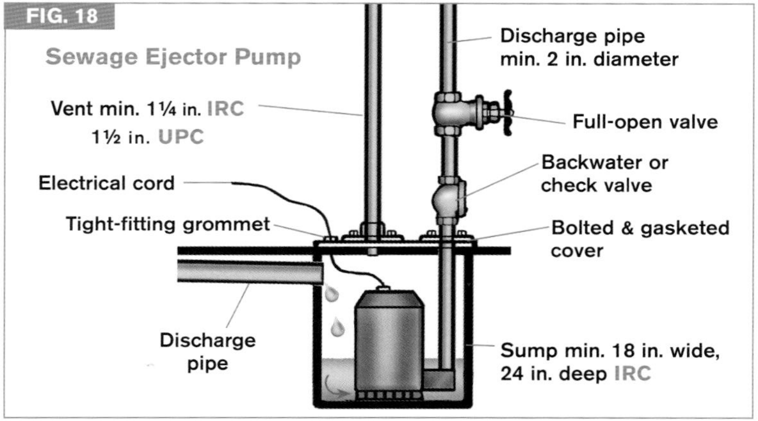

TABLE 11 — MINIMUM EJECTOR PUMP CAPACITY [T3007.6]

Discharge Pipe Diameter	Pump Capacity
2 in.	21 gpm
2½ in.	30 gpm
3 in.	46 gpm

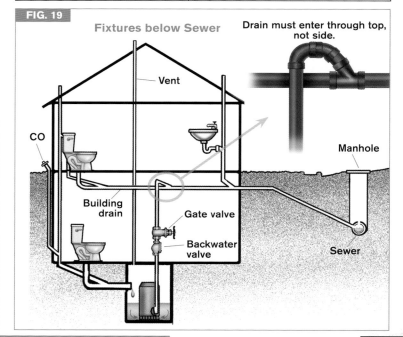

FIG. 19

Fixtures below Sewer

Drain must enter through top, not side.

Vent

CO

Manhole

Building drain

Gate valve

Backwater valve

Sewer

FIG. 18

Sewage Ejector Pump

Discharge pipe min. 2 in. diameter

Vent min. 1¼ in. IRC
1½ in. UPC

Full-open valve

Electrical cord

Backwater or check valve

Tight-fitting grommet

Bolted & gasketed cover

Discharge pipe

Sump min. 18 in. wide, 24 in. deep IRC

FIXTURES BELOW MANHOLE COVER OR SEWER

ON-SITE SEWAGE DISPOSAL SYSTEMS

In addition to codes enforced by building departments, on-site sewage disposal systems may be regulated by the local department of environmental health services. Septic tanks and leach fields must be sized and situated so they do not have an adverse impact on local water supply systems. The minimum areas shown in **T12B** are a general planning guide, and local percolation test data and the area will be followed. The area will vary based on the results of soil percolation test data and the type of construction. The IRC does not directly deal with on-site systems, and the ICC has a separate code for these, the IPSDC. The UPC has an appendix chapter K for such systems.

Septic Tanks   09 IPSDC   09 UPC

☐ Tank in flood hazard area anchored in place   [303.2]   {K1.0D}
☐ Min capacity per T12   [802.7.1]   {K2.0}
☐ Tank min 5 ft. from building   [802.8]   {n/a}

TABLE 12A	SEPTIC TANK CAPACITY [T802.7.1] {T K-2}		
Bedrooms	**Min. Tank Size ISPDC**	**Min. Tank Size UPC**	**Max UPC DFUs**
1-2	750	750	15
3	1,000	1,000	20
4	1,200	1,200	25
5	1,425	1,500	33
6	1,650	1,500	33

TABLE 12B	MIN. ABSORPTION AREA [T603.1]		
Percolation Class	**Percolation Rate**[A]	**Seepage Trenches**[B]	**Seepage Beds**[B]
1	0 to < 10	165	205
2	10 to < 30	250	315
3	30 to < 45	300	375
4	45 to 60	330	415

A. Minutes req'd for water to fall 1 in.
B. Sq. ft. per bedroom.

TRAPS & TAILPIECES

Traps prevent sewer gases, vermin, and other contaminants from entering the dwelling. The trap seal must be a sufficient depth (2 in.) to maintain a seal and not so deep (4 in. max) as to become blocked with sludge or create a siphoning effect. Trap arms (fixture drains) must be vented, otherwise the negative pressure created by water moving down the pipe will cause air to be sucked through the trap seal. The maintenance of proper trap seals is the underlying principle behind the code rules for drainage, traps, and venting.

General   09 IRC   09 UPC

☐ Each fixture reqs separate trap EXC   [3201.6]   {1001.1}
  • Fixtures with integral traps (WCs)   [3201.6X1]   {1001.1}
  • 2 or 3 lavs, LTs, or sinks of same type OK on
    1 center trap in same room   [3201.6X2]   {1001.2}
  • Laundry trap may drain to CW standpipe   [3201.6X3]   {n/a}
  • Fixtures sharing trap max 30 in. apart horizontal   [3201.6X2]   {1001.2}
☐ Trap seal min 2 in., max 4 in. F20   [3201.2]   {1005.0}
☐ Set traps level & protect from freezing   [3201.3]   {1005.0}
☐ No S traps, bell traps, drum traps, traps with moving parts,
   or traps with interior partitions EXC F21   [3201.5]   {1004.0}
  • Lav traps with plastic or stainless partitions.   [3201.5]   {1004.0}
☐ Size per T6   [3201.7]   {1003.3}
☐ Trap size 2 fixture outlet size   [3201.7]   {1003.3}
☐ No double traps (in series)   [3201.6]   {1004.0}

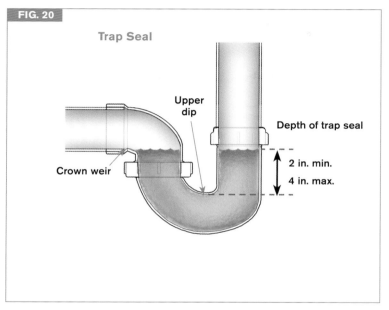

FIG. 20

Trap Seal

Upper dip

Crown weir

Depth of trap seal

2 in. min.

4 in. max.

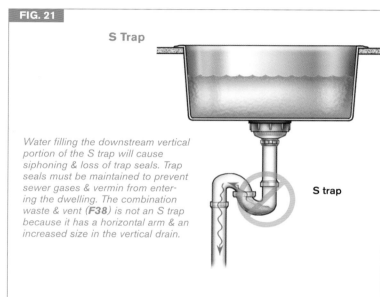

FIG. 21

S Trap

*Water filling the downstream vertical portion of the S trap will cause siphoning & loss of trap seals. Trap seals must be maintained to prevent sewer gases & vermin from entering the dwelling. The combination waste & vent (**F38**) is not an S trap because it has a horizontal arm & an increased size in the vertical drain.*

S trap

TRAPS & TAILPIECES

PLUMBING

Fixture Tailpieces

	09 IRC	09 UPC
☐ Fixture tailpiece max 24 in. vertical distance EXC F22	[3201.6]	{1001.4}
• CW standpipes 18–42 in. {18–30 in. UPC} F62	[2706.2]	{804.1}
☐ IRC: Max 30 in. horizontal distance F22	[3201.6]	{Ø}
☐ UPC: Max 24 in. total developed length F22	[n/a]	{1001.4}
☐ Directional fittings req'd for continuous wastes from disposer or DW (i.e., wyes, combos, or tees with baffles) F22,23	[2707.1]	{404.4}

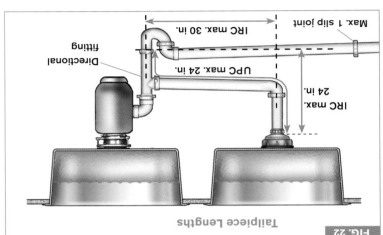

FIG. 22

Tailpiece Lengths

Max 1 slip joint

IRC max. 30 in.

Directional fitting

UPC max. 24 in.

IRC max. 24 in.

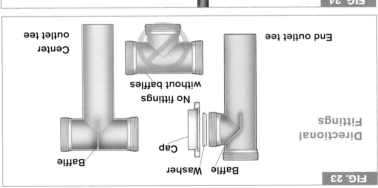

FIG. 23

Directional Fittings

Center outlet tee

End outlet tee

No fittings without baffles

Baffle

Cap

Washer

Baffle

FIG. 24

Slip Joints & Access

Slip joints

Overflow drain

An access opening at least 12 in. × 12 in. is req'd for repair or replacement of concealed slip joints. The opening can be in a ceiling or a wall.

Trap Arms

	09 IRC	09 UPC
☐ Trap same size as trap arm_____	[3201.7]	{1003.3}
☐ Trap arm length min 2× trap arm diameter F27 _____	[3105.3]	{1002.2}
☐ Trap arm length & slope per table EXC T13_____	[3105.1]	{1002.2}
• Trap arm length from WC unlimited (6 ft. in UPC) ___	[3105.1X]	{T10-1}
☐ Trap arms < 3 in. diameter min slope ¼ in./ft. _____	[3005.3]	{708.0}
☐ Total fall of trap arm max 1 pipe diameter F25 _____	[3105.2]	{n/a}
☐ Only 1 trap permitted on trap arm EXC _____	[3201.6]	{1001.1}
• 2 trap arms allowed to join through double-wye fitting to common vent F34_____	[3107.1&2]	{Ø}
☐ Tubing traps req trap adapter F26_____	[n/a]	{1003.2}
☐ Max 1 slip joint allowed on outlet side of trap F22 _____	[n/a]	{1003.2}
☐ CO req'd if direction change > 90° in < 3 in. arm_____	[n/a]	{1002.3}
☐ Slip joints req'd to be accessible F24 _____	[3201.1]	{404.2}

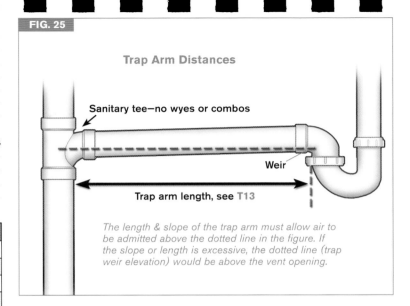

FIG. 25

Trap Arm Distances

Sanitary tee—no wyes or combos

Weir

Trap arm length, see **T13**

The length & slope of the trap arm must allow air to be admitted above the dotted line in the figure. If the slope or length is excessive, the dotted line (trap weir elevation) would be above the vent opening.

TABLE 13	TRAP ARM DISTANCE TO VENT [T3015.1] {T10-1}		
Trap Arm Diameter	**Min.**	**IRC Max.**	**UPC Max.**
1¼ in.	2½ in.	5 ft.	2 ft. 6 in.
1½ in.	3 in.	6 ft.	3 ft. 6 in.
2 in.	4 in.	8 ft.	5 ft.
3 in.[A]	6 in.	12 ft.	6 ft.
4 in. or larger[A]	8 in.	16 ft.	10 ft.[B]

A. In the IRC, these arms can have ⅛ in./ft. slope. In the UPC, all arms must slope ¼ in./ft.
B. The maximum length from a WC to the vent is 6 ft. in the UPC & unlimited in the IRC.

VENTS

Vents prevent atmospheric pressure differences across traps and are essential to maintaining the trap seal. Without vents, the water in the seal could be sucked out, leaving the occupants unprotected from contaminants downstream of the trap. The IRC and UPC have very different approaches to venting.

General	09 IRC	09 UPC
☐ All fixture traps req venting	[3101.2.1]	{901.0}
☐ Vent system not to be used for any other purposes	[3101.3]	{n/a}
☐ No flat dry vents (take off above horizontal centerline) F30	[3104.3]	{905.2}
☐ Slope vents to drain to soil or waste piping	[3104.2]	{905.1}
☐ Change direction with appropriate fittings F29	[3104.2]	{903.3}
☐ No vent opening below trap weir except WCs F25	[3105.2]	{905.5}
☐ No crown vents: min 2 pipe diameters from trap F27	[3105.3]	{1002.2}
☐ Horizontal dry vents min 6 in. above FLR F28	[3104.4]	{905.3}
☐ Horizontal branch vents min 6 in. above FLR F28	[3104.5]	{905.3}
☐ Piping < 6 in. above FLR req's drainage type fittings	[n/a]	{905.3}

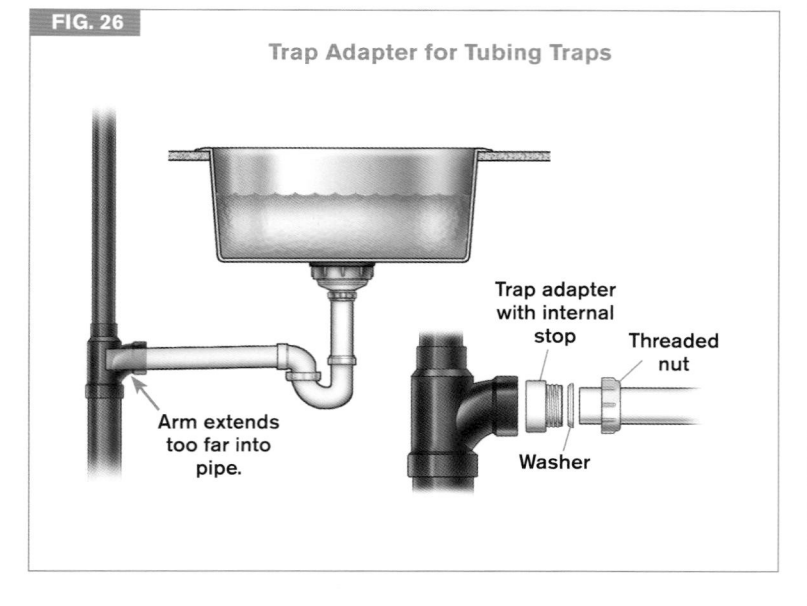

FIG. 26

Trap Adapter for Tubing Traps

Trap adapter with internal stop

Threaded nut

Washer

Arm extends too far into pipe.

FIG. 27

Crown Venting

Too close, must be at least 2 pipe diameters

Improper application of sanitary tee; cannot be placed on back

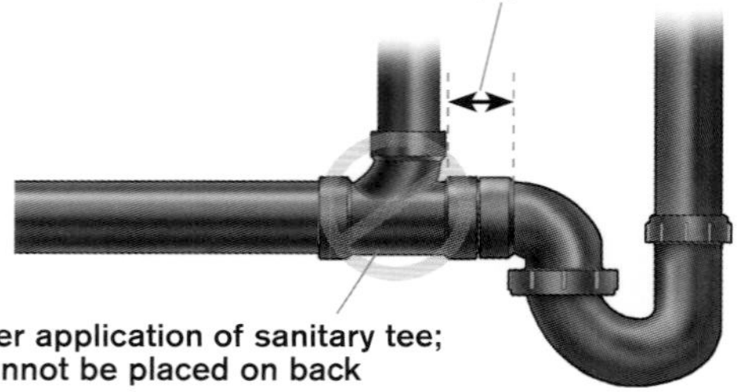

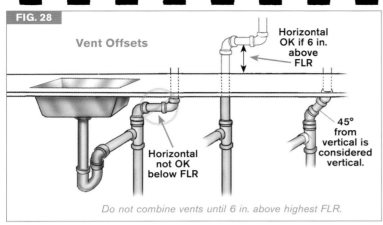

FIG. 28

Vent Offsets

Horizontal OK if 6 in. above FLR

Horizontal not OK below FLR

45° from vertical is considered vertical.

Do not combine vents until 6 in. above highest FLR.

FIG. 30

Vent Takeoff above Center Line

≥45°

Vent takeoffs must be at least 45° from horizontal. A direct horizontal takeoff could be blocked by effluent in the pipe, which would make the vent ineffective.

FIG. 29

Branch Vents

¼ bend

Sanitary tee

*Two or more individual vents may combine to form a branch vent, thereby reducing the number of penetrations through the roof. Horizontal piping must still have ¼ in./ft. of slope, & fittings must be oriented in the direction of water flow. Refer to **T14** for sizing of the branch vent.*

TABLE 14	UPC VENT SIZE AND LENGTH {T7-5}					
Pipe size (in.)	1¼	1½	2	2½	3	4
Max. DFUs	1	8	24	48	84	256
Max. length (ft.)^A	45	60	120	180	212	300

A. Max. horizontal length is ⅓ of total length. If pipes are increased 1 size, the length limitations of this table do not apply.

VENTS

PLUMBING

Size

	09 IRC	09 UPC
☐ Min size 1¼ in.	[3113.1]	{904.1}
☐ Vents min ½ size of drain served	[3113.1]	{904.1}
☐ Size per no. of DFUs served & length of vent **T14**	[n/a]	{904.1}
☐ Increase 1 pipe size if developed vent length > 40 ft.	[3113.1]	{n/a}
☐ Increase 1 pipe size if > ⅓ of vent is horizontal	[n/a]	{904.2X}
☐ Waste stack vent same size as waste stack **F16**	[3109.3]	{n/a}
☐ Total area of vents ≥ to size of building sewer **T15**	[n/a]	{904.1}
☐ Vents for fixtures discharging through pumps, ejectors, or backwater valves do not meet area requirement	[n/a]	{904.1}

Vertical Wet Venting

	09 IRC	09 UPC
☐ All wet vented fixtures to be on same story	[3108.4]	{908.1}
☐ Size wet vent per DFUs of upper drains **T16**	[3108.3]	{n/a}
☐ WCs at same elevation & below other drains	[3108.4]	{n/a}
☐ Each fixture drain connect independently to wet vent	[3108.4]	{n/a}
☐ Limited to trap arms of 1-DFU & 2-DFU fixtures	[n/a]	{908.1}
☐ Max 4 fixtures	[n/a]	{908.1}
☐ 6 ft. max developed length of wet vent **F32**	[n/a]	{908.1}
☐ Wet vent min 1 pipe size larger than req'd waste & min 2 in.	[n/a]	{908.2}

TABLE 15	**UPC VENT AREA FILL-IN TABLE** {904.1}

Vent Size (in.)	Area (sq. in.)	No. of Vents	Net Vent Area
1¼	1.23		
1½	1.77		
2	3.14		
3	7.07		
4	12.57		
	TOTAL		

Example:
4 in. building drain = 12.57 sq. in.
One 1¼ in. vent = 1.23 sq. in.
Three 2 in. vents = 9.42 sq. in.
1.23 + 9.42 = 10.65 sq. in.
Thus more venting would be req'd.

FIG. 31
Common Venting IRC
1¼ in.
KS
1½ in.
Size per **T16** based on DFUs of upper fixture
2 in.
CW
2 in.

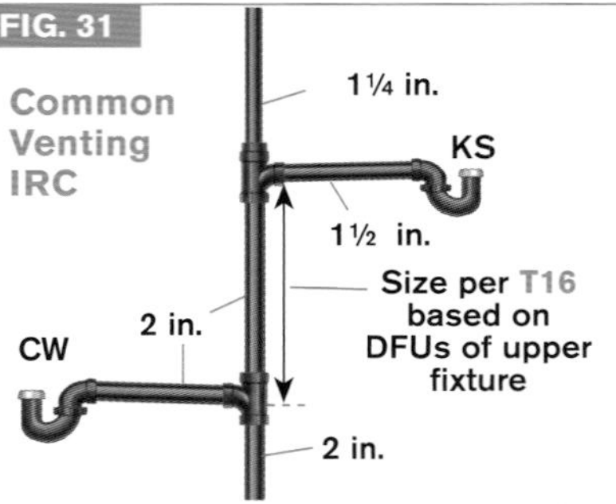

FIG. 32
Wet Venting UPC
1½ in.
Lav
2 in.
1 pipe size larger than req'd drain
Max. 6 ft.
Shower
2 in.

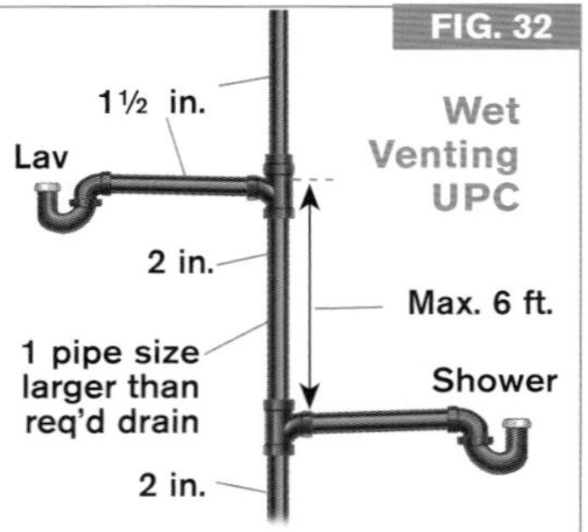

FIG. 33
Common Venting UPC
Back-to-back

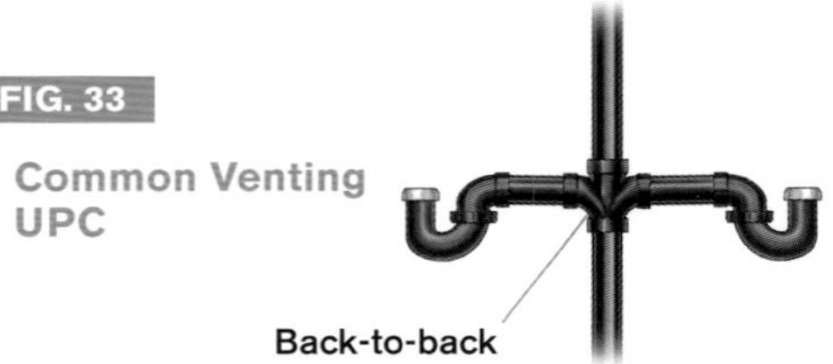

*UPC common vented fixtures must enter through back-to-back sanitary tee that has a barrel size 2 sizes larger than inlets— see **F9**.*

Common Vent

	09 IRC	09 UPC
☐ OK only for fixtures on same floor level **F31,33**	[3107.1]	{905.6}
☐ Max 2 fixtures to vertical common vent	[3107.1]	{905.6}
☐ UPC must connect at approved double fitting **F9**	[n/a]	{905.6}
☐ Connect fixtures to vertical vent at same level **F33** EXC	[3107.2]	{905.6}
• Vent connection downstream OK in IRC **F34**	[3107.2]	{Ø}
☐ Size common vent per DFUs of upper drain **T16**	[3107.3]	{n/a}
☐ Upper fixture cannot be WC	[3107.3]	{908.1}

TABLE 16 — IRC COMMON VENT SIZES [T3107.3]

Pipe Size (in.)	Max. Discharge from Upper Fixture Drain (Fixture Units)
1½	1
2	4
2½ to 3	6

Vent Termination

	09 IRC	09 UPC
☐ At least 1 vent must extend outdoors (all vents UPC)	[3102.1]	{906.1}
☐ Req'd IRC exterior vent must be dry vent	[3102.2]	{n/a}
☐ Req'd IRC exterior vent not an island fixture vent F36	[3102.2]	{n/a}
☐ Req'd IRC exterior vent min ½ size of building drain	[3102.3]	{n/a}
☐ Vents through roof min 6 in. above roof	[3103.1]	{906.1&3}
☐ Min 12 in. horiz from adjacent vertical surfaces	[n/a]	{906.1}
☐ Min 7 ft. above roof used as deck	[3103.1]	{906.3}
☐ Min 7 ft. above roof if within 10 ft. horizontal of roof deck	[n/a]	{906.3}
☐ Min 2 ft. (3 ft. UPC) above openings within 10 ft. EXC	[3103.5]	{906.2}
• OK if 4 ft. below building openings	[3103.5]	{Ø}
☐ Min 3 ft. distance from PL	[n/a]	{906.2}
☐ Provide flashing for roof penetrations	[3103.3]	{906.5}
☐ Vent pipes secured in approved manner when anchoring flagpoles, aerials, or similar items (OK only when roof used for other than weather protection in UPC)	[3103.4]	{906.3}

Snow or Frost Closure

	09 IRC	09 UPC
☐ Applies when min design temperature < 0°F	[3103.2]	{906.7}
☐ Min 3 in. diameter (2 in. UPC) termination	[3103.2]	{906.7}
☐ Transition to larger diameter min 1 ft. below roof	[3103.2]	{906.7}
☐ 6 in. above snow line (UPC 10 in. above roof or per AHJ)	[3103.1]	{906.7}

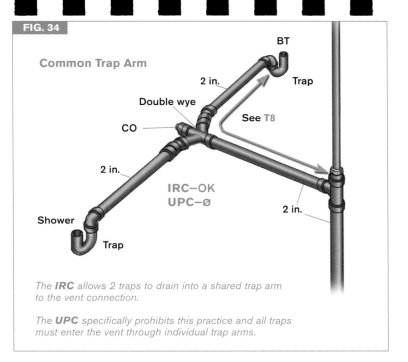

FIG. 34

Common Trap Arm

BT · Trap · 2 in. · Double wye · See T8 · CO · 2 in. · IRC–OK · UPC–ø · 2 in. · Shower · Trap

*The **IRC** allows 2 traps to drain into a shared trap arm to the vent connection.*

*The **UPC** specifically prohibits this practice and all traps must enter the vent through individual trap arms.*

SPECIAL VENTING SYSTEMS

The IRC offers a number of options for island sinks, including air admittance valves, combination waste and vent, and the loop vents shown in **F35 & F36**. The UPC allows only the method shown in **F35** unless specific approval is obtained from the AHJ.

Island Sinks	09 IRC	09 UPC
☐ Island venting limited to sinks (disposer OK) & lavs ____ [3112.1]		{909.0}
☐ Island vented with drainage pattern fittings only **F35,36**__[3112.3]		{909.0}
☐ Island vent above fixture drain outlet **F36** (UPC: as high as possible **F35**) before returning downward _____ [3112.2]		{909.0}
☐ Lowest part of island vent shall connect full size to vertical drain or top half of a horizontal drain **F36**_____ [3112.3]		{Ø}
☐ COs req'd in island vents & drains **F35,36** _____ [3112.3]		{909.0}
☐ Connect island vent downstream of fixture drain **F36**_____ [n/a]		{909.0}
☐ Foot vent req'd through wye branch off below-floor vent **F35** [n/a]		{909.0}
☐ CO req'd in vertical section of foot vent **F35**_____ [n/a]		{909.0}
☐ No upstream fixtures on drain serving island _____[n/a]		{909.0}

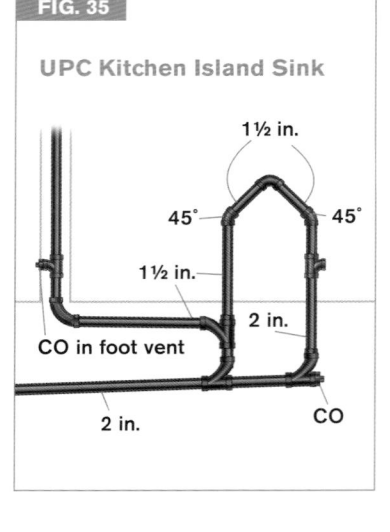

FIG. 35

UPC Kitchen Island Sink

1½ in.
45°   45°
1½ in.
CO in foot vent
2 in.
2 in.   CO

FIG. 36

IRC Kitchen Island Sink

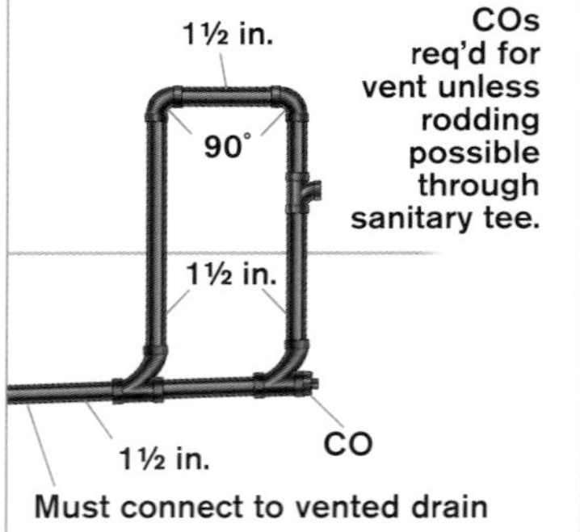

1½ in.
90°
COs req'd for vent unless rodding possible through sanitary tee.
1½ in.
1½ in.   CO
Must connect to vented drain

Combination Waste & Vent (CW&V)	09 IRC	09 UPC
☐ UPC CW&V reqs specific approval from AHJ _____[n/a]		{910.2}
☐ Only sinks (no disposers), lavs & floor drains_____ [3111.1]		{Ø}
☐ 1 vertical pipe (max 8 ft. length) allowed between fixture drain & horizontal CW&V pipe _____ [3111.2]		{Ø}
☐ Max slope of CW&V piping ½ in./ft., min slope ¼ in./ft. [3111.2.1]		{n/a}
☐ CW&V must connect to vented horizontal drain or have vent connected to CW&V pipe **F38** _____ [3111.2.2]		{n/a}
☐ Vent connected to CW&V pipe must rise 6 in. min above fixture FLR before horizontal offsets _____ [3111.2.2]		{n/a}

Horizontal Wet Venting

	09 IRC	09 UPC

☐ Horizontal wet venting OK for any combination of fixtures
within 1 or 2 bathroom groups on same floor **F37** _____ [3108.1] {908.2.1}

☐ Fixtures within bathroom groups only ones allowed
on horizontal wet vent_____ [3108.1] {908.2.1}

☐ Other vented fixtures OK downstream of wet vent _____ [3108.1] {908.2.1}

☐ UPC WC must be downstream of other drain connections __[n/a] {908.2.1}

☐ Dry vent connection to wet vent must be individual
or common vent connected to non-WC fixture EXC __ [3108.2.1] {908.2.2}

• OK for horizontal WC drain to be wet vent _____ [3108.2.1][5] {Ø}

☐ Max trap arm length from trap weir to wet vent **T13** _____ [3108.5] {908.2.1}

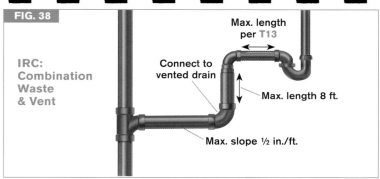

FIG. 38

IRC: Combination Waste & Vent

Max. length per **T13**

Connect to vented drain

Max. length 8 ft.

Max. slope ½ in./ft.

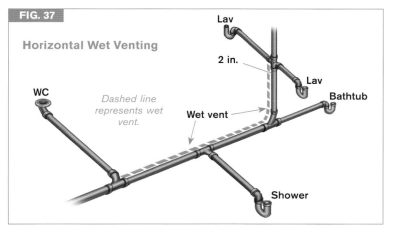

FIG. 37

Horizontal Wet Venting

Lav

2 in.

Lav

WC

Bathtub

Dashed line represents wet vent.

Wet vent

Shower

TABLE 17	MAX. DFUs FOR WET VENTS & CW&V [T3108.3 & 3111.3] & {908.2 & 908.4.3}			
Pipe Size (in.)	Wet Vents		CWV [IRC]	
	IRC	UPC	To Branch	To Building Drain
1½	1	Ø	Ø	Ø
2	4	4	3	4
2½	6	8[A]	6	26
3	12	8[B]	12	31
4	32	8[B]	20	50

A. For horizontal wet vents, max load is 4 DFUs on 2½ in. pipe.
B. More than 8 possible for horizontal wet vents.

SPECIAL VENTING SYSTEMS

PLUMBING

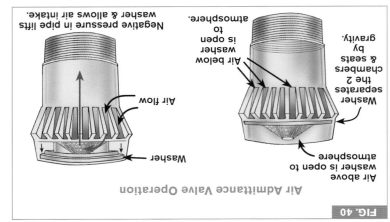

AIR ADMITTANCE VALVES

Air admittance valves (AAVs) operate by gravity, as shown in **F39 & F40**, and have no metal or rubber parts that could corrode or deform. In the IRC they can be used for individual fixtures or for branches. The UPC does not explicitly approve AAVs, although some jurisdictions might accept them under the provisions for Alternate Materials and Methods in 301.2. If the UPC is the code in your area, be sure to check with the local building department before installing AAVs.

General

IRC 09

- [ ] Install after DWV leak test [3114.2]
- [ ] OK at individual, branch, circuit & stack vents **F39** [3114.3]
- [ ] Individual & branch type AAV to vent only fixtures on same floor level [3114.3]⁶
- [ ] & that connect to a horizontal branch drain
- [ ] Individual fixture AAV min 4 in. above fixture drain **F39** [3114.4]
- [ ] Stack-type AAV min 6 in. above FLR of highest fixture [3114.4]
- [ ] AAV within same max distance as conventional vent **T13** [3114.4]
- [ ] AAVs terminating in attic min 6 in. above insulation [3114.4]
- [ ] AAVs must be accessible [3114.5]
- [ ] Space containing AAV must be ventilated [3114.5]
- [ ] Min 1 vent to outdoors [3114.7]
- [ ] Not OK for sumps or tanks without an engineered design [3114.8]⁷

FIG. 39

Air Admittance Valve

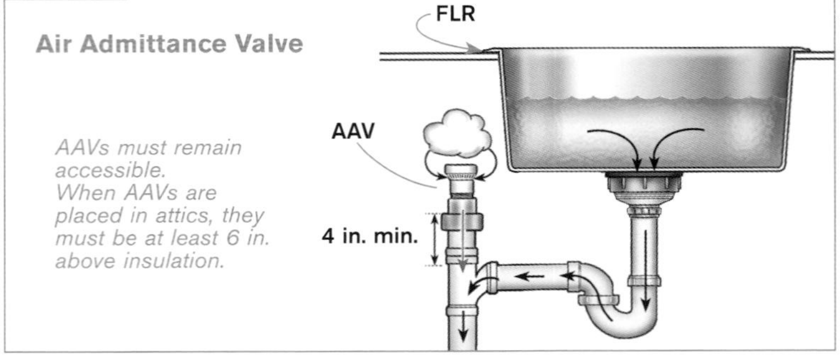

FLR

AAV

AAVs must remain accessible. When AAVs are placed in attics, they must be at least 6 in. above insulation.

4 in. min.

FIG. 40

Air Admittance Valve Operation

Air above washer is open to atmosphere

Washer separates the 2 chambers & seats by gravity.

Air below washer is open to atmosphere.

Washer

Air flow

Negative pressure in pipe lifts washer & allows air intake.

WATER SUPPLY & DISTRIBUTION

Water supply piping must provide an adequate flow of clean potable water and be free of any cross-connections that would introduce contaminants. Piping systems must be protected against damage and movement. Modern plumbing systems often use plastic pipe and tubing, and in some systems the branch piping originates from a central manifold, rather than a traditional series system with a main line and branches.

General

	09 IRC	09 UPC
☐ Use only approved materials **T18**	[2905.4&5]	{604.1}
☐ Proper installation & support req'd **T3,4**	[2605.1]	{314.0}
☐ Min water service ¾ in.	[2903.7]	{610.8}
☐ Pressure tank & pump req'd if supplied pressure < 40psi (15psi in UPC)	[2903.3]	{608.1}
☐ Regulator req'd if pressure > 80psi **F46**	[2903.3.1]	{608.2}
☐ Mechanical water hammer arrestors (not air chambers) req'd near quick-close valves (DW or CW)	[2903.5]	{609.10}
☐ Arrestors to conform to ASSE 1010	[2903.5]	{609.10}
☐ No pipes previously used for other than potable water	[2608.5]	{604.7}

PEX Tubing

	09 IRC	09 UPC
☐ Bend radius AMI **T19**	[manu]	{manu}
☐ Not allowed in 1st 18 in. of WH connections	[manu]	{604.11.2}
☐ ⅜ in. tubing limited to 60 ft. developed length	[2903.8.2]	{manu}
☐ WH fed from end of cold water manifold req's manifold 1 size larger than WH feed	[2903.8.2]	{manu}

TABLE 18	WATER PIPE MATERIALS [T2905.4&5] {T6-4}			
Material	**IRC**		**UPC**	
	Service	Distribution	Service	Distribution
ABS	✔	∅	∅	∅
Brass pipe	✔	✔	✔	✔
CPVC	✔	✔	✔	✔
Cu tubing	✔	✔	✔	✔
Ductile iron	✔	∅	✔	✔
Galvanized steel	✔	✔	✔	✔
PE	✔	∅	✔	∅
PE-AL-PE	✔	✔	✔	✔
PEX	✔	✔	✔	✔
PEX-AL-PEX	✔	✔	✔	✔
PEX-AL-HDPE	✔	✔	∅	∅
PP	✔	✔	∅	∅
PVC	✔	∅	✔	∅
Stainless steel	✔	✔	✔	✔

TABLE 19	RECOMMENDED MIN. BEND RADIUS FOR PEX	
Tubing Size (in. nominal)	**Tubing outer diameter (in.)**	**Bend Radius (in.)[A]**
⅜	½	4
½	⅝	5
¾	⅞	7
1	1⅛	9

A. As recommended by the Plastic Pipe & Fitting Association.

Joints & Connections

	09 IRC	09 UPC
☐ Cu to galvanized steel req's brass or dielectric fitting _[2905.17.1]		{316.2.1}
☐ Cu joints in or under concrete slab on grade within building req brazed wrought-Cu fittings _____[n/a]		{609.3.2}
☐ Slip joints only at exposed fixture supply _____ [2905.6]		{606.2.3}
☐ Unions req'd (within 12 in. in UPC) in WH hookups ___ [2801.3]		{609.5}
☐ Unions req'd 12 in. of softeners, filters, regulators, etc. ____[n/a]		{609.5}

Prohibited Joints

	09 IRC	09 UPC
☐ Joints between different types of plastic req adapter _[2905.17.2]		{316.2.3}
☐ No female threaded PVC fittings_____ [local]		{606.2.2}

Required Valves

	09 IRC	09 UPC
☐ Accessible main valves req'd near entrance_____ [2903.9.1]		{605.2}
☐ Main & WH valve must be full-open type **F41,42** __ [2903.9.1&2]		{605.2}
☐ Throttling valves not OK for main & WH **F43**_____ [2903.9.1&2]		{605.2}
☐ Main valve must be on discharge side of water meter_____ [local]		{605.2}
☐ Main valve must have bleed orifice or separate drain _ [2903.9.1]		{n/a}
☐ Valves req'd on fixture supply except tub & shower___ [2903.9.3]		{605.5}
☐ Valves only at distribution manifold OK if labeled ____ [2903.8.5]		{605.5}
☐ Hose bibbs subject to freezing req valve with drain (stop-and-waste-type) EXC _____ [2903.10]		{313.6}
• Frostproof hose bibbs with stem through insulation _[2903.10X]		{313.6}
☐ Valves req'd on cold water supply at each WH _____ [2903.9.2]		{605.2}
☐ All shutoffs req access _____ [2903.9.3]		{605.6}

FIG. 41 — Gate Valve
FIG. 42 — Ball Valve
FIG. 43 — Globe Valve

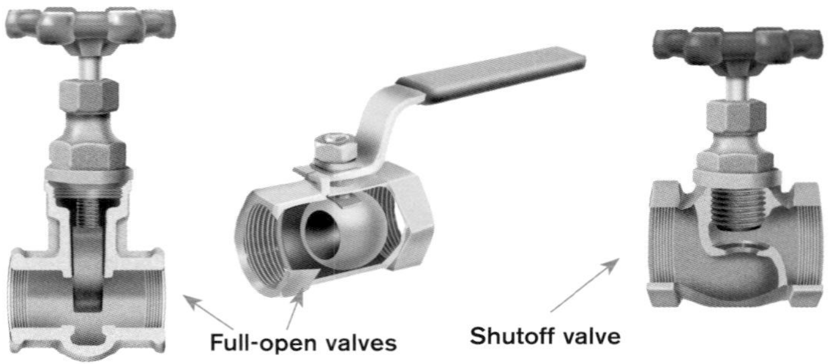

Full-open valves

Shutoff valve

CROSS-CONNECTION CONTROL

Backflow preventers protect water systems from backup and contamination. Vacuum breakers prevent contaminants from entering through systems such as lawn sprinklers. A physical separation in the form of an air gap prevents contamination at waste receptors, such as sinks.

Protection of Potable Water

	09 IRC	09 UPC
☐ Prevent contamination of potable water supply _____ [2902.1]		{602.1}
☐ Connections for private to public water supply prohibited [2902.1]		{602.4}
☐ Reduced pressure principle backflow preventers OK for:		
• Boilers with conditioning chemicals _____ [2902.5.1]		{T6-2}
• Fire-sprinkler systems with additives_____[2902.5.4.1]		{T6-2}
• Lawn irrigation systems with chemical injectors ____ [2902.5.3]		{T6-2}
• Solar heating piping with additives _____ [2902.5.5]		{T6-2}

Protection of Potable Water (cont.)

	09 IRC	09 UPC

☐ Atmospheric vacuum breakers OK for:
- Hose connections (not needed for tank drain valves) [2902.4.3] {603.4.7}
- Swimming pool inlets without an air gap _____ [2902.1] {603.4.5}
- Irrigation system, 6 in. above highest head **F63** ____ [2902.5.3] {T6-2}

☐ Integral air gaps in fixtures to recognized standards OK for:
- Reverse osmosis drinking water treatment units _____ [2908.2] {603.4.13}
- DWs_____ [2717.1] {n/a}
- Pullout spouts & sprayers with integral backflow AMI [2902.4.2] {603.3.7}
- Pull-out or separate shower spray wands _____ [2902.4.2] {603.3.7}
- Flush tank fill valves _____ [2902.4.1] {603.4.2}

☐ Fixture outlet receptor air gaps:
- Min 2× diameter of outlet & per table **F44, T20** ____ [2902.3.1] {603.2.1}

TABLE 20	MIN. REQUIRED AIR GAPS [T2902.3.1] & {T6-3}			
Opening Diameter & Typical Fixtures (in.)	**Not Affected by Side Walls (in.)**		**Affected by Side Walls^A (in.)**	
	IRC	UPC	IRC	UPC
≤½ (lav)	1	1	1½	1½
≤ ¾ (LT)	1½	1½	2½	2¼
≤ 1 (BT)	2	2	3	3
> 1 (pool)	2× diameter	3× diameter	2× diameter	3× diameter

A. Affected by side walls = any time the distance from the spout to the wall is < 3× the diameter of the effective opening, or < 4× the diameter for 2 intersecting walls.

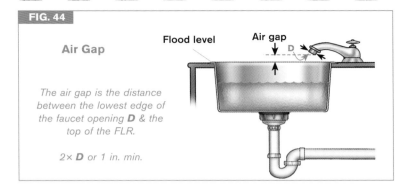

FIG. 44

Air Gap

*The air gap is the distance between the lowest edge of the faucet opening **D** & the top of the FLR.*

*2× **D** or 1 in. min.*

Flood level — Air gap — D

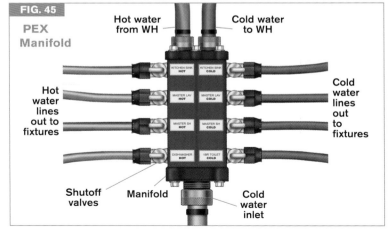

FIG. 45

PEX Manifold

Hot water from WH — Cold water to WH

Hot water lines out to fixtures

Cold water lines out to fixtures

KITCHEN SINK HOT / KITCHEN SINK COLD
MASTER LAV HOT / MASTER LAV COLD
MASTER SH HOT / MASTER SH COLD
DISHWASHER HOT / 1BR TOILET COLD

Shutoff valves — Manifold — Cold water inlet

WATER SUPPLY SIZING

Modern plastic supply systems typically use a method of parallel distribution from a central manifold. Traditional water supply systems are installed with a large main supply & a series of progressively smaller pipes toward the final fixture outlet. That method is still used in the UPC. The IRC no longer acknowledges that method & instead simply requires that the system be designed to provide the required capacities of T21. In addition to these major code changes, many jurisdictions now have ordinances that mandate water efficiency and conservation.

IRC Water Supply Design   09 IRC

☐ Design system to provide capacity of T21 under peak demand [2903.1]
☐ Max consumption limited to rates in T25 [2903.2]
☐ Mains, branches & risers sized according to supply demand, available pressure, friction loss of water meter & developed length of pipe including equivalent length of fittings F45 [2903.7]

TABLE 21	MIN. CAPACITIES AT FIXTURE SUPPLY OUTLETS [2903.1]	
Fixture Outlet	**Flow Rate (gpm)**	**Flow Pressure (psi)**
Bathtub	4	20
Bidet	2	20
DW	2.75	8
LT	4	8
Lav	2	8
Shower (pressure-balancing or thermostatic mixing)	3	20
Shower (temperature controlled)	3	20
Hose bibb	5	8
KS	2.5	8
WC (tank type)	3	20
WC (flushometer)	1.6	20
WC (one piece)	6	20

TABLE 22	WATER SUPPLY FIXTURE UNITS FILL-IN TABLE [2903.6] & (T6-4)					
	IRC			UPC		
Fixture	**Hot**	**Cold**	**Comb.**	**Comb.**	**No.**	**Extension**
BT	1	1	1.4	4		
CW	1	1	1.4	4		
DW	1.4	–	1.4	1.5		
Hose bibb	–	2.5	2.5	2.5		
KS	1	1	1.4	1.5		
Lav	0.5	0.5	0.7	1		
LT	1	1	1.4	1.5		
Shower (per head)	1	1	1.4	2.0		
WC	–	2.2	2.2	2.5		
Laundry group^A	1.8	1.8	2.5	–		
Kitchen group^B	1.9	1	2.5	–		
Half-bath group^C	0.5	2.5	2.6	–		
Full-bath group^D	1.5	2.7	3.6	–		
Total Demand						

A. Laundry group = CW & LT.
B. Kitchen group = DW & sink with or without garbage grinder.
C. Half-bath group = WC & lav.
D. Full-bath group = WC & lav & bathtub (with or without shower) or shower stall.

TABLE 23	MANIFOLD SIZING [2903.8.1]	
Inlet Pipe Size (in.)	**Max. GPM Plastic^A**	**Max. GPM Metal^B**
¾	17	11
1	29	20
1¼	46	31
1½	66	44

A. Based on velocity limitation of 12 ft./second.
B. Based on velocity limitation of 8 ft./second.

TABLE 24 — CONVERTING WSFUs to GPM [T2903.6(1)]

Load (WSFUs)	Demand (gpm)	Load (WSFUs)	Demand (gpm)
1	3.0	14	17.0
2	5.0	15	17.5
3	6.5	16	18.0
4	8.0	17	18.4
5	9.4	18	18.8
6	10.7	19	19.2
7	11.8	20	19.6
8	12.8	25	21.5
9	13.7	30	23.3
10	14.6	35	24.9
11	15.4	40	26.3
12	16.0	45	27.7
13	16.5	50	29.1

TABLE 25 — MAXIMUM FLOW RATES FOR PLUMBING FIXTURES [T2903.2]

Plumbing Fixture or Fixture Fitting	Max. Flow Rate
Lav faucet	2.2 gpm at 60 psi
Showerhead[A,B]	2.5 gpm at 80 psi
Sink faucet	2.2 gpm at 60 psi
WC	1.6 gal per flush

A. Handheld shower sprays are also considered showerheads
B. Individual states have different policies on whether the max. flow rate applies to each individual head in a multiple showerhead assembly or to the overall assembly. Check with your local jurisdiction.

TABLE 26 — UPC WATER SIZING TABLE {T6-6}

Meter	Supply	Units Allowed per Lengths of Pipe					
30–45 psi		40 ft.	60 ft.	80 ft.	100 ft.	150 ft.	200 ft.
3/4 in.	1/2 in.[A]	6	5	4	3	2	1
3/4 in.	3/4 in.	16	16	14	12	9	6
3/4 in.	1 in.	29	25	23	21	17	15
1 in.	1 in.	36	31	27	25	20	17
1 in.	1 1/4 in.	54	47	42	38	32	28
46–60 psi		40 ft.	60 ft.	80 ft.	100 ft.	150 ft.	200 ft.
3/4 in.	1/2 in.[A]	7	7	6	5	4	3
3/4 in.	3/4 in.	20	20	19	17	14	11
3/4 in.	1 in.	39	39	36	33	28	23
1 in.	1 in.	39	39	39	36	30	25
1 in.	1 1/4 in.	78	78	76	67	52	44
> 60 psi		40 ft.	60 ft.	80 ft.	100 ft.	150 ft.	200 ft.
3/4 in.	1/2 in.[A]	7	7	7	6	5	4
3/4 in.	3/4 in.	20	20	20	20	17	13
3/4 in.	1 in.	39	39	39	39	35	30
1 in.	1 in.	39	39	39	39	38	32
1 in.	1 1/4 in.	78	78	78	78	74	62

A. Min. building supply is 3/4 in.

WATER SUPPLY SIZING

PLUMBING

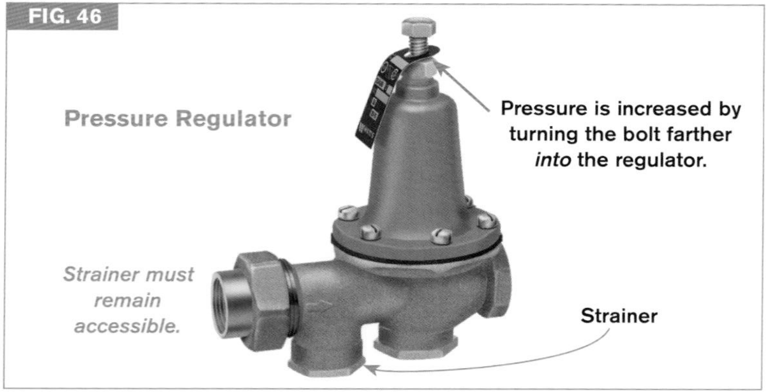

PRESSURE REGULATORS

Excessive pressure increases the risk of leaks and scalding. When supply pressure exceeds 80 psi, a regulator is required. A screen in the regulator can prevent it from clogging with sediment. Regulators without an integral bypass feature create a closed system downstream from the regulator. As the water heater recovers heat, pressure rises. To prevent excessive pressure, expansion tanks are then required & are usually placed in the cold water line just above the water heater.

General	IRC 09	UPC 09
☐ Req'd when water pressure at building > 80psi	[2903.3.1]	{608.2}
☐ Strainer req'd ahead of regulator F46	[n/a]	{608.2}
☐ Regulator & strainer accessible without removing piping	[manu]	{608.2}
☐ Regulated pressure computed at 80% of setting	[n/a][10]	{608.2}
☐ Expansion tank req'd for closed systems F47	[2903.4.1]	{608.3}
☐ Expansion tank req'd for systems with supply check valves	[2903.4.2]	{608.3}

FIG. 47

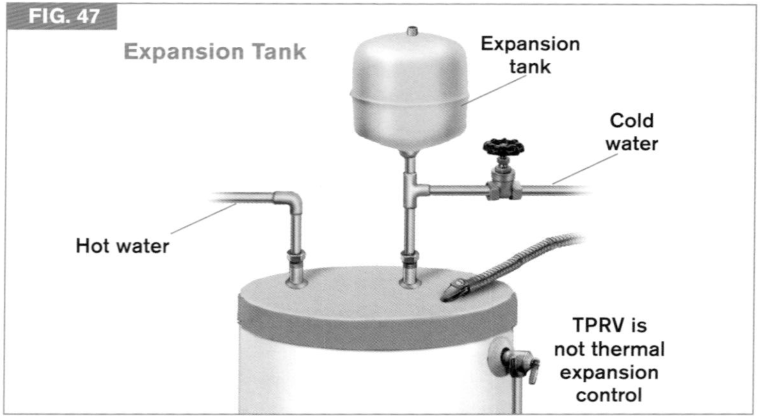

Expansion Tank

TPRV is not thermal expansion control

Hot water

Cold water

Expansion tank

TABLE 27	UPC WATER SIZING WORKSHEET* {610.8}	
1. Determine fixture unit demand (total from T22)		
2. Min. daily static pressure at meter or source		
3. Subtract ½ lb. pressure per ft. of rise		
4. Deduct pressure losses for filters, regulators, etc.		
5. Find pressure range group in T26		
6. Find column for developed length to most remote fixture		
7. Find row meeting fixture unit demand (total from T26)		
8. Find req'd meter & pipe size in left column of T26		

A. The same procedure can be used for branches.

FIG. 46

Pressure Regulator

Strainer

Strainer must remain accessible.

Pressure is increased by turning the bolt farther into the regulator.

GAS PIPING

Gas pipe sizes depend upon the appliance demand, gas pressure, Btus per cubic foot of gas, and length of run. Corrugated stainless steel tubing systems (CSST) can be run at medium pressure (approximately 2 psig) to a central manifold where the pressure is then reduced to the operating pressure used by the appliances, typically not more than 0.5 psig. This method allows smaller main runs and uses less tubing than a system operating at utilization pressure.

General	09 IRC	09 UPC
☐ Nonsteel pipe req's yellow label marked "gas" in black letters at 5 ft. intervals EXC	[2412.5]	{n/a}
• When located in same room as appliance served	[2412.5]	{n/a}
☐ Meters identified to indicate which premises served	[2412.7]	{1209.6.5}
☐ LPG storage per NFPA 58 (see **p.152**)	[2412.2]	{1213.0}
☐ No piping in circulating air duct, chimney or gas vent, ventilating duct, or elevator shaft	[2415.1]	{1211.2.5}
☐ No concealed piping in solid partition except in chase	[2415.2]	{1211.3.3}

Joints & Fittings	09 IRC	09 UPC
☐ Joints threaded, flanged, brazed, or welded	[2414.10.1]	{1209.5.8.1}
☐ Clear fittings of burrs, brush, blow out chips & scales	[2414.7]	{1209.5.5}
☐ No unions, tubing fittings, bushings, right-left couplings, or compression couplings in concealed locations	[2415.3]	{1211.3.2}

Materials	09 IRC	09 UPC
☐ Steel or black pipe min schedule 40	[2414.4.2]	{1209.5.2.2}
☐ Cu tubing type K or L	[2414.5.2]	{1209.5.3.2}
☐ No Cu or brass if > 0.3 grains H_2S per 100 cu. ft. gas	[2414.5.2]	{1209.5.3.2}
☐ No pipe repair – pipe with defects must be replaced	[2414.7]	{1209.5.5}
☐ CSST AMI (UPC: comply with CSA LC-1)	[2414.5.3]	{1209.5.3.4}

TABLE 28	TYPICAL GAS APPLIANCE DEMAND [T2413.2] {T12-1}			
Appliance	Typical kBtu/hr.	Actual kBtu/hr.	Typical cu.ft./hr.[A]	Actual cu.ft./hr.
FAU or hydronic boiler	100		91	
Space & water heating units	120		109	
Instantaneous WH 2 gpm	143		130	
Instantaneous WH 4 gpm	285		259	
Storage tank WH 30–40 gal.	35		32	
Storage tank WH 50 gal.	50		45	
Built-in oven	25		22	
Built-in cooktop	40		36	
Freestanding range	65		59	
Barbecue	40		36	
Clothes dryer	35		32	
Direct-vent fireplace	40		36	
Gas log	80		73	
Total cu. ft./hr. max. gas demand				

A. Based on 1100 Btu/cu. ft.–consult local gas provider for actual values.

Underground   **09 IRC**   **09 UPC**

☐ Protect piping subject to corrosion from soil or moisture [2415.9] {1211.1.3}
☐ Zi coatings not sufficient protection underground _____ [2415.9] {1211.1.3}
☐ Coatings & wrappings factory-applied EXC _____ [2415.9.2] {1211.1.3}
  • Nipples & fittings where field coating OK if AMI ___ [2415.9.2X] {1211.1.3}
☐ Min cover depth 12 in. (UPC 18 in. EXC) _____ [2415.10] {1211.1.2A}
  • UPC OK at 12 in. if external damage not likely ____ {1211.1.2A}
☐ Pipe trenches to have firm continuous bearing _____ [2415.11] {1211.1.2B}
☐ No pipe penetrating foundation walls below grade ___ [2415.4][11] {n/a}
☐ UPC allows below ground penetration if sleeved _____[Ø] {1211.1.5}
☐ Plastic OK only underground & outdoors EXC _____ [2415.15.1] {1211.1.7}
  • Anodeless risers & wall head adapters _____ [2415.15.1X] {1211.1.7X}
  • Regulator vent connections to exterior _____ [[2414.6.3] {n/a}
☐ Provide yellow insulated tracer wire min 18 AWG (14 AWG in UPC) along plastic pipe & terminating above ground [2415.15.3] {1211.1.7C}

Gas Piping in or below Slab   **09 IRC**   **09 UPC**

☐ Piping may not penetrate foundation below grade ____ [2415.4][11] {n/a}
☐ Piping in slab req's protected channel or conduit _____ [2415.6][12] {n/a}
☐ Conduit under slab req's protective conduit **F48** ____ [2415.12][12] {1211.1.6}
☐ Conduit with both ends terminating in building should not have ends sealed & both must be accessible _ [2415.6.2&2415.12.2][13] {n/a}
☐ Conduit with one end on exterior & one on interior: **F48**
  • Seal pipe to conduit in interior _____ [2415.6.1&2415.12.1] {1211.1.6}
  • Exterior pipe min 4 in. outside building __ [2415.6.1&2415.12.1] {1211.1.6}
  • Conduit vented above grade _____ [2415.6.1&2415.12.1] {1211.1.6}
  • Conduit to prevent water & insect entry _ [2415.6.1&2415.12.1] {1211.1.6}

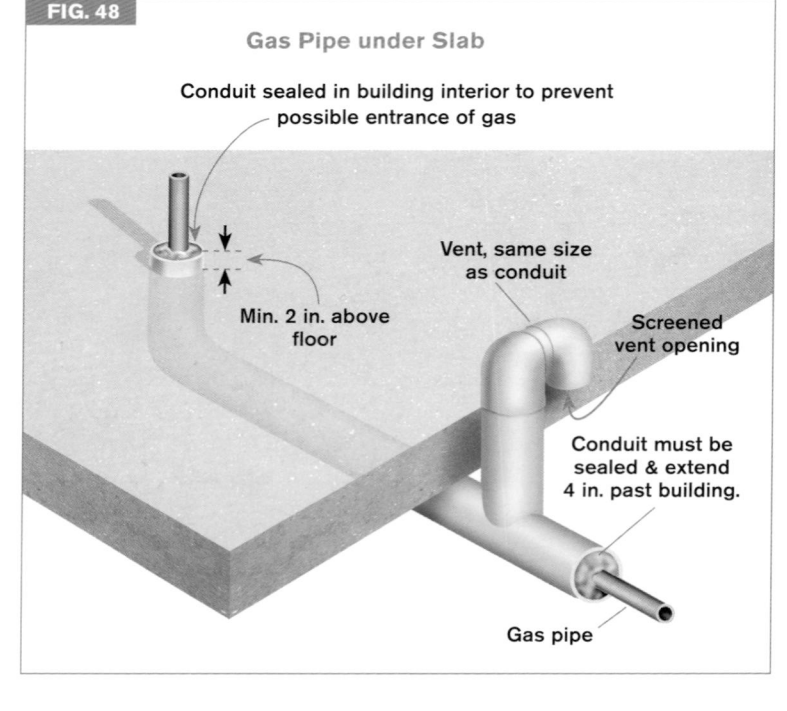

FIG. 48

Gas Pipe under Slab

Conduit sealed in building interior to prevent possible entrance of gas

Min. 2 in. above floor

Vent, same size as conduit

Screened vent opening

Conduit must be sealed & extend 4 in. past building.

Gas pipe

Protection & Installation 09 IRC 09 UPC

- ☐ Outdoor piping min 3½ in. above ground or roof surface [2415.7] {n/a}
- ☐ Shield plates 4 in. past edge of framing members for other than black or Zi steel pipe < 1½ in. from face of framing **F50** _____ [2415.5] {manu}

Piping Support 09 IRC 09 UPC

- ☐ Max support intervals for gas pipe **T29**_____ [2424.1] {1211.2.6B}
- ☐ Hangers to dampen excessive vibration _____ [2418.2] {1211.2.6A}
- ☐ Hangers must allow for expansion & contraction of pipe [2418.2] {1211.2.6C}

TABLE 29	GAS PIPING & TUBING SUPPORT [T2424.1] {T12-3}		
Steel Pipe Nominal Size (in.)	Max. Support Spacing (ft.)	Smooth-Wall Tubing Nominal Size (in.)	Max. Support Spacing (ft.)
½	6	½	4
¾ or 1	8	⅝ or ¾	6
≥1¼ (horizontal)	10	⅞ or 1 (horizontal)	8
≥1¼ (vertical)	Every floor level	⅞ or 1 (vertical)	Every floor level

Electrical 09 IRC 09 UPC

- ☐ Electrical bond req'd for above-ground gas piping _____ [2411.1] {1211.15.1}
- ☐ EGC (**p.188**) supplying equipment OK as bond EXC __ [2411.1] {1211.15.1}
  - • CSST req's 6AWG bond to service at building entry [2411.1.1] {1211.15.2}
- ☐ Gas piping not OK as grounding electrode in earth _____ [2410.1] {1211.15.3}

Drips & Sediment Traps 09 IRC 09 UPC

- ☐ Slope piping ¼ in./15 ft. for other than dry gas _____ [2419.1] {n/a}
- ☐ Nondry gas: Accessible drips at meter & as needed____ [2419.2] {1211.8.1}
- ☐ Sediment traps req'd as close as practical to appliance inlets **F49** EXC _____ [2419.4] {1212.7}
  - • Ranges, dryers, gas lights, fireplaces & outdoor grills__ [2419.4] {1212.7}

Valves, Shutoffs & Appliance Connections 09 IRC 09 UPC

- ☐ All valves accessible & protected from damage_____ [2420.1.3] {1212.5}
- ☐ Valve within 6 ft. of appliance & in same room EXC __ [2420.5.1] {1212.5}
  - • Valve can be at manifold within 50 ft. of appliance _ [2420.5.3][14] {Ø}
- ☐ Valve upstream from union at appliance _____ [2420.5.1] {1212.5}
- ☐ Multiple buildings req shutoff for each building _____ [2420.3] {1211.11.2}
- ☐ Valve ahead of each MP regulator **F51**_____ [2420.4] {1211.11.1}
- ☐ Cap any unused outlets _____ [2415.13] {1211.9.2A}
- ☐ Connector can be rigid pipe, CSST AMI, ≤6 ft., or L&L connectors entirely in same room as appliance _____ [2422.1] {1212.1}
  - *(Note: flexible connectors may be req'd in seismically active areas.)*
- ☐ Connector max 6 ft. length EXC _____ [2422.1.2.1][15] {n/a}
  - • Developed length of rigid pipe OK > 6 ft. if sized as pipe (not as connector) & if valve within 6 ft. of appliance_____ [2422.1.2.1X] {n/a}

FIG. 49

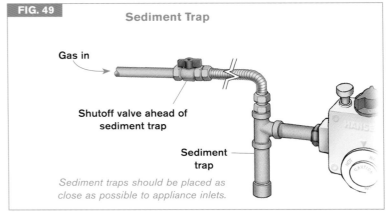

Sediment Trap

Gas in

Shutoff valve ahead of sediment trap

Sediment trap

Sediment traps should be placed as close as possible to appliance inlets.

GAS PIPING

PLUMBING

99

CORRUGATED STAINLESS STEEL TUBING (CSST)

The standard for CSST is ANSI/IAS LC 1-2005, which includes a requirement that workers be certified before installing it. The manufacturers offer training & certification courses. Concerns over damage to CSST from indirect lightning strikes have caused some jurisdictions to restrict its use, and have led to the bonding requirements on **p.99.**

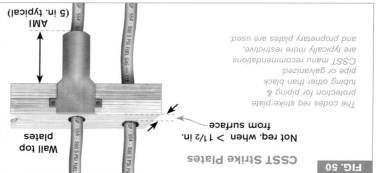

FIG. 50

CSST Strike Plates

Not req. when > 1¹⁄₂ in. from surface →

Wall top plates

↕ AMI (5 in. typical)

The codes req strike-plate protection for piping & tubing other than black pipe or galvanized. CSST manu recommendations are typically more restrictive, and proprietary plates are used.

CSST–Typical manufacturer recommendations:

- Support per manu tables
- Size per manu tables
- Bending radius per manu tables
- No direct burial—routing through conduit OK
- Striker plates per manu F50
- Avoid kinking, twisting, or contact with sharp objects
- Protect where passing through sheet metal
- Regulators in vented area or with vent limiters

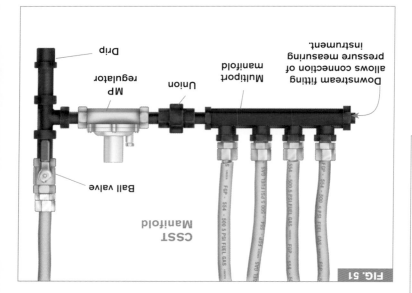

FIG. 51

Drip

MP regulator

Union

Multiport manifold allows connection of pressure measuring instrument.

Downstream fitting

Ball valve

CSST Manifold

Medium Pressure (MP) Regulators F51

	09 IRC	09 UPC
☐ MP regulators must be accessible	[1209.7.3]	
☐ MP regulators req tee between shutoff & regulator	[2421.2]	{manu}
☐ Capped tee fitting req'd downstream	[2421.2]	{manu}
☐ Vented regulators must be vented to outdoors EXC	[2421.3]	{1209.7.5B}
• If equipped with approved vent-limiting device	[2421.3X]	{1209.7.5B}
☐ Vent piping must run independently to outdoors	[2421.3.1]	{n/a}

GAS PIPE SIZE

Gas piping systems in series (**F52**) can be sized by either the longest length or the branch length method. Systems with MP regulators are sized by the hybrid pressure method.

General

	09 IRC	09 UPC
☐ Size per max demand based on appliance ratings	[2413.2]	{1209.4.1}
☐ Assume all appliances operating simultaneously	[2413.2]	{1209.4.2}
☐ Size AMI or per tables **T30–32**	[2413.3]	{1209.4.3}[16]

TABLE 30	PROCEDURES FOR SIZING GAS PIPE [2413.3] {1217.1}
1. Determine Btu/cu. ft. from local gas provider
2. Determine cu. ft./hr. demand for each appliance
3. Sketch layout with piping lengths to each appliance (**F52**)
4. Determine total cu. ft./hr. demand on each pipe section
5. Determine length to most remote appliance
6A. (Longest length method) use row of **T32** for that length for all appliances.
6B. (Branch length method) use same row for all sections in series with most remote appliance. For other branches, use actual length of each branch.

The **Longest Length** method is more conservative & compensates for pressure losses throughout the system. The **Branch Length** method has less leeway & consideration should be given to the lengths of pipe fittings. The codes accept both methods. Systems with MP regulators use the "hybrid pressure" method, where the pipe sizes before the regulator are determined separately, each by the longest length method.

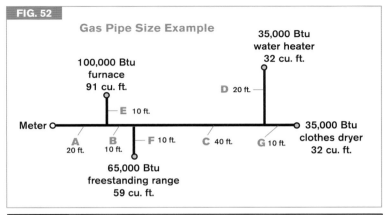

FIG. 52 — Gas Pipe Size Example

TABLE 31		GAS PIPE SIZE EXAMPLE			
Pipe Section	Total cu. ft./hr.[A]	Longest Length	Longest-Length Method	Actual Lengths	Branch-Length Method
A	214	90 ft.	1 1/4 in.	90 ft.	1 1/4 in.
B	129	90 ft.	1 in.	90 ft.	1 in.
C	64	90 ft.	3/4 in.	90 ft.	3/4 in.
D	32	90 ft.	1/2 in.	90 ft.	1/2 in.
E	91	90 ft.	3/4 in.	30 ft.	1/2 in.
F	59	90 ft.	3/4 in.	40 ft.	1/2 in.
G	32	90 ft.	1/2 in.	80 ft.	1/2 in.

A. Based on 1,100 Btu/cu. ft.–Contact local provider for actual values.

TABLE 32	CUBIC FEET CAPACITY OF SCHEDULE 40 METALLIC GAS PIPE[A] [T2413.4(1)] {T12-8}								
Pipe Length (ft.)	Nominal Pipe Size								
	Demand Capacity (in cu.ft./hr)								
	¹/₂	³/₄	1	1¹/₄	1¹/₂	2	2¹/₂	3	4
10	172	360	678	1,390	2,090	4,020	6,400	11,300	23,100
20	118	247	466	957	1,430	2,760	4,400	7,780	15,900
30	96	199	374	768	1,150	2,220	3,530	6,250	12,700
40	81	170	320	657	985	1,900	3,020	5,350	10,900
50	72	151	284	583	873	1,680	2,680	4,740	9,660
60	65	137	257	528	791	1,520	2,430	4,290	8,760
70	60	126	237	486	728	1,400	2,230	3,950	8,050
80	56	117	220	452	677	1,300	2,080	3,670	7,490
90	52	110	207	424	635	1,220	1,950	3,450	7,030
100	50	104	195	400	600	1,160	1,840	3,260	6,640
125	44	92	173	355	532	1,020	1,630	2,890	5,890
150	40	83	157	322	482	928	1,480	2,610	5,330
175	37	77	144	296	443	854	1,360	2,410	4,910
200	34	71	134	275	412	794	1,270	2,240	4,560
250	30	63	119	244	366	704	1,120	1,980	4,050
300	27	57	108	221	331	638	1,020	1,800	3,670
350	25	53	99	203	305	587	935	1,650	3,370
400	23	49	92	189	283	546	870	1,540	3,140
450	22	46	86	177	266	512	816	1,440	2,940
500	21	43	82	168	251	484	771	1,360	2,780

A. Based on inlet pressure < 2 psi, pressure drop 0.5 in. water column, specific gravity 0.60.

GAS PIPE SIZE EXAMPLE FILL-IN

Pipe Section	Total cu. ft./hr.[A]	Longest Length	Longest-Length Method	Actual Lengths	Branch-Length Method
A					
B					
C					
D					
E					
F					
G					

A. Btu/cu. ft. (from gas supplier).

WATER HEATERS

Water heaters should be maintained at as low a temperature as comfortably practical to reduce the risk of scalding. An undersize water heater is more likely to be turned to a dangerously high setting. Aside from the water heater thermostat control, other means of protection against scalding are recommended. These include tempering valves at the water heater or at individual fixtures. Temperatures at the point of use in excess of 120°F are considered a hazard.

Tankless water heaters are becoming more popular, including hybrid systems that contain a small storage tank and circulating line. Water heaters that are part of a boiler system are discussed on **p.138**.

General   09 IRC   09 UPC

☐ Size to meet demand **T33** _____ [2801.1]   {501.0}
☐ Replacement WHs req permits _____ [105.1]   {503.0}
☐ Valve req'd on cold water supply at or near WH _____ [2903.9.2]   {605.2}
☐ Valve must be full open type **F41,42**_____ [2903.9.2]   {605.2}
☐ WH also used for space heating must be L&L for both _ [2448.2]   {n/a}
☐ Systems also used for space heating req master mixing
     valve to temper domestic water to ≤140°F **F88**_____ [2802.2]   {n/a}
☐ Unions req'd within 12 in. to allow removal **F47** _____[n/a]   {609.5}
☐ Electric WH req's in-sight or lockable disconnect **F59**_ [T4101.5]   {506.1}

For information on combustion air and venting of gas-fired water heaters, see pp.134–135.

TABLE 33	WATER HEATER MIN. CAPACITY[A] {T5-1}	
No. of Bathrooms	No. of Bedrooms	1st hr. Rating[B]
1 to 1½	1	42
	2 to 3	54
2 to 2½	2	54
	3 to 4	67
	5	80
3 to 3½	3	67
	4 to 6	80

A. Based upon the 1st-hr. rating found on the "Energy Guide" label. This number is approximately equal to storage size plus hourly recovery rate.
B. This table can also be used to size tankless WHs.

Special Locations   09 IRC   09 UPC

☐ Fuel-fired WH prohibited in storage closets_____ [2005.2]   {n/a}
☐ Not in bedrooms, bathrooms, or their closets EXC ____ [2005.2]   {505.1}
   • Direct-vent WH OK without enclosure_____ [2406.2]   {505.1}
   • WH OK in dedicated enclosure with solid, weatherstripped,
     self-close door & all combustion air from exterior _____ [2005.2]   {505.1}
☐ Ignition source 18 in. above garage floor EXC **F53** ____ [2801.6]   {508.14}
   • Flammable vapor ignition-resistant (FVIR) WHs **F54** _[2408.2X]   {508.14}
   • WH in separate enclosed space accessible from outside
     the garage & no combustion air from garage_____ [2406.2]   {508.14}
☐ Min 18 in. above floor in area where flammables stored
     (basements) unless FVIR **F53,54** _____ [2801.6]   {508.13}
☐ SDC D$_0$, D$_1$ & D$_2$ seismic bracing req'd **F53**__[1307.2 & 2801.7]   {508.2}
☐ Barrier or elevation req'd in garage or carport **F53** ____ [1307.3.1]   {508.14}
☐ Min. 3 in. concrete pedestal if supported on ground_[1305.1.4.1]   {508.3}

FIG. 53 Water Heater in Garage

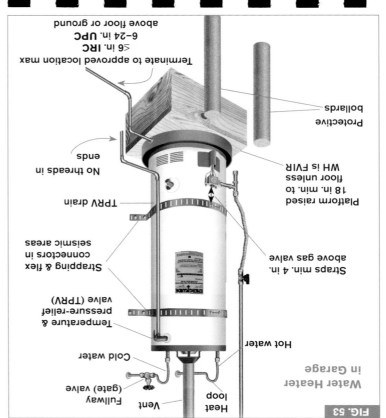

Terminate to approved location max
≤6 in. IRC
6–24 in. UPC
above floor or ground

Protective bollards

No threads in
spreads in

Platform raised
18 in. min. to
floor unless
WH is FVIR

Straps min. 4 in.
above gas valve

TPRV drain

Strapping & flex
connectors in
seismic areas

Temperature &
pressure-relief
valve (TPRV)

Cold water

Hot water

Fullway
(gate) valve

Vent

Heat loop

FIG. 54 FVIR Water Heater

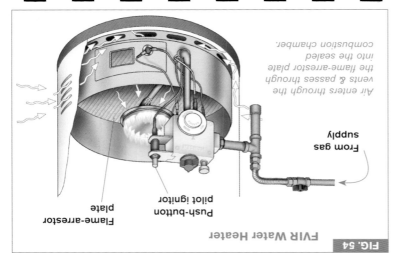

Flame-arrestor plate

Push-button pilot ignitor

From gas supply

Air enters through the vents & passes through the flame-arrestor plate into the sealed combustion chamber.

Access & Working Space

	09 IRC	09 UPC
☐ Clearances to combustibles AMI	[2408.5]	{505.3.1}
☐ Remain accessible for service, inspection & removal	[2801.3]	{505.3.1}
☐ Appliance must fit through attic door	[2005.1,1305.1.3]	{509.4.1}
☐ Attic hatch/door min 22 in. wide × 30 in. high	[2005.1,1305.1.3]	{509.4.1}
☐ Attic min 24 in. passageway, solid floor to WH	[2005.1,1305.1.3]	{509.4.3}
☐ Max 20 ft. from attic access if ceiling < 6ft	[2005.1,1305.1.3]	{509.4.2}
☐ Min 30 × 30 in. level working platform req'd EXC.	[1305.1.3]	{509.4.4}
• Platform not req'd if it can be serviced from opening	[1305.1.3X1]	{Ø}
☐ Attic req's light & receptacle near WH	[2005.1,1305.1.3.1]	{509.4.5}
☐ Light switch req'd at entrance to attic	[2005.1,1305.1.3.1 & 3903.4]	{509.4.5}

Tankless Water Heaters

	09 IRC	09 UPC
☐ Type III vent typically req'd AMI **F55** _____	[2427.3.1]	{510.1.2}
☐ PRV AMI _____	[2005.1]	{505.6}
☐ Size gas line to max Btu rating **F52** _____	[2413.2]	{1209.4.1}

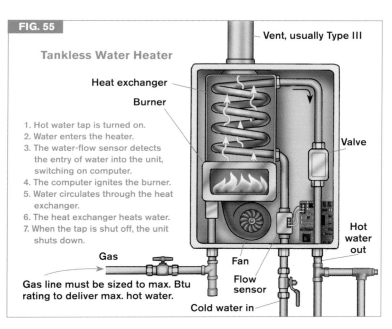

FIG. 55

Tankless Water Heater

Vent, usually Type III

Heat exchanger

Burner

1. Hot water tap is turned on.
2. Water enters the heater.
3. The water-flow sensor detects the entry of water into the unit, switching on computer.
4. The computer ignites the burner.
5. Water circulates through the heat exchanger.
6. The heat exchanger heats water.
7. When the tap is shut off, the unit shuts down.

Valve

Hot water out

Gas

Fan

Flow sensor

Gas line must be sized to max. Btu rating to deliver max. hot water.

Cold water in

Temperature & Pressure Relief Valves

	09 IRC	09 UPC
☐ All WHs req pressure relief device **F56,58** _____	[2803.1]	{505.4}
☐ All WHs req temperature limiting device **F56, F57** _____	[2803.1]	{505.5}
☐ Devices may be combination TPRV (mandatory for storage-tank WHs in UPC) **F56** _____	[2803.5]	{608.3}
☐ Temperature probe top 6 in. of tank (AMI in UPC) **F53, 59**	[2803.4]	{505.6}
☐ Settings not > 150psi OR 210°F _____	[2803.3&4]	{608.4}
☐ Watts 210 also req's PRV **F57,58** _____	[2803.1]	{505.6}

TPRV Drain Piping

	09 IRC	09 UPC
☐ No shutoff valves before or downstream of TPRV _____	[2803.6]	{505.6}
☐ Piping may not be shared with condensate drain or relief valves of other systems _____	[2803.6.1]	{608.4}
☐ Drain must end outside building or at other approved location (IRC allows floor, pan, exterior, or indirect waste _____	[2803.6.1]	{608.5}
☐ TPRV may discharge into pan (not allowed in UPC) _	[2803.6.1][17]	{Ø508.5}
☐ End ≤6 in. (6–24 in. UPC) from ground or receptor __	[2803.6.1]	{608.5}
☐ Drain size at least same as outlet of valve _____	[2803.6.1]	{608.5}
☐ Must drain by gravity; cannot run uphill or be trapped	[2803.6.1]	{608.5}
☐ No kinks or restrictions in pipe _____	[2803.6.1]	{608.5}
☐ End of pipe visible & no threads on end **F53,59** _____	[2803.6.1]	{608.5}
☐ Material can be any allowed for water distribution (only Zi steel, CPVC, hard-drawn Cu, or listed TPRV drain in UPC) **F53,59** _____	[2803.6.1]	{608.5}
☐ Protect from freezing (terminate through air gap to indirect receptor located in a heated space) _____	[2803.6.1]	{608.5}
☐ May not drain to crawl space _____	[2803.6.1]	{608.5}

FIG. 56

Temperature & Pressure-Relief Valve

When the WH is in a basement or below grade, it may not be possible to arrange for a gravity drain of the TPRV valve. A Watts 210 valve (F57) might be an allowable option. The temperature-sensing bulb of the valve goes in the upper portion of the tank & the gas piping runs through the valve. The Watts 210 shuts off the gas if the temperature is excessive. In addition, a separate water pressure–relief valve (F58) must be installed in the piping & must drain by gravity to an approved location. Check with the local AHJ to see if this method is accepted in your area.

FIG. 57

FIG. 58

Pressure-Relief Valve

Watts 210 Gas Shutoff Valve

Required Pans & Drain

	09 IRC	09 UPC
☐ Watertight corrosion-resistant pan req'd for WHs in attics or where leakage could cause damage **F53,59**	[2801.5][18]	{508.4}
☐ Pan 24-gage Zi or listed corrosion-resistant material ___	[2801.5]	{508.4}
☐ Pan drain size min ¾ in. **F53,59**	[2801.5.1]	{508.4}
☐ Pan drain req'd to end in indirect waste or outdoors 6 to 24 in. above grade (to any approved location in UPC)	[2801.5.2]	{508.4}
☐ Pan min 1½ in. deep **F53,59**	[2801.5.1]	{n/a}

FIG. 59

TPRV Discharge Pipes

TPRV discharge pipe:
- *No threads on end*
- *Not trapped*
- *No smaller than relief valve outlet*
- *No valves or fittings*
- *Discharge to readily observable location*
- *Max. 6 in. above receptor in IRC*
- *Not to drain to pan in UPC*

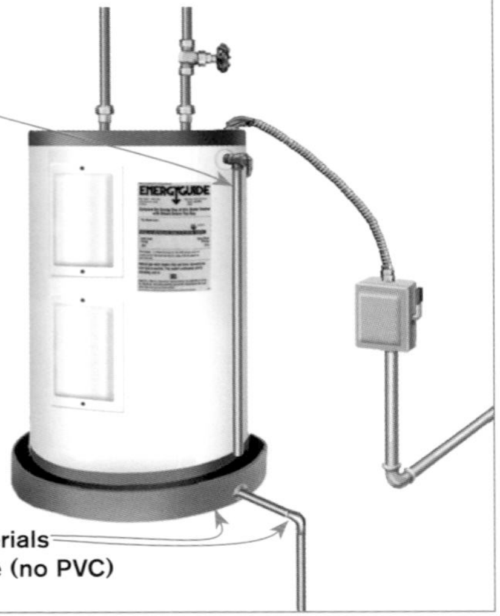

Pan & TPR drain only with materials approved for interior water pipe (no PVC)

FIXTURES

Fixtures include faucets, showers, sinks, toilets, hose bibbs & similar equipment. They must be arranged so as to prevent cross-connections between the supply and waste. Fixtures must be constructed to recognized standards to maintain a sanitary condition. Fixtures such as toilets and showers must also conform to local water conservation regulations.

General
	09 IRC	09 UPC
☐ Fixtures req'd to be smooth, impervious & free from concealed fouling areas F63	[2701.1]	{401.1}
☐ Watertight seal req'd between fixtures & walls or floors (caulk base of WC)	[2705.1]	{407.2}
☐ Separate accessible shutoff req'd at each fixture EXC	[2903.9.3]	{605.5}
• Tubs & showers	[2903.9.3]	{605.5}
☐ Shutoffs can be at manifolds or at fixture	[2903.8.5]	{605.5}
☐ Shutoffs at manifolds must be labeled F45	[2903.8.5]	{605.5}
☐ Hot on left, cold on right when facing outlet	[2722.2]	{415.0}
☐ Drain strainers req'd except WCs & urinals	[2702.1]	{404.1}
☐ Tailpiece min 1¼ in. for lavs & bidets	[2703.1]	{404.3}
☐ Tailpiece min 1½ in. for other fixtures	[2703.1]	{404.3}
☐ Floor drains req removable strainers	[2719]	{411.1}
☐ Floor drains not OK under or restricted by appliances	[2719][19]	{n/a}

Kitchens
	09 IRC	09 UPC
☐ Sink min outlet 1½ in diameter	[2714]	{T7-3}
☐ OK for sink, DW & disposer on same 1½ in. trap	[2717.3]	{T7-3}
☐ 2 in. drain req'd for sink downstream of trap	[n/a]	{T7-3}
☐ DW supply req's air gap or integral backflow device	[2717.1]	{301.1.1}
☐ May discharge directly to a trap, trapped fixture, branch wye tailpiece on kitchen sink, or directly to disposer	[2717.2&3]	{Ø}
☐ Secure drain hose to underside of counter F61	[2717.2&3]	{n/a}

Kitchens (cont.)
	09 IRC	09 UPC
☐ Air gap fitting above sink flood level req'd for DW drain F60	[n/a]	{807.4}
☐ No connection to discharge side of disposer F60	[n/a]	{404.4}
☐ Reverse osmosis systems to recognized standards	[2908.1]	{603.4.13}
☐ Reverse osmosis systems req air gap	[2908.2]	{603.4.13}
☐ No saddle fittings or tapping/drilling of drain line	[3003.2]	{311.2}

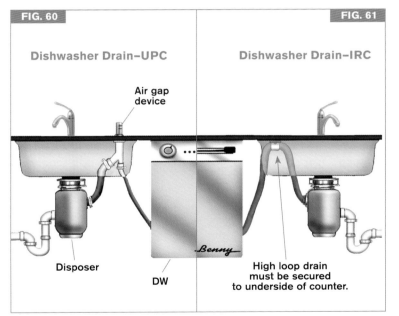

FIG. 60 — Dishwasher Drain–UPC

FIG. 61 — Dishwasher Drain–IRC

Air gap device

Disposer

DW

High loop drain must be secured to underside of counter.

Benny

Laundry

	09 IRC	09 UPC
☐ Standpipe 18-42 in. (18-30 in. UPC) above trap F62	[2706.2]	{804.1}
☐ Must drain through air break (no pressurized waste)	[2718.1]	{805.0}
☐ No trap below floor	[n/a]	{804.1}
☐ Trap 6-18 in. above floor F62	[n/a]	{804.1}
☐ CW may drain directly into LT	[2706.3X2)]	{T7-3}
☐ LT may drain into washer standpipe within 30 in. if standpipe min 30 in. above weir & above FLR of LT	[2706.2.1][20]	{∅}

FIG. 62

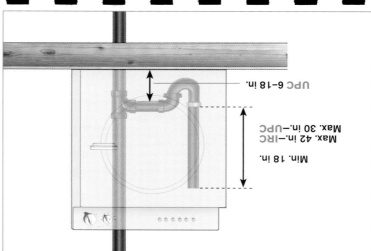

Laundry Standpipe

UPC 6-18 in.

Max. 42 in.—IRC
Max. 30 in.—UPC

Min. 18 in.

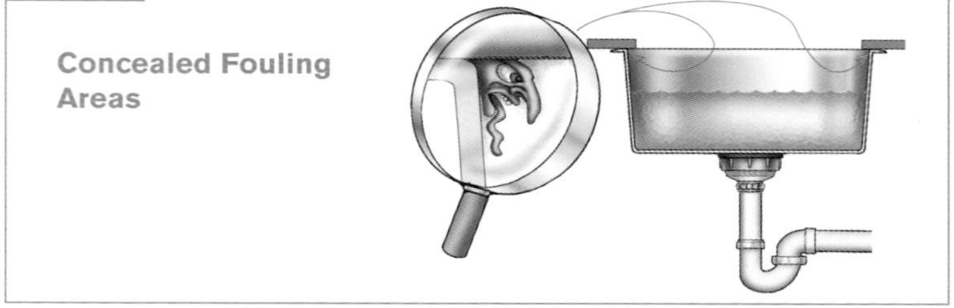

Outdoors & Irrigation Systems F64

	09 IRC	09 UPC
☐ Hose bibbs req backflow preventer or vacuum breaker	[2902.4.3]	{603.4.7}
☐ Irrigation vacuum breakers above highest head	[2902.5.3]	{T6-2}

FIG. 64

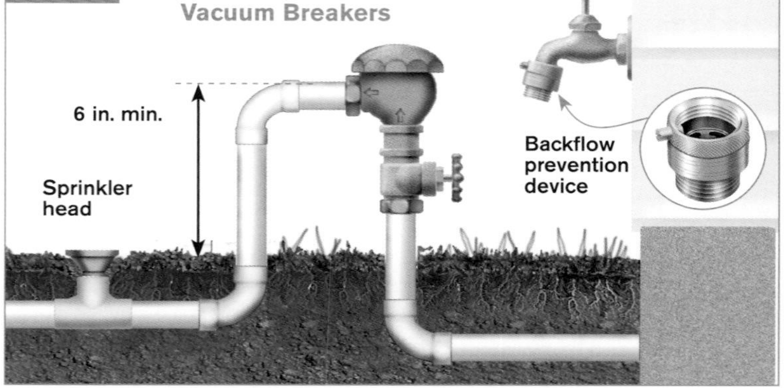

Vacuum Breakers

Backflow prevention device

Sprinkler head

6 in. min.

FIG. 63

Concealed Fouling Areas

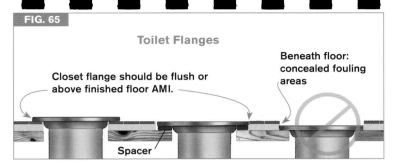

FIG. 65

Toilet Flanges

Closet flange should be flush or above finished floor AMI.

Beneath floor: concealed fouling areas

Spacer

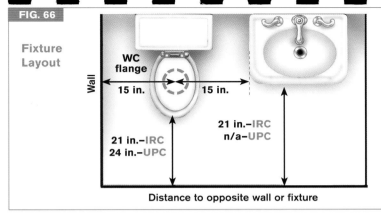

FIG. 66

Fixture Layout

Wall

WC flange

15 in. | 15 in.

21 in.–IRC
24 in.–UPC

21 in.–IRC
n/a–UPC

Distance to opposite wall or fixture

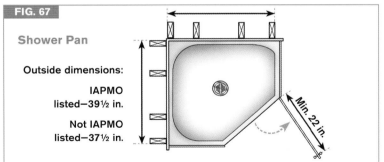

FIG. 67

Shower Pan

Outside dimensions:

IAPMO
listed—39½ in.

Not IAPMO
listed—37½ in.

Min. 22 in.

Toilets & Bidets	09 IRC	09 UPC
☐ Floor flanges req'd for floor outlets **F65** _____	[2705.1]	{408.3}
☐ Secure floor flange with corrosion-resistant fasteners __	[2705.1]	{408.3}
☐ WC or bidet req's min 15 in. clearance from center to side walls or outer rim of adjacent fixtures or partitions or vanity **F66** __	[2705.1][21]	{407.5}
☐ Min 21 in. (24 in. UPC) front clearance **F66** _____	[2705.1]	{407.5}
☐ No offset or reducing floor flanges _____	[3002.3.1]	{408.3}
☐ Max distance from closet ring to vent 6 ft. UPC (unlimited IRC) _____	[3105.1X]	{T10-1}
☐ Ballcock critical level ≥ 1 in. above overflow pipe _____	[2712.4]	{603.4.2}

Tubs	09 IRC	09 UPC
☐ Slip joints accessible, min 12 in. × 12 in. door **F24** ____	[2704.1]	{404.2}
☐ Over-rim bath spout–min air gap 2 in. from FLR ____	[T2902.3.1]	{T6-3}
☐ Overflow min 1½ in. diameter _____	[2713.1]	{404.3}
☐ Tub or whirlpool max water temperature 120°F _____	[2713.3]	{414.5}

Showers   09 IRC   09 UPC

☐ Min area 900 sq. in. (1024 sq. in. UPC) & min diameter 30 in.
  measured from finished wall **F67**_____ [2708.1]   {411.7}
☐ Min shower area to be maintained to 70 in. above drain_ [2708.1]   {411.7}
☐ Showerheads, valves, grab bars & soap dishes allowed to
  protrude into req'd min. space_____ [2708.1]   {411.7}
☐ Shower walls nonabsorbent to min 72 in. above drain ___ [307.2]   {IS-4.6}
☐ Finished threshold height min 1 in. below rest of shower
  receptor & 2–9 in. above top of drain **F68** _____ [2709.1]   {411.6}
☐ Door must open outward **F67**_____ [2708.1]   {411.6}
☐ Door min 22 in. wide **F67**_____ [2708.1.1]   {411.6}
☐ Finished floor slope ¼ in. to ½ in./ft._____ [2709.1]   {411.6}
☐ Secure shower valve, head/riser to permanent structure [2708.2]   {411.11}
☐ Showerhead not discharging directly at door_____[n/a]   {411.10}
☐ Listed anti-scald/pressure balance valve req'd 120°F max [2708.3]   {418.0}

Shower Pan & Liner   09 IRC   09 UPC

☐ Min rough pan 900 sq. in. (1024 sq. in. UPC)_____ [2708.1]   {411.7}
☐ Must conform to approved standards _____ [2709.2]   {411.8}
☐ Slope underlayment ¼ in./ft. **F68** _____ [2709.3]   {411.8}
  • Liner min 3 in. above dam **F68** _____ [2709.2]   {411.8}
☐ Pan liner plastic or 3 layers hot mop type 15 felt_____ [2709.2]   {411.8}
☐ Special attention to hot mop corner installation; extend
  4 in. all directions from corner _____ [2709.2.3]   {411.8}
☐ PVC & CPE sheet lining AMI_____ [2709.2.1&2]   {411.8}
☐ Weep holes at drain req'd & must remain clear **F68** ____ [2709.4]   {411.8}
☐ No fasteners < 1 in. above finished threshold _____[2709. 3]   {411.8}
☐ Roll over top of rough threshold (no penetrations through top)
  & fasten to outside edge **F68** _____ [2709.3]   {411.8}

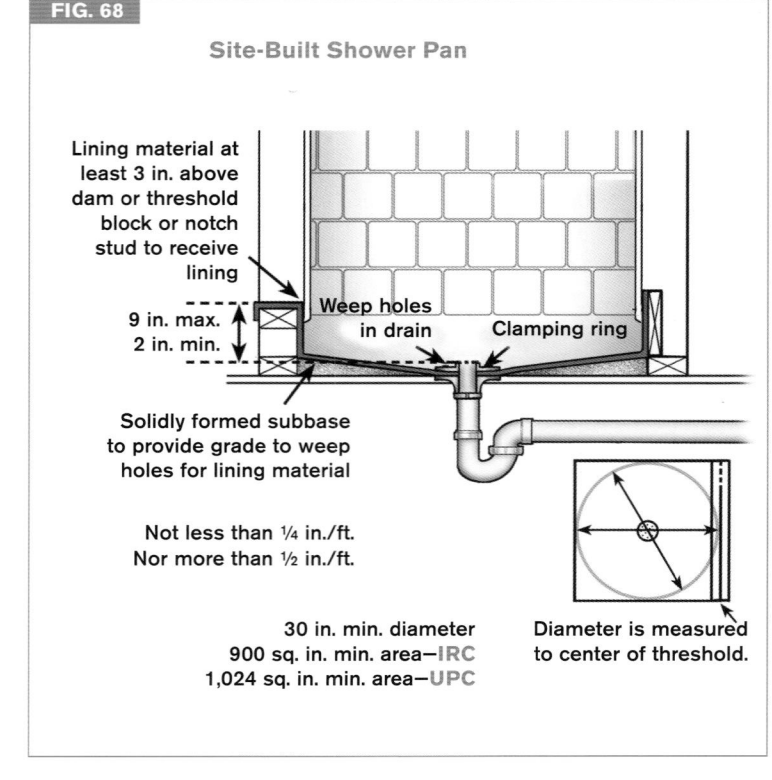

FIG. 68

Site-Built Shower Pan

Lining material at
least 3 in. above
dam or threshold
block or notch
stud to receive
lining

9 in. max.
2 in. min.

Weep holes
in drain   Clamping ring

Solidly formed subbase
to provide grade to weep
holes for lining material

Not less than ¼ in./ft.
Nor more than ½ in./ft.

30 in. min. diameter
900 sq. in. min. area—IRC
1,024 sq. in. min. area—UPC

Diameter is measured
to center of threshold.

INSPECTIONS

General

	09 IRC	09 UPC
☐ Nothing concealed until inspected & approved	[2503.2]	{103.5.1}
☐ Testing to be conducted in presence of AHJ	[2503.1]	{105.5.3.1}

Water Supply

	09 IRC	09 UPC
☐ Test all piping before cover or concealment	[2503.2]	{103.5.1.1}
☐ Water pipe test at working pressure 15 minutes EXC	[2503.7]	{609.4}
• 50 psi air for other than plastic pipe	[2503.7]	{609.4}
☐ Water for testing must be from potable water source	[2503.7]	{609.4}
☐ Reduced-pressure-principle backflow devices tested at installation & annually	[2503.8.2]	{603.3.3}
☐ Test gauges req'd to have increments of:	[2503.9]	{319.1-3}
• 0.1 psi up to 10 psi test pressure		
• 1 psi up to 100 psi test pressure		
• 2 psi up if over 100 psi test pressure		

DWV Systems

	09 IRC	09 UPC
☐ Rough-in plumbing, one of the following:		
• Water test–10 ft. head for 15 minutes OR	[2503.5.1]	{712.2}
• Air test–5 psi (10 in. mercury column) 15 minutes	[2503.5.1	{712.3}
☐ Building sewer 10 ft. head 15 minutes (or air test UPC)	[2503.4]	{723.0}
☐ UPC no air testing for plastic pipe	[n.a]	{712.1}
☐ Finished plumbing: fill each drain, inspect traps EXC	[2503.5.2]	{712.1}
• BO may req gas test with smoke 15 minutes @ 1 in. water column or peppermint test 2 ounces in 10 quarts water	[2503.5.2]	{n/a}

Gas

	09 IRC	09 UPC
☐ Leave all joints exposed until tested	[2417.3]	{1214.2.1}
☐ Test pressure min 1½ × working pressure, min 3 psig	[2417.4.1]	{1214.3.2}
☐ Test time min 10 minutes	[2417.4.2]	{1214.3.3}

Gas (cont.)

	09 IRC	09 UPC
☐ Test medium air, nitrogen, or CO_2 (not oxygen)	[2417.2]	{1214.1.7}
☐ Test gauge scale not > 5× test pressure	[2417.4.1]	{1214.3.1}
☐ Cap outlets before pressure test	[2417.3.4]	{1214.2.5}
☐ Inspect for open fittings or valves before turning on gas	[2417.6.2]	{1214.5.2}
☐ Check for leakage immediately after turning on gas	[2417.6.3]	{1214.5.3}
☐ Soapy water or gas detector OK for locating leaks	[2417.5.1]	{1214.4.2}
☐ Matches not OK for locating leaks	[2417.5.1]	{1214.4.2}
☐ Purge appliances before placing in operation	[2417.7.4]	{1214.6.4}

Note: Air must be safely displaced from the fuel lines before they are placed in use. An addendum to several fuel gas codes specifies the purging procedures to be used. Further information is available at: http://www.nfpa.org/Assets/files/AboutTheCodes/54/TIA54-09-3.pdf/ Fireblocking: Purpose & Materials.

STRUCTURAL

Pipe Protection in Framing F69

	09 IRC	09 UPC
☐ Steel-plate protection for pipes other than galvanized & cast iron in notches or holes < 1½ in. (1 in. UPC) of face of framing	[2603.2.1]	{313.9}
☐ Protection min 2 in. above sole plates & below top plates	[2603.2.1]	{n/a}
☐ Protection min 1½ in. beyond outside of pipe	[n/a]	{313.9*}
☐ Protective plates min 16 gage (18 gage UPC)	[2603.2.1]	{{313.9}

Structural Modifications: General

	09 IRC	09 UPC
☐ No structural member weakened or impaired	[2603.1]	{313.2}
☐ Drilling & notching per T34, 35	[2603.2]	{313.2}

Fireblocking

	09 IRC
☐ Purpose is to cut off concealed draft openings	[302.11]
☐ At openings around vents, ducts, pipes & cables at ceilings & floors (see p.9)	[302.11]
☐ Caulking does not have to be fire-rated	[302.11]

FIG. 69 Piping Protection

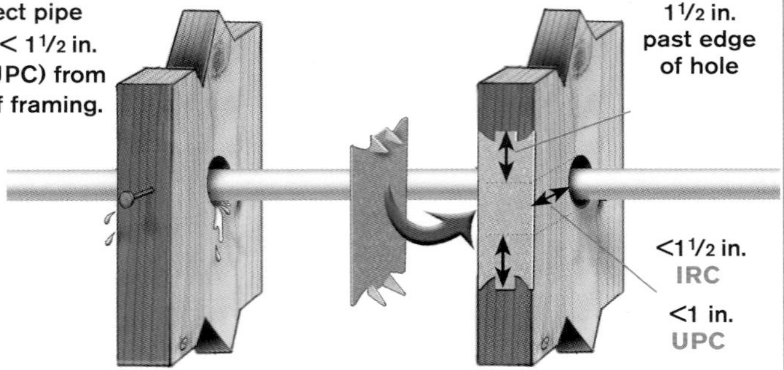

Protect pipe when < 1½ in. (1 in. UPC) from face of framing.

Min. extension 1½ in. past edge of hole

<1½ in. **IRC**

<1 in. **UPC**

FIG. 70

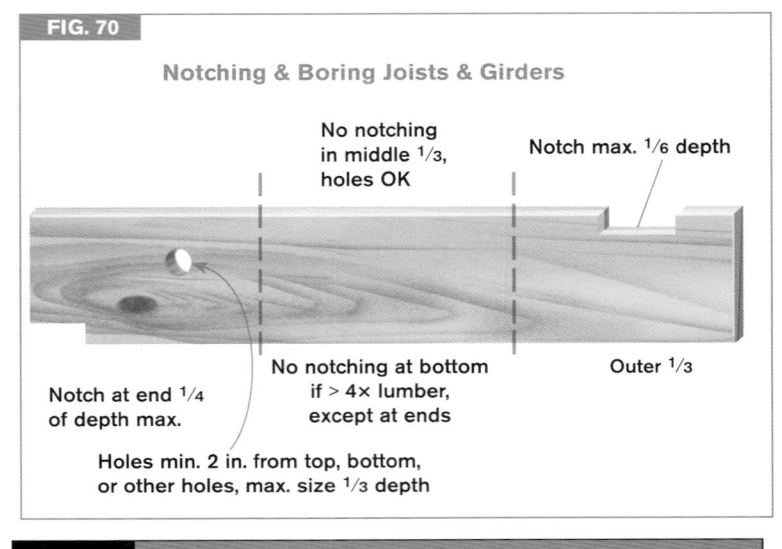

Notching & Boring Joists & Girders

No notching in middle ⅓, holes OK

Notch max. ⅙ depth

Notch at end ¼ of depth max.

No notching at bottom if > 4× lumber, except at ends

Outer ⅓

Holes min. 2 in. from top, bottom, or other holes, max. size ⅓ depth

Notches in Joists, Girders & Rafters　　　**09 IRC**

- ☐ Notches in sawn lumber max ⅙ depth of member **F70, T34** _____ [502.8.1]
- ☐ Max length of notches in sawn lumber ⅓ depth of member **F70** ___ [502.8.1]
- ☐ Notches in sawn lumber not in middle ⅓ **F70** _____ [502.8.1]
- ☐ End notches max ¼ depth of member **F70** _____ [502.8.1]
- ☐ No notches in tension side of members ≥ 4 in. except at ends _____ [502.8.1]
- ☐ Holes 2 in. min to top or bottom or notch or other hole **F70** _____ [502.8.1]
- ☐ Engineered product (TJI®'s) notches, holes, or cuts AMI only _____ [502.8.2]
- ☐ No field modification of trusses, e.g., notching, cutting _____ [802.10.4]
- ☐ No alteration or loading of trusses without written concurrence of design professional _____ [802.10.4 & 2405.2]

TABLE 34	NOTCHING & BORING JOISTS [502.8.1]			
Nominal[A] Dimension Joist or Girder	Max. Diameter Bored Hole	Max. Notch Length	Max. Notch Depth Outer ⅓	Max. Depth End Notch
6	1½ in.	1¾ in.	⅞ in.	1⅜ in.
8	2⅜ in.	2⅜ in.	1³⁄₁₆ in.	1⅞ in.
10	3¹⁄₁₆ in.	3¹⁄₁₆ in.	1½ in.	2⅜ in.
12	3¾ in.	3¾ in.	1⅞ in.	2⅞ in.

A. Table numbers based on actual dimensions: typically 5½, 7¼, 9¼ & 11¼.

Stud Notching & Boring

- ☐ Notching 25% max in bearing wall, 40% nonbearing **F71, T35** _____ [602.6]
- ☐ Bored holes min 5/8 in. from face of stud_____ [602.6]
- ☐ Boring 40% max in bearing wall, 60% nonbearing EXC **F71, T35**____ [602.6]
  - • 2 successive doubled bearing studs 60% OK **F71** _____ [602.6]
- ☐ Holes not in same stud section as cuts or notches _____ [602.6]
- ☐ If top plate > 50% removed for pipes, attach galvanized 16-gage metal tie 1½ in. wide with min 8-10d nails each side of notch EXC _____ [602.6.1]
  - • If entire side of wall with notch covered with wood sheathing ___ [602.6.1X]

TABLE 35	STUD NOTCHING & BORING [602.3(5)]		
Bearing Walls:	**2×4**	**3×4**	**2×6**
Max. notch depth	7/8 in.	7/8 in.	1 3/8 in.
Max. diameter bored hole	1 3/8 in.	1 3/8 in.	2 3/16t in.
Max. diameter bored hole doubled studs	2 in.	2 in.	3 1/4 in.
Nonbearing Walls:	**2×3**	**2×4**	**3×4**	**2×6**
Max. notch depth	1 in.	1 3/8 in.	1 3/8 in.	2 3/16 in.
Max. diameter bored hole	1 1/2 in.	2 in.	2 in.	3 1/4 in.

FIG. 71

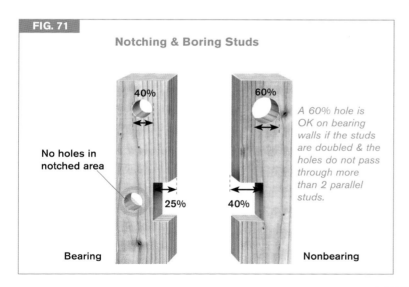

Notching & Boring Studs

40%

No holes in notched area

25%

Bearing

60%

40%

Nonbearing

A 60% hole is OK on bearing walls if the studs are doubled & the holes do not pass through more than 2 parallel studs.

In 1752, Benjamin Franklin brought the first bathtub to America. After designing a more comfortable model, he took it with him on his travels to Europe.

STRUCTURAL

PLUMBING

GLOSSARY

A

AAV (air admittance valve): One-way valve designed to admit air into the plumbing system during periods of relative negative pressure in the vent system to protect traps from siphonage and to remain sealed under zero or positive pressure. F39,40

ABS (acrylonitrile-butadiene-styrene): A plastic pipe that is usually black and used for DWV. It can also be used for building water supply and for vent piping of high-efficiency condensing appliances.

Accessible: Capable of being exposed without damage to the building or component structure or finishes, and that may require removal of access doors or fasteners requiring tools.

AHJ (Authority Having Jurisdiction): An organization responsible for enforcing the code, typically the building department and its authorized representatives.

Airbreak: A physical separation in which a discharge pipe from a fixture, appliance, or device drains indirectly into a receptor and enters below the flood level rim of a receptor, such as a clothes washer standpipe.

Air chamber: A pressure surge-absorbing device operating through the compressibility of air. Expansion tanks have air chambers. Mechanical water hammer arresters have replaced air chambers in piping systems.

Air gap (for drainage systems): Unobstructed vertical distance through free atmosphere between the outlet of a waste pipe and the flood level rim of the receptacle into which the waste pipe is discharging. F44

Air gap (for water distribution systems): An unobstructed vertical distance through the free atmosphere between the lowest opening from any pipe or faucet supplying water to a tank, plumbing fixture, or other device and the flood level rim of the receptacle.

Antisiphon valves: Valves or devices that prevent siphoning, typically by an opening to the atmosphere.

Approved: Accepted by the AHJ. UL and other testing laboratories do not approve materials; they test products and determine their conformity to published standards. Only the AHJ can approve them.

B

Backflow: A flow of water or other liquids, mixtures, or substances into the distributing pipes of a potable supply of water from any source other than its intended source.

Backflow connection: Any arrangement whereby backflow can occur.

Backflow preventer: A device or means to prevent backflow into the potable water system. These include air gaps, vacuum breakers, pressure principle backflow prevention assemblies, and reduced pressure principle backflow prevention assemblies.

Backpressure: A potential backflow condition, created by pressure in the potential backflow source higher than the pressure in the water main.

Backsiphonage: Backflow caused by a loss of supply pressure.

Backwater valve: A device that prevents the backflow of sewage. F17-19

Ball cock: A valve in a toilet tank to control the supply of water into the tank.

Bathroom: In ASHRAE, a bathroom is a room containing a tub, shower, spa, or other source of moisture. A half bath contains only a water closet and lavatory & is not considered a bathroom. In the NEC, a bathroom is a room containing a basin and another plumbing fixture.

Bathroom group: A group of fixtures including a water closet, one or two lavatories, and bathtub, combination bath/shower, or a shower. The group can also include a urinal or bidet and an emergency floor drain.

Bonding jumper: A conductor installed to electrically connect metal gas piping to the grounding electrode system.

Branch: Any part of the drains except a main, riser, or stack.

Branch interval: A vertical measurement of distance, at least 8 ft., between connections of horizontal branches to a drainage stack, measured down from the highest horizontal branch connection. F16

Branch vent: A vent connecting two or more individual vents with a vent stack or stack vent. F29

Btu (British thermal unit): The quantity of heat necessary to raise the temperature of 1 lb. of water 1°F.

Building drain: The lowest piping of a drainage system that conveys discharge of soil, waste, and drainage pipes in the building to the building sewer beginning 30 in. (24 in. in UPC) outside the building wall.

Building sewer: Horizontal piping from a drainage system extending from the building drain to the public or private sewer or private sewage disposal system.

C

Check valve: A device used to prevent the flow of liquids in a direction not intended in the design of the system. Check valves are not backflow preventers. They are often used in solar systems.

CO (cleanout): An opening for cleaning the drainage system. *See p.75.*

Common vent: A pipe venting two trap arms on the same branch, either back-to-back or one above another. F31, 33

Concealed: Not exposed to view without removal of building surfaces or finishes.

Contamination: Impairment of potable water quality that creates a health hazard.

Continuous waste: A drain from 2 or more adjacent fixtures connecting the compartments of a set of fixtures to a trap or connecting other permitted fixtures to a common trap. An example is a double kitchen sink with each side connecting together and then to the trap. F22, 23

CPVC (chlorinated polyvinyl chloride): Plastic pipe designed for hot and cold water. Water distribution pipe is typically cream-colored, and orange CPVC is used for automatic fire sprinkler piping.

Cross-connection: A backflow connection or other arrangement whereby a potable water system may be contaminated.

Crown weir (trap weir): Highest point of the inside portion of the bottom surface of the horizontal waterway at the exit of a trap. Water flows into the trap arm once it rises above the crown weir. F20

DWV (drain, waste, and vent): The system of piping and fittings that carries drainage, waste, and sewer gases and that equalizes atmospheric pressure at traps to protect the occupants from the contaminated gases in the system.

E

Effective opening: The cross-sectional area of a water outlet, expressed in terms of the diameter of an equivalent circle. In the case of a faucet, measured at the smallest orifice or in the supply piping.

F

Fireblock: Building materials installed to resist the free passage of flame to other areas through small concealed spaces of the building.

Firestop: Until the early 1990s, this term was used for what today is called fireblocking. A penetration firestop assembly is a group of materials installed to resist free passage of flame through an assembly, typically around a duct, vent, or chimney passing through a rated ceiling, floor, or wall.

Fixture branch drain: A drain serving two or more fixtures that discharges to another drain or to a stack.

Fixture branch supply: A water supply pipe between the fixture supply pipe & the water distributing pipe.

Fixture drain: A drain from a fixture trap to a junction with any other drain; also called a trap arm. F25

Fixture unit: A unit of measure for the drain (DFU) or water supply (WSFU) load from different fixtures.

CW&V (combination waste drain & vent system): A system employing horizontal and/or vertical pipes functioning as vents and drains for sinks and floors by providing for free movement of air above the water line in the horizontal pipe. CW&Vs are not self-scouring and their use is restricted. F38

D

Developed length: The distance measured along the centerline of a pipe.

(DFU) Drainage fixture unit: A value used to calculate a fixture's load on the drainage system. T6

Directional fittings: Drainage fittings designed to join two pipes and direct their flow, such as wyes, combos, or tees with baffles. F22, 23

Discharge pipe: A pipe that conveys the discharge from a fixture or plumbing appliance.

Draft hood: A nonadjustable device integral to an appliance or made part of the appliance connector. It provides for the escape of flue gases from the appliance in the event of insufficient draft, backdraft, or stoppage. A draft diverter (typical on water heaters) prevents backdraft from entering the appliance and neutralizes the stack effect on the operation of the appliance.

Drain: A pipe carrying soil and water-borne waste.

Drainage system: Includes all the piping within public or private premises, which conveys sewage to a legal point of disposal.

FLR (flood level rim): The level at which water overflows from a fixture or the surface to which it is fastened. **F28**

G

Gas connector: Tubing or piping that connects the gas supply piping to the appliance.

Grade: The slope or fall of a line of pipe in reference to a horizontal plane. In drainage, it is usually expressed as the fall in vertical units compared to horizontal units or a fraction of an inch per foot length of pipe, such as ¼ in. per ft.

H

Hangers: *See "Supports."*

Horizontal: Any pipe or vent that is less than 45° from horizontal.

Horizontal branch drain: A drainage pipe that extends from a stack or drain and that serves two or more fixtures.

Hot water: Water at a temperature of 110°F (120°F in the UPC).

I

In sight: *See "Within sight".*

Indirect-fired water heater: A water heater with a storage tank with a heat exchanger used to transfer heat from an external source to heat potable water. The storage tank could derive its heat source from an external source, such as solar or a boiler, or an internal source.

Indirect waste pipe: A waste discharge into the drainage system through an airbreak into a trap, fixture, or receptor, such as a clothes washer standpipe. **F62**

Individual vent: A pipe that vents a fixture trap. **F25**

J

Joint: Connection between two pipes:

Brazed joint: Any joint obtained by joining metal parts with alloys that melt at temperatures above 840°F (449°C), but lower than the melting temperature of the parts to be joined.

Expansion joint: A loop, return bend, or return offset that accommodates pipe expansion and contraction.

Flexible joint: A joint that allows movement of one pipe without deflecting the other pipe.

Mechanical joint: A joint that uses compression to seal the joint.

Slip joint: A joint that incorporates a washer or special packing material to create a seal. **F22, 24**

Soldered joint: A joint obtained by joining of metal parts with metallic mixtures or alloys that melt at a temperature less than 800°F (427°C) & above 300°F (149°C).

Welded joint or seam: Any joint or seam obtained by the joining of metal parts in the plastic molten state.

L

Label: A marking applied on a product that identifies the manufacturer, the function or designation of the product, and the agency that has evaluated a sample of that product.

Labeled: Equipment, materials, or products affixed with a label or other identifying mark to attest that the product complies with identified standards or has been found suitable for a specific purpose. See "Listed."

Liquefied petroleum (LP) gas: LP or propane gas is composed primarily of propane, propylenes, butanes, or butylenes or mixtures thereof that are gaseous under normal atmospheric conditions but capable of being liquefied under moderate pressure at normal temperatures. LP gas is typically stored in tanks on site. Unlike natural gas (CH_4), LP gas (C_3H_8) is heavier than air.

Liquid waste: A discharge from any fixture or appliance that does not receive fecal matter.

Listed: Equipment or materials on a list published by an approved organization that is concerned with product evaluation and that maintains periodic inspection of production of listed equipment or materials. The listing will state that the product meets specified standards or has been found suitable for a specific purpose.

Low-pressure hot-water heating boiler: A boiler furnishing hot water at pressures not exceeding 160psi or temperatures not exceeding 250°F.

Luminaire: A complete lighting fixture including the lamp(s), mounting assembly, and cover.

M

Main: A principal artery of any system of continuous piping to which branches may be connected.

MP regulator: A line pressure regulator that reduces gas pressure from a medium pressure to a range at which appliances can use it (typically 0.5 psig). These are often found in CSST systems with a central manifold.

N

Nationally Recognized Testing Laboratory (NRTL): A testing facility recognized by OSHA as qualified to provide testing and certification of products and services. Examples of NRTLs are CSA, NDF, and UL.

Natural gas: Gas, usually odorized methane, supplied to the site by a gas utility company and metered at the site. Natural gas is lighter than air.

O

Offset: A combination of elbows or bends in a line of piping that brings a section of the pipe out of line, but into a line parallel with the other section. F28

P

PEX tubing: Water-supply tubing made of cross-linked polyethylene, typically found in parallel distribution systems. See p.91. PEX-AL-PEX has a thin layer of aluminum sandwiched between layers of PEX. The aluminum serves as an oxygen barrier and helps overcome the bending memory of the PEX.

Plumbing system: The water-supply system, drainage system, storm drains, sewers, connected fixtures, supports, appurtenances, and appliances.

Potable water: Water fit for human consumption.

Pressure-balancing valve: A mixing valve that senses incoming hot and cold water pressures and compensates for fluctuations in either to stabilize outlet temperature. They can be built into shower and tub mixers or added as separate devices ahead of other individual fixtures.

Pressure-relief valve (PRV): A device designed to protect against high pressure and to function as a relief mechanism.

Public sewer: A sewer controlled by a public authority.

Public water main: Water supply pipe controlled by public authority.

R

Readily accessible: Access that does not require removing a panel or door. For electrical equipment, this also means not having to resort to use of a ladder.

Reduced-pressure principle backflow preventer (RPPBP): An RPPBP consists of two independently acting check valves, internally pressure-forced to the normally open position. These checks are separated by an intermediate chamber that is equipped with a means of automatic relief. Should there be a reversal of flow, the downstream liquid will drain instead of placing back pressure on the supply liquid. RPPBPs have a means of field testing.

Relief vent: A vent providing air circulation between vent and drainage systems.

Rim: An unobstructed open edge of a fixture. **F28**

Riser: A water or gas supply pipe that extends one or more stories.

Rough-in: Part of the plumbing system that is installed in a structure before fixture installation.

S

Sewage: Liquid waste that contains chemicals or animal or vegetable matter.

Sewage ejector: A device for lifting sewage at high velocity with air or water. **F18,19**

Sewage pump: A device, other than an ejector, for lifting sewage from a sump.

Shielded coupling: An approved elastomeric sealing gasket with an approved outer shield and a tightening mechanism (i.e., no hub coupling).

Slope: Fall or pitch along a line of pipe.

Soil pipe: A pipe that conveys waste including fecal matter.

Stack: A vertical drain line that extends one or more stories. **F16**

Stack vent: A vent that extends from a stack.

Stack venting: A method of venting fixtures through the soil or waste stack. **F16**

Static pressure: The pressure existing in a system without any flow.

Sump: A tank or pit that receives waste and is discharged by mechanical means. **F19**

Supports: Devices used to support or secure pipes, fixtures, or equipment. **F1**

T

Tailpiece: Pipe or tubing connecting the outlet of a plumbing fixture to a trap. F22

Temperature & pressure-relief valve (TPRV): A device designed to protect against high pressure or temperature and to function as a relief for either. F56

Top dip (of trap): *See "Upper dip".*

Trap: A fitting or device that employs a liquid seal to prevent the escape of sewer gas from the plumbing system. F20

Trap arm: A horizontal pipe between a trap and the connection to the drain and vent system—also called a fixture drain. F25

Trap seal: Vertical distance between the crown weir and the upper dip of the trap. F20

Tubular brass: Traps, waste bends, and tailpieces with slip-joint connections.

U

Upper dip: Highest point in the internal cross section of the trap at the lowest part of the bend. By contrast, the bottom dip is the lowest point in the internal cross section.

V

Vent (plumbing): A pipe or device for introduction of air into the plumbing system to equalize pressure, allow drainage, and prevent siphoning of trap seals.

Vent stack: A vertical vent pipe that provides air circulation for the drainage system.

Vertical: Any pipe or vent that is 45° or more from horizontal.

W

Waste: Liquid-borne waste free of fecal matter.

Waste pipe: A pipe that conveys only waste.

Water main: A water supply controlled by a public entity or utility.

Water pipe: A pipe that conveys water to fixtures and outlets.

Water supply system: Pipes, valves, fittings & supports that supply water to and throughout a residence & its accessories, such as sprinkler piping.

Weir: *See "Crown weir".* F20

Wet vent: A vent that also serves as drain. F32,37

Whirlpool bathtub (hydromassage tub): A bathtub fixture equipped & fitted with a pump and circulating piping & that is drained after each use. A spa is not a whirlpool tub.

Within sight: Within 50 ft. & with an unobstructed line of sight.

WSFU (water supply fixture unit): A measure of the estimated normal demand on the water supply by various types of plumbing fixtures. T22

No.	Page	Code No.	Description
1	75	IRC 3005.2.6	The 2006 IRC allowed cleanouts outside within 3 ft. of the building wall as an alternative.
2	78	UPC 710.1	Drain cleanouts through backwater valves require permanent labeling: "backwater valve downstream."
3	79	IRC 3007.3.5	Though this rule was implied by other rules, the 2006 IRC did not specifically require connection through a wye at the top of the gravity drain.
4	79	IRC 3007.6	The 2006 IRC specified the velocity, not the gpm, of the pump. T11 is new in the 2009 IRC.
5	89	IRC 3108.2.1	A water closet is now permitted to be located upstream of the dry vent connection to the horizontal wet vent.
6	90	IRC 3114.3	Branch or stack type AAVs can only vent fixtures on the same floor level & which connect to horizontal drains.
7	90	IRC 3114.8	AAVs may not vent sumps or tanks without an engineered design.
8	94	IRC T2903.1	The min. dynamic flow pressure at bathtubs, showers, and WCs was increased in 2009.
9	94	IRC 2903.7	The IRC WSFU method previously used was similar to the UPC method, and is now located in Appendix P, which typically requires local adoption to be enforceable. The new method is based on conversion of WSFUs to GPM flow rates. IRC Appendix P provides examples.
10	96	IRC 2903.7	The 2006 IRC applied an 80% multiplier to the minimum daily supply pressure used as the basis of pipe size calculations. The rule is now in Appendix P in the 2009 IRC.
11	98	IRC 2415.4	The 2009 IRC does not allow any gas pipe to penetrate a foundation below grade, including sleeved underground penetrations to basements.
12	98	IRC 2415.6 & 2415.12	The 2009 IRC distinguishes between gas piping inside a concrete slab and piping below the slab.
13	98	IRC 2415.6.2 2415.12.2	In the 2006 IRC a conduit beneath or encased in a slab with both ends terminating indoors would have both ends sealed. In 2009 neither is sealed.
14	99	IRC 2420.5.3	Appliance shutoff valves no longer have to be in the same room and within 6 ft. of the appliance if they are clearly labeled and located at a manifold within 50 ft. and serving no other purpose.
15	99	IRC 2422.1.2.1	In the 2006 IRC the length was 3 ft. except for dryers and ranges & rigid pipe.
16	101	UPC 1209.4.3	The 2006 UPC & UMC restricted the "branch length" method to 250 cu. ft./hr. and did not acknowledge the "longest length" method.
17	105	IRC 2803.6.1	The IRC now explicitly allows the TPR discharge piping to drain to the water heater pan.
18	106	IRC 2801.5	All water heaters (including tankless) req pan in 2009 IRC if leakage could cause damage.
19	107	IRC 2719.1	Appliances may not restrict access to floor drains or be located over floor drains.
20	108	IRC 2706.2.1	A standpipe receiving continuous waste of a laundry tray must be above the laundry tray FLR.
21	109	IRC 2705.1	Toilets and bidets now require 15 in. clearance from their centerline to adjacent vanity or partition.

PLUMBING CODE CHANGES

PLUMBING

In colonial America, most homes were warmed by building a fire in a fireplace. This method resulted in sending most of the heat up the chimney, using a lot of wood and causing many house fires. In 1742, Ben Franklin invented an iron furnace stove, equipped with loosely fitting iron plates through which air circulated & warmed before passing into the room. It warmed homes more efficiently, less dangerously & with less wood—resulting in less air pollution. He named this furnace stove the "Pennsylvania Fireplace," although today it is known as the "Franklin Stove."

Code ☑Check® Mechanical Fourth Edition

By DOUGLAS HANSEN & REDWOOD KARDON
Illustrations & Layout by Paddy Morrissey

For more information on the building, electrical, and mechanical codes, valuable resources, and why Benjamin Franklin is featured in the Code Check series, visit www.codecheck.com.

C*ode Check Mechanical 4th edition* is an illustrated guide to common code questions in residential heating, ventilation & air conditioning systems. The book emphasizes the safety principles that are at the heart of the codes for these systems.

The primary code used in this book is the 2009 edition of the *International Residential Code for One- & Two-Family Dwellings*, published by the International Code Council (ICC). It is the most widely used residential code in the United States. The other major code referenced here is the *2009 Uniform Mechanical Code*, published by the International Association of Plumbing & Mechanical Officials (IAPMO). For most topics, these different codes are in agreement. Each of these codes also references standards, many of which are maintained by the organizations in Table 2 (**T2**).

Additional codes for specialized items are listed in **T1**. The National Fire Protection Association (NFPA) publishes several of these.

The 2009 cycle of codes is likely to remain in effect in most areas for at least 3 or 4 years after the cover date. Energy codes vary greatly from one area to another & may modify or overrule the code requirements shown in this book. Before beginning any project, check with your local building department to determine the codes that apply in your area.

Thanks to Hamid Naderi of ICC for his editorial input.

TABLE 1		CODES & STANDARDS USED IN THIS BOOK
Organization	**Edition**	**Code**
ASHRAE	2010	ASHRAE 62.2 Ventilation & Acceptable Indoor Air Quality in Low-Rise Residential Buildings
IAPMO	2009	Uniform Mechanical Code
ICC	2009	International Residential Code
NFPA	2011	NFPA 31 Standard for Installation of Oil-Burning Equipment
NFPA	2011	NFPA 58 Liquefied Petroleum Gas Code
NFPA	2010	NFPA 211 Standard for Chimneys, Fireplaces, Vents & Solid Fuel-Burning Appliances
NFPA	2011	NFPA 70 National Electrical Code

KEY TO USING CODE CHECK

Code Check Mechanical condenses large amounts of code information by using several "shorthand" conventions that are explained here. Each rule described in Code Check begins with a checkbox and ends with the code citations. When only one code is shown, the code citation is inside of brackets, and when two codes are shown, the second code is shown inside of braces, as in the following example from p. 128.

☐ Maintain accessibility for service of gas appliances _____ [1305.1] {304.1}

This line is stating that all appliances must be accessible for service. The rule is found in section 1305.1 of the IRC & section 304.1 of the UMC.

TABLE 2	ORGANIZATIONS
Acronym	Name
ACCA	Air Conditioning Contractors of America
ASHRAE	American Society of Heating, Air Conditioning & Refrigeration Engineers
ASME	American Society of Mechanical Engineers
ASTM	ASTM International (formerly the American Society for Testing & Materials)
CSA	Canadian Standards Association
IAPMO	International Association of Plumbing & Mechanical Officials
ICC	International Code Council
NFPA	National Fire Protection Association
UL	Underwriters Laboratories

These lines are stating that combustion air is required for natural draft appliances except direct-vent appliances installed in accordance with the manufacturer's instructions. The abbreviations C.A. & AMI are explained in the list on p. 125.

☐ C.A. req'd for natural draft appliances EXC [2407.1] {701.1.1}
• Direct-vent appliances installed AMI [2407.1] {701.1X1}

A line ending in EXC means that an exception to the rule is contained in the line that follows. The following example is from p.134.

Single-Wall Vents   09 IRC   09 UMC

☐ Not allowed in dwellings _____ [n/a] {802.7.4.1}⁹

This line is saying that single-wall gas appliance vents are not allowed in dwellings. The IRC does not have such a rule, so the citation there is "n/a." In the UMC, the rule is in section 802.7.4.1 & it is a change from the 2006 edition. The change is explained further on p. 168.

A change from the previous code edition is shown by a code citation in a different color. The superscript blue number after the code citation refers to the table on p.168, where more information about the change is found. The following example is from p.144.

Gas Vent Terminations   09 IRC   09 UMC

☐ B vents ≤ 12 in. diameter per F31, T8 if >8 ft. from wall _____ [2426.4] {802.6.2}

This line is stating that B vents more than 8 feet from a wall are to be installed in accordance with rules shown in Figure 31 & Table 8.

References to figures & tables are preceded by an F or a T as in the following example from p.147.

ABBREVIATIONS

AC	=	air conditioning
ACH	=	air changes per hour
AHJ	=	Authority Having Jurisdiction
AMI	=	in accordance with manufacturer's instructions
ASTM	=	ASTM International, formerly the American Society for Testing & Materials
AWG	=	American Wire Gage
BO	=	building official
Btu	=	British thermal unit
C.A.	=	combustion air
cfm	=	cubic feet per minute
CPVC	=	chlorinated PVC pipe
cu.	=	cubic, as in cubic feet
Cu	=	copper
EGV	=	equipment grounding conductor
EXC	=	exception to rule will follow in the next line
FAU	=	forced-air unit (central heater)
ft.	=	foot/feet
gal	=	gallon(s)
gpm	=	gallons per minute
HP	=	heat pump
hr	=	hour(s)
in.	=	inch(es)
kBtu	=	1,000 Btu

L&L	=	listed & labeled
lb.	=	pound(s)
LP	=	liquefied petroleum (LP gas)
manu	=	manufacturer, manufacturer's
max	=	maximum
min	=	minimum
no.	=	number
o.c.	=	on center
PE	=	polyethylene tubing
PEX	=	crossed-link polyethylene tubing
PL	=	property line
PRV	=	pressure relief valve
psf	=	pounds per square foot
psi	=	pounds per square inch
psig	=	pounds per square inch gage
PVC	=	polyvinyl chloride pipe
req	=	require
req'd	=	required
req's	=	requires
SDC	=	Seismic Design Category
sq.	=	square, as in sq. ft.
TPRV	=	temperature & pressure relief valve
V	=	volt(s)
WH	=	water heater(s)

TABLE OF CONTENTS

GENERAL MECHANICAL SYSTEM REQUIREMENTS

The IRC & UMC have many sections that are identical except for their numbering. The model code for systems using natural gas is *NFPA 54–The National Fuel Gas Code* & many provisions of the more commonly adopted IRC & UMC are taken directly from it.

Administration
	09 IRC	09 UMC
□ New installations, alterations & repairs req permits EXC	[105.1]	{11.1}
• Portable equipment & minor replacement parts	[105.2]	{11.2}
• Components (including piping) within L&L equipment	[105.2]	{11.2}
□ BO or AHJ may accept alternative materials, methods & equipment	[104.11]	{105.0}
□ Installations not covered by IRC must comply with International Mechanical Code or International Fuel Gas Code	[1301.1]	{n/a}
□ Continued use of existing installations OK if safe & compliant with code at time of construction	[1202.2]	{104.2}

Listing & Labeling
	09 IRC	09 UMC
□ All appliances L&L	[1302.1]	{302.1}
□ Install listed appliances AMI & per listing	[1307.1 & 1401.1]	{303.1}
□ Attach installation & operating instructions to appliance	[1307.1]	{303.1}
□ Fuel-fired factory-applied nameplates must include:	[1303.1]	{306.1}

• Manu name, model & serial number
• Operating & maintenance instructions
• Hourly Btu rating
• Type of fuel
• Req'd clearances from combustibles

Appliance Maintenance
	09 IRC	09 UMC
□ Maintain system in proper working order	[1202.3]	{104.4}
□ BO may order reinspections to determine compliance	[1202.3]	{104.4}

Minimum heating requirements
	09 IRC	09 UMC
□ Habitable rooms req installed heat source capable of min 68°F at 3 ft. above floor & 2 ft. from exterior wall	[303.8]	{n/a}
□ Portable space heater not OK to meet above rule	[303.8]	{n/a}

APPLIANCE ACCESS & LOCATION

Appliances must remain accessible for inspection, service, repair & replacement without the need to remove permanent construction. Appliances must be located where they are not subject to flooding or damage & with adequate clearances from combustible surfaces. Appliances in under-floor areas & attics have specific rules. The IRC does not address appliances on roofs, though it does state that in the absence of specific rules, those from the International Mechanical Code can apply.

Flood Elevation
	09 IRC	09 UMC
□ Locate above design flood elevation EXC	[1301.1.1]	[3072]
• If designed to prevent water entry & resist buoyancy	[322.1.6X]	[3072]
□ No equipment mounted on breakaway flood walls	[322.3.4]	[3072.1]

Appliance Access
	09 IRC	09 UMC
□ Maintain accessibility for service of gas appliances	[1305.1]	{304.1}
□ Min 30 x 30 in. level work space on control side EXC	[1305.1]	{304.0X}
• 18 in. deep space OK for room heaters	[1305.1]	{304.0X}
□ Equipment room door min 24 in. wide F32	[1305.1.2]	{304.0}
□ Equipment must fit through door F32	[1305.1.2]	{304.0}
□ Oil-burning FAUs in alcoves per F32 EXC	[1305.1.1]	{n/a}
• Replacement appliances clearances AMI	[1305.1.1X]	{n/a}

Appliances under Floors | 09 IRC | 09 UMC

☐ Access opening min rough-framed size 22 × 30 in. __ [1305.1.4] {904.11.1}
☐ Appliance must fit through opening_____ [1305.1.3] {904.11.1}
☐ Passageway min 22 in. wide × 30 in. high_____ [1305.1.4] {904.11.1}
☐ Passageway max 20 ft. long EXC _____ [1305.1.4] {904.11.2}
  • Passageway ≥ 6 ft. high OK for unlimited length__ [1305.1.4X2] {904.11.2}
☐ Min 30 × 30 in. level space on service side **F1** _____ [1305.1.4] {904.11.4}
☐ Support on concrete slab above ground or suspend from
  floor min 6 in. above ground **F1**_____[1305.1.4.1] {n/a}
☐ Excavations min 6 in. below appliance, 12 in. on sides,
  30 in. on control side **F1** _____[1305.1.4.2] {n/a}
☐ If excavation > 12 in. below adjacent grade, line with concrete
  extending 4 in. above adjacent grade **F1** _____ [1305.1.4] {n/a}
☐ Luminaire & receptacle outlet req'd near appliance__[1305.1.4.3] {904.11.5}
☐ Switch for luminaire req'd at passageway entrance _[1305.1.4.3] {904.11.5}

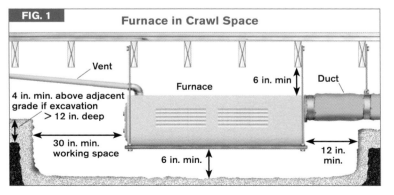

FIG. 1

Furnace in Crawl Space

Vent

4 in. min. above adjacent grade if excavation > 12 in. deep

30 in. min. working space

Furnace

6 in. min.

6 in. min.

Duct

12 in. min.

Heating & Cooling Appliances in Garages | 09 IRC | 09 UMC

☐ Ignition source min 18 in. above floor EXC _____ [1307.3] {307.1}
  • Appliances in separate enclosed space accessible from
    outside garage & no combustion air from garage _____ [1307.3] {307.1}
☐ Protect appliances from impact_____[1307.3.1] {307.1}
☐ Ducts through common wall to house min 26-gage steel [302.5.2] {n/a}
☐ No duct openings into garage _____ [302.5.2] {n/a}
☐ Openings around duct penetrations through common wall
  sealed with approved materials _____ [302.5.3] {n/a}

Heating & Cooling Appliances in Attics | 09 IRC | 09 UMC

☐ Rough-framed access opening min 22 × 30 in._____ [1305.1.3] {904.11.1}
☐ Appliance must fit through opening_____ [1305.1.3] {904.11.1}
☐ Solid floor min 24 in. wide from entrance to appliance [1305.1.3] {904.11.3}
☐ Max 20 ft. from access opening to appliance EXC ___ [1305.1.3] {904.11.2}
  • 50 ft. OK if passageway ≥ 6 ft. high
    (unlimited length OK in UMC)_____ [1305.1.3X2] {904.11.2}
☐ Min 30 × 30 in. level work platform EXC _____ [1305.1.3] {904.11.4}
  • OK to omit if service possible from access opening [1305.1.3X1] {∅}[1]
☐ Luminaire & receptacle req'd near appliance **F17** ___[1305.1.3.1] {904.11.5}
☐ Switch for luminaire req'd at entrance **F17**_____[1305.1.3.1] {904.11.5}

Clearances from Combustibles | 09 IRC | 09 UMC

☐ Install with clearances AMI_____[1306.1 & 2409.3.1] {303.1}
☐ Appliances in rooms "large in comparison to size of
  equipment" (not in alcove) clearance reduction
  allowed **T9** _____ [1306.2& 2409.3.3] {303.2}
☐ Appliances installed in alcoves or closets must be L&L for
  same & no clearance reduction allowed_____ [2409.3.2] {303.2X}

AIR-CONDITIONING & HEAT PUMPS

Central air-conditioning efficiency is measured in seasonal energy efficiency rating (SEER) & heat pumps are measured in heating seasonal performance factor (HSPF). The numbers are higher as efficiency is increased. Min efficiency standards for heat pumps & air conditioners have increased in recent years. Tax credits have been available for consumers upgrading their equipment to meet or exceed certain standards. As of this writing, the Department of Energy min SEER is 13, the min for an Energy Star rating is 14 & the min for a tax credit is 16. When replacing an existing system, both the indoor & outdoor units will likely have to be replaced to qualify.

Air-Conditioning (AC) & Heat Pumps (HPs)	09 IRC	09 UMC
☐ Factory-applied nameplates must include: _____	[1303.1]	{306.3}
• Label with manu name, model & serial number		
• Operating & maintenance instructions or publication number of manual		
• Rating in volts, amperes & Btus or watts, no. of phases if > 1		
• Req'd clearances		
☐ HP return air duct min 6 sq. in. per kBtu output _____	[1403.1]	{manu}
☐ Outdoor heat pump unit on min 3 in. raised pad **F2** ___	[1403.2][2]	{1106.2}
☐ Furnace with cooling coil reqs pressure capacity min 0.5 in. water column or L&L for cooling _____	[1411.2]	{904.8A&B}
☐ Cooling coil downstream from heat exchangers unless L&L for upstream (stainless steel heat exchanger) _____	[1411.2]	{904.8C}
☐ Central AC req's air filter **F2** _____	[1401.1]	{311.2}
☐ Condenser not near clothes dryer vent_____	[manu]	{manu}
☐ Refrigerant vapor (suction) lines insulated min R4 **F2**___	[1411.5]	{manu}

Refrigerant tubing should be secured, supported & protected from damage as recommended by the manu. Underground tubing must be protected from corrosion.

Window & Through-Wall Units	09 IRC	11 NEC
☐ Must have equipment grounding conductor; no adapters to existing 2-slot receptacles _____	[3908.1]	{440.61}
☐ Max cord length 10 ft. if 120V, 6 ft. if 240V _____	[manu]	{440.64}
☐ Cord plug OK as disconnect if controls ≤ 6 ft. of floor ___	[manu]	{440.63}
☐ Arc fault circuit interrupter (AFCI) or leakage current detection interrupter (LCDI) req'd in attachment plug _____	[manu]	{440.65}
☐ Max load rating 80% of individual circuit _____	[3602.12.1]	{440.62B}
☐ Max load rating 50% of shared circuit _____	[3602.12.2]	{440.62C}

Condensate Disposal	09 IRC	09 UMC
☐ Condensate may not drain to public way _____	[1411.3]	{309.1}
☐ Drainpipe min ¾ in. with min ⅛ in./ft. slope **F2**_____	[1411.3]	{309.1}
☐ Threaded female PVC fittings only on plastic male _____	[manu]	{309.5}
☐ May drain to indirect waste (lavatory tailpiece, tub overflow) **F2** _____	[1411.3]	{309.1}
☐ No direct connection to waste or vent pipe _____	[1411.3]	{309.1}
☐ If condensate stoppage would damage building components, install one of the following methods: **F2** _____	[1411.3.1]	{309.2}
• Secondary drain to conspicuous point of disposal __	[1411.3.1]	{309.2}
• Auxiliary drain pan with drain to conspicuous point__	[1411.3.1]	{309.2}
• Auxiliary drain pan with detector & drain fitting _____	[1411.3.1][3]	{n/a}
• Water level detection in primary with interlocked cutout	[1411.3.1]	{n/a}
☐ Down-flow units with no secondary & no way to install auxiliary drain pan req internal blockage detector with interlocked cutout _____	[1411.3.1.1]	{n/a}
☐ No drilling (saddle fittings) of drain, waste, or vent (DWV) pipes to accept condensate drain _____	[3003.2]	{UPC}

FIG. 2

Heat Pump Operating in Cooling Mode

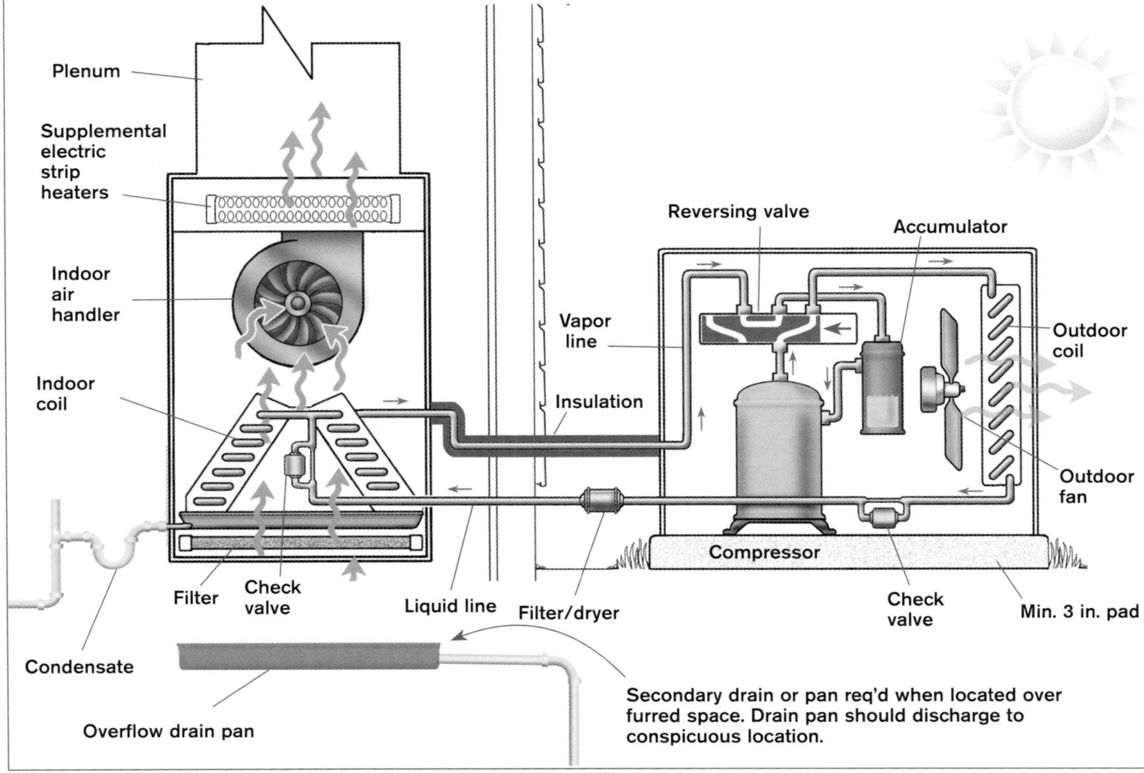

Plenum

Supplemental electric strip heaters

Indoor air handler

Indoor coil

Filter

Check valve

Condensate

Overflow drain pan

Vapor line

Insulation

Liquid line

Filter/dryer

Reversing valve

Accumulator

Outdoor coil

Outdoor fan

Outdoor fan

Check valve

Min. 3 in. pad

Compressor

Secondary drain or pan req'd when located over furred space. Drain pan should discharge to conspicuous location.

FIG. 3

Reversing Valve in Heating Mode

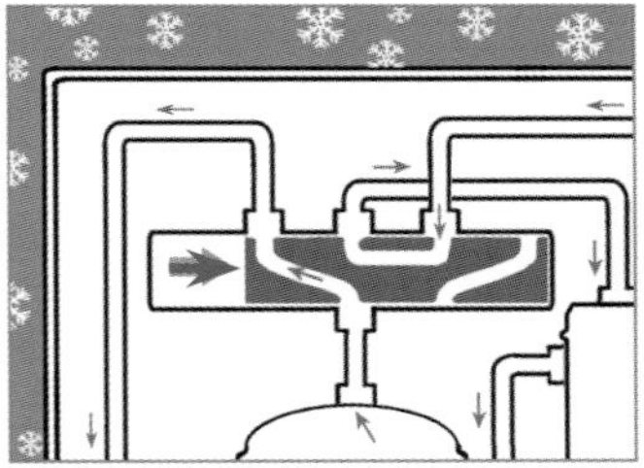

Heat pumps can be used for both heating & cooling.

*A **reversing valve** determines the direction of refrigerant flow.*

*In **heating mode,** the outdoor coil extracts heat from the atmosphere & the indoor coil gives up that heat to the interior space.*

When the outdoor temperature is below the balance point, supplementary electric strip heaters are activated in the indoor air handler.

*In **cooling mode,** the system operates as in **F2**.*

MECHANICAL   **AIR-CONDITIONING & HEAT PUMPS**   131

DUCTS

General · 09 IRC · 09 UMC

☐ Factory-made ducts L&L & installed AMI _____ [1601.2] {602.3}

☐ Max 2 stories for vertical riser on factory-made duct _____ [n/a] {604.3}

☐ Fireblocking around duct penetrations between floors _[302.11] {n/a}

Return Air · 09 IRC · 09 UMC

☐ Must be open to min 25% area served _____ [1602.2 & 2442.5] {311.3}[4]

☐ Not from bathroom, kitchen, mechanical room, closet, garage, or separate dwelling unit _____ [1602.2 & 2442.5] {311.3&4}[4]

☐ Sole return can be from room with fuel-burning equipment if supply air provided, return min 10 ft. from equipment & room volume min 100 cu. ft./kBtu of equipment _____ [1602.2, 2442.5] {311.3}[4]

☐ Duct min size 2 sq. in./kBtu output rating _____ [2442.2] {n/a}

Installation · 09 IRC · 09 UMC

☐ Ground clearance min 4 in. _____ [1601.4.7] {604.2&3}

☐ 18 in. vertical clearance where needed under duct to prevent cutting off access to crawl space _____ [n/a] {604.1A}

☐ Min 2 in. encasement for ducts under/in concrete ___ [1601.1.2] {604.2}

☐ Duct joints sealed, taped, or gasketed airtight _____ [1601.4.1] {602.4}

☐ Flex ducts req UL-181B pressure-sensitive tape **F4,6** _____ [1601.4.1] {602.4}

☐ Flex duct per L&L (1 1/2 in. strap every 4 ft.) **F5** _____ [1601.4.3] {T6-10}

☐ Round metal horizontal duct support: min 1/2 in. metal straps or 12-gauge wire every 10 ft. **F5** _____ [1601.4.3] {T6-7}

☐ Stud cavities prohibited as supply air duct [1103.2.3 & 1601.1.1] {n/a}

☐ Stud cavities & joist spaces OK as returns if no condensation [not conveying air from one floor to another] _____ [1601.1.1] {602.1}

Insulation in Unconditioned Space · 09 IRC · 09 UMC

☐ R-6 min in floor trusses, R-8 elsewhere EXC _____ [1103.3] {605.0}

• Ducts completely inside building thermal envelope [1103.2.1X] {605.0XB}

Note: Energy codes may take precedence over mechanical code insulation reqs.

FIG. 4

Stretch Flex Duct

Manufactured ducts must be supported & stretched so they are as straight as possible without kinks that obstruct airflow.

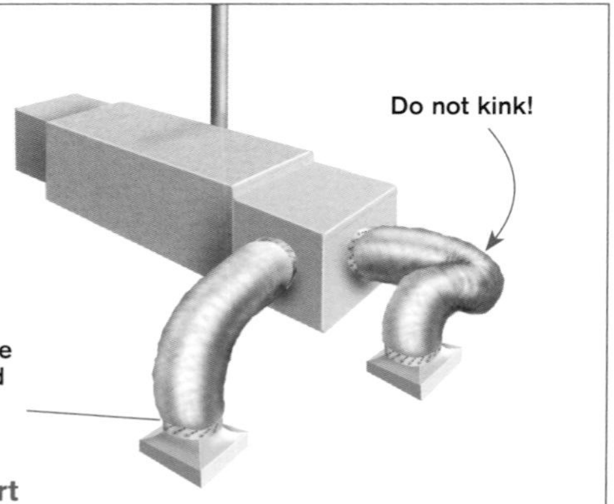

Do not kink!

Tape must be marked UL181.

FIG. 5

Flex Duct Support

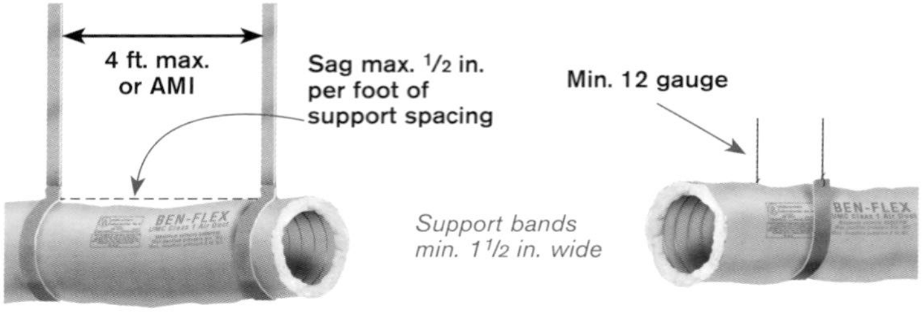

4 ft. max. or AMI

Sag max. 1/2 in. per foot of support spacing

Min. 12 gauge

Support bands min. 1 1/2 in. wide

FIG. 6 — Duct Splices

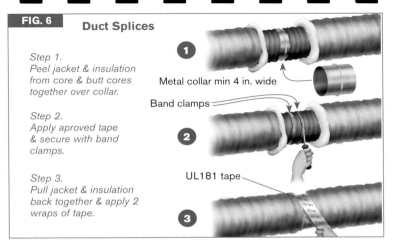

Step 1.
Peel jacket & insulation from core & butt cores together over collar.

Step 2.
Apply aproved tape & secure with band clamps.

Step 3.
Pull jacket & insulation back together & apply 2 wraps of tape.

Metal collar min 4 in. wide

Band clamps

UL181 tape

EVAPORATIVE (SWAMP) COOLERS

In dry climates evaporative coolers can reduce the sensible temperature & provide fresh air to the building interior. Care must be taken in locating these to ensure that objectionable odors are not brought into the building. They are also used to provide makeup air to commercial kitchens.

General F7	09 IRC	09 UMC
☐ Install AMI	[1413.1]	{304.1}
☐ Ground-mounted units secured in place & level base min 3 in. above adjoining ground	[1413.1]	{405.3}
☐ Platform-mounted unit min 6 in. above adjoining ground	[n/a]	{405.3}
☐ Provide flashing at openings into building	[1413.1]	{405.3}
☐ OFF switch or disconnect in sight if motor > ⅛hp	[4101.5]	{308.0}

General F7 (cont.)	09 IRC	09 UMC
☐ Backflow protection on supply (internal air gap OK)	[1413.2]	{UPC}
☐ Min 10 ft. horizontal clearance to plumbing or gas vents	[303.4.1]	{UPC}
☐ Electrical receptacle within 25 ft. (UPC) not req'd IRC	[3801.11X]	{308.0}

FIG. 7 — Evaporative (Swamp) Cooler

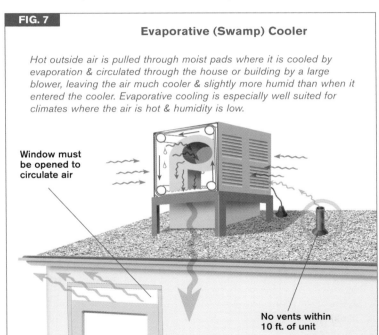

Hot outside air is pulled through moist pads where it is cooled by evaporation & circulated through the house or building by a large blower, leaving the air much cooler & slightly more humid than when it entered the cooler. Evaporative cooling is especially well suited for climates where the air is hot & humidity is low.

Window must be opened to circulate air

No vents within 10 ft. of unit

GAS APPLIANCE COMBUSTION AIR

General | 09 IRC | 09 UMC

Note: The IRC & UMC address combustion air only for gas-burning appliances. Oil-fired appliances are governed by NFPA 31 (see p.149).

- ☐ C.A. req'd for natural draft appliances EXC _____ [2407.1] {701.1.1}
  - Direct-vent appliances installed AMI _____ [2407.1] {701.1X1}
- ☐ Draft hood in same space as appliance_____ [2407.3] {701.1.3}
- ☐ Provide make-up air to offset exhaust fans (kitchen, bath)[2407.4] {701.1.4}

Mechanically Supplied Combustion Air | 09 IRC | 09 UMC

- ☐ Mechanical C.A. supply min 0.35 cu. ft./minute/kBtu__ [2407.9] {701.7}
- ☐ Appliance interlock req'd if mechanically supplied C.A. [2407.9.2] {701.8.2}

Openings | 09 IRC | 09 UMC

- ☐ Outside air openings req screens with mesh ≥ 1/4 in._ [2407.10] {701.9B}
- ☐ No screens allowed on ducts terminating in attic ____ [2407.11] {701.10}
- ☐ Net free area of louvers 75% for metal, 25% for wood [2407.10] {701.9A}
- ☐ Motorized louvers/dampers req appliance interlock __ [2407.10] {701.9C}

Ducts | 09 IRC | 09 UMC

- ☐ Duct galvanized metal or material of equivalent performance _____ [2407.11] {701.10}
- ☐ Ducts to outdoors min dimension 3 in. _____ [2407.6] {701.4}
- ☐ No manual dampers in C.A. ducts _____ [n/a] {701.11}
- ☐ Joist/stud space as C.A. duct ≤ 1 fireblock removed [2407.11X] {701.10X}
- ☐ Exterior openings min 12 in. above finished ground **F11, 12** _____ [2407.11] {701.10}
- ☐ Ducts may serve only 1 enclosure or appliance space [2407.11] {701.10}
- ☐ Horizontal ducts to upper part of enclosure may not slope down to source (upper duct not to originate from below) **F12** _ [2407.11] {701.10}
- ☐ Upper & lower ducts remain separate to source____ [2407.11] {701.10}

FIG. 8 — Vertical Ducts to Outdoors

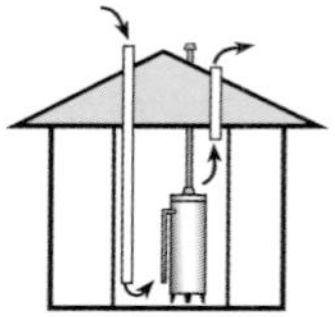

2 vertical ducts min. 1 sq. in./4 kBtu each

FIG. 9 — Vertical Ducts to Attic

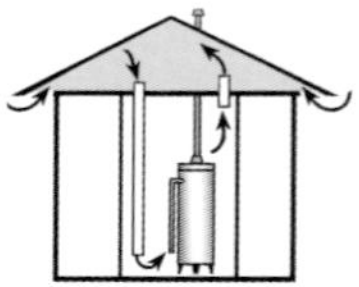

2 openings to ventilated attic min. 1 sq. in./4 kBtu each & sleeved min 6 in. above joist

Single-Opening Method | 09 IRC | 09 UMC

- ☐ Single direct exterior opening OK in upper 12 in. of enclosure min 1 sq. in./3kBtu & ≥ sum of vent connectors **F10**____ [2407.6.2] {701.4.2}
- ☐ Single opening can be to ventilated attic_____ [F2407.6.2] {F7-5}

Two-Opening Method | 09 IRC | 09 UMC

- ☐ 2 openings in upper & lower 12 in. **F11, 12** _____ [2407.6.1] {701.4.1}
- ☐ 2 direct exterior openings min 1 sq. in./4 kBtu **T3, F10** [2407.6.1] {701.4.1}
- ☐ 2 vertical ducts min 1 sq. in./4 kBtu **T3, F8, F9** ____ [2407.6.1] {701.4.1}
- ☐ 2 horizontal ducts min 1 sq. in./2 kBtu **T3, F12** ____ [2407.6.1] {701.4.1}

FIG. 10 — Single-Opening Method

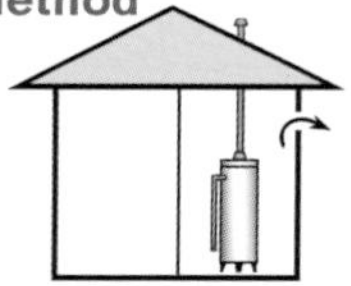

1 opening in upper 12 in. of exterior wall min. 1 sq. in./3 kBtu

FIG. 11 — 2 Direct Exterior Openings

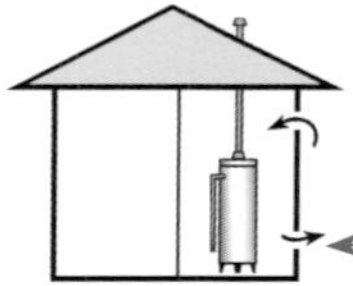

2 openings in exterior wall min. 1 sq. in./4 kBtu each

FIG. 12 — Horizontal Ducts

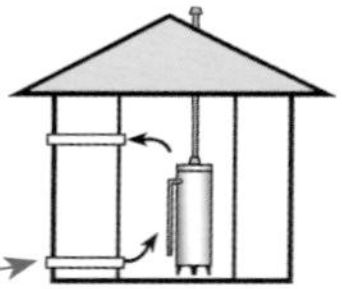

≥ 12 in. above finished ground

2 openings in exterior wall min. 1 sq. in./2 kBtu each

Attic & Crawl-Space Sources

09 IRC   09 UMC

☐ Ventilated attics & crawl spaces considered
   equivalent to outdoors **F9,13**_____[F2407.6.1(1&2)]   {F7-2&3}
☐ Crawl space only for lower C.A., not upper **F14** _____ [2407.11]   {701.10}

FIG. 13

Crawl-Space & Attic Openings

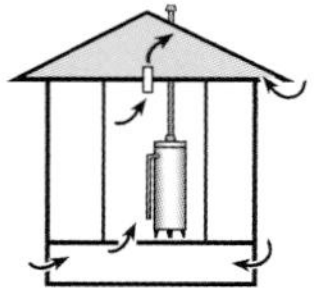

Attic & crawl space min. 1 sq. in./kBtu each

FIG. 14

Crawl Space Cannot Be Upper Air Source

Crawl space may not provide upper combustion air source

Indoor Air Source

09 IRC   09 UMC

☐ ACH = air changes per hour_____ [2407.5.2]   {701.2.1.1}
☐ Indoor air source alone only OK if infiltration > .40 ACH [2407.5]   {701.2}
☐ Min volume of space 50 cu. ft./1 kBtu/hr. **T3, F15**___ [2407.5.1]   {701.2.1}
☐ Indoor air volume includes rooms directly communicating
   with appliance space **F16** _____ [2407.5]   {701.2}
☐ Openings connecting indoor spaces req'd to be located in
   upper & lower 12 in. of appliance space **F15** _____ [2407.5.3.1]   {701.3.1}
☐ Openings connecting indoor spaces min 100 sq. in.
   each & min 1 sq. in./kBtu if on same level, 2 sq. in. if on
   different levels **T3**_____ [2407.5.3]   {701.3.1}
☐ If ACH < 0.40, min volumes for known air infiltration method:
   • Non fan-assisted appliance (21 cu. ft./ACH) per kBtu[2407.5.2]   {701.2.2}
   • Fan-assisted appliance (15 cu. ft./ACH) per kBtu _ [2407.5.2]   {701.2.2}

FIG. 15

Confined Space Indoors

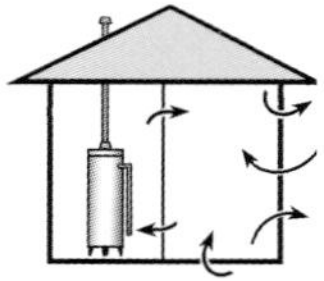

Openings from enclosed appliance space to building interior min. 100 sq. in. each & per **T3**. One in upper 12 in. & 1 in. lower 12 in. of enclosed space.

FIG. 16

All Air from Indoors

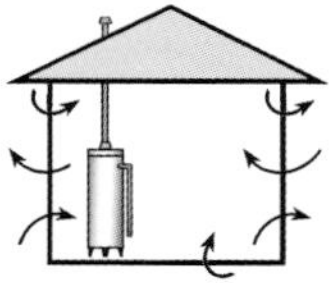

Space with > 0.40 ACH sufficient if volume ≥ 50 cu. ft./kBtu.

TABLE 3	COMBUSTION AIR OPENING SIZES			
	Indoor Air[A]		Outdoor Air Openings	
Btu	Opening size[B]	cu. ft. min. (sq. ft.[C])	1 in./2kBtu/hr.	1 in./4kBtu/hr.
30k	100 sq. in.	1,500 (188)	15 sq. in.	7.5 sq. in.
40k	100 sq. in.	2,000 (250)	20 sq. in.	10sq. in.
50k	100 sq. in.	2,500 (313)	25 sq. in.	12.5 sq. in.
60k	100 sq. in.	3,000 (375)	30 sq. in.	15 sq. in.
80k	100 sq. in.	4,000 (500)	40 sq. in.	20 sq. in.
100k	100 sq. in.	5,000 (625)	50 sq. in.	25 sq. in.
125k	125 sq. in.	6,250 (781)	62.5 sq. in.	31.3 sq. in.
150k	150 sq. in.	7,500 (938)	75 sq. in.	37.5 sq. in.

A. For construction with known air infiltration rate > 0.40/hr.
B. Req'd opening between confined space (< 50 cu. ft. per kBtu) & unconfined space.
C. Example: sq. ft. for 8 ft. ceiling—use actual room volume.

FORCED-AIR FURNACES

General Rules & Clearances   09 IRC   09 UMC

- ☐ Prohibited in bedroom, bathroom, or their closets EXC [2406.2]   {904.1}
  - Direct-vent type installed AMI _____ [2406.2]   {904.1}
  - Separated by weather-stripped self-closing door & all combustion air from exterior _____ [2406.2]   {904.1}
- ☐ Equipment room door large enough to remove appliance [min 24 in. wide in IRC] _____ [1305.1.2]   {304.0}
- ☐ Work space min 30 in. deep & wide in front of appliance [1305.1]   {304.0}
- ☐ FAUs in alcoves or closets must be L&L for alcove _ [2409.3.2]   {904.2B}
- ☐ FAUs clearance AMI EXC _____ [2409.3.2, 2409.4.2]   {904.2}
  - Clearance reduction OK if room large in comparison with size of equipment _____ [2409.3.3, 2409.4.3]   {904.2A&B}
- ☐ Install above design flood elevation _____ [1401.5]   {307.2}
- ☐ Air filter req'd AMI _____ [2442.1]   {311.2}

Electrical Requirements   09 IRC   11 NEC

- ☐ Receptacle within 25 ft. of appliance _ [1305.1.3.1 & 1305.1.4.3]   {210.63}
- ☐ Crawl space furnace req's light with switch at access [1305.1.4.3]   {210.70A3}
- ☐ Attic furnace req's light with switch at access _____ [1305.1.3.1]   {210.70A3}
- ☐ Individual circuit req'd for central heating _____ [3703.1]   {422.12}
- ☐ No other equipment on central heating circuit EXC _____ [3703.1]   {422.12}
  - Associated pumps, humidifiers, air cleaners & AC _____ [3703.1]   {422.12X}

Underfloor   09 IRC   09 UMC

- ☐ Equipment support on grade req's min 3 in. pad _____ [2408.4][5]   {904.3.1.1}[5]
- ☐ Suspended equipment min 6 in. above grade _____ [2408.4]   {904.3.1.2}[6]
- ☐ Passageway min 22 in. wide × 30 in. high _____ [1305.1.4]   {904.11.1}
- ☐ Passageway max distance 20 ft. to equipment EXC_ [1305.1.4]   {904.11.2}
  - Unlimited length if passageway 6 ft. high & 22 in. wide _____ [1305.1.4X2]   {904.11.2}

Attic   09 IRC   09 UMC

- ☐ Opening min 20 in. wide {22 in. UMC} & > appliance size _____ [1305.1.3]   {904.11.1}
- ☐ Passageway min 22 in. wide × 30 in. high _____ [1305.1.3]   {904.11.1}
- ☐ Max distance from opening 20 ft. EXC _____ [1305.1.3]   {904.11.2}
  - [IRC: 50 ft. if ≥ 6 ft. high] {UMC: Unlimited if ≥ 6 ft.} [1305.1.3X]   {904.11.2}
- ☐ Solid floor min 24 in. wide to equipment _____ [1305.1.3]   {904.11.3}
- ☐ Min 30 × 30 in. platform in front of firebox EXC **F17** _ [1305.1.3]   {904.11.4}
  - Not req'd if equipment can be serviced from opening [1305.1.3X1]   {∅}[1]

FIG. 17

Attic Furnace

Min. 5 ft. from flue collar to termination

No added loads on trusses except per design.

Min. 30 in. platform on control side

Light switch

Platform noncombustible or AMI

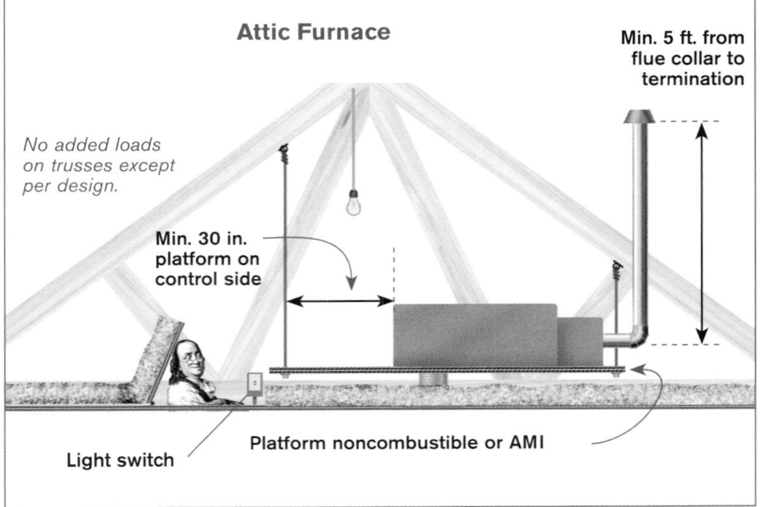

Garage

		09 IRC	09 UMC
☐	Protect appliance from impact_____[1307.3.1 & 2408.3X]	{307.1}
☐	Ignition source min 18 in. above floor EXC ___[1307.3 & 2408.2]	{307.1}
	• Flammable vapor ignition resistant (FVIR) appliances [2408.2X]	{307.1}
	• Separate space not communicating with garage air ___ [2408.2]	{307.1}
☐	Ducts & penetrations min 26-gage steel_____ [302.5.2]	{n/a}
☐	No duct openings into garage _____ [302.5.2]	{n/a}

Condensing Furnaces (Category IV) F18

		09 IRC	09 UMC
☐	Size venting AMI _____ [2426.5]	{802.6.3.2}
☐	Install vent & support AMI _____ [2426.5]	{802.6.1.1}
☐	Positive-pressure systems req'd to be gas tight _____[2427.3.3]	{802.3.4.3}
☐	No mixing of natural to forced draft connectors or vents[2427.3.3]	[802.3.4.4]
☐	Burner interlock req'd to forced-vent fan _____[2427.3.3]	[802.3.4.5]
☐	Furnaces with combustion air piping terminating in same location as vent piping typically considered direct-vent (see **p.142**) _____ [2427.8X1] {802.8.1X1}
☐	Terminate min 3 ft. above forced-air inlets within 10 ft. __ [2427.8]	{802.8.1}
☐	Termination clearances from building openings min 4 ft. below, 4 ft. horizontal & 1 ft. above _____ [2427.8]	{802.8.2}
☐	Terminate min 12 in. above grade _____ [2427.8]	{802.8.2}
☐	No vent termination where vapor would be a nuisance__ [2427.8]	{802.8.4}

Condensate Disposal

		09 IRC	09 UMC
☐	Provide means to collect & dispose of condensate_____ [2427.9]	{310.1}
☐	Condensate drains AMI _____ [2427.8][7]	{802.8.4}[7]
☐	Auxiliary drain pan req'd if condensate stoppage could damage any building component EXC _____ [1411.4 & 2404.10]	{n/a}
	• Automatic cutout installed in drain system_ [1411.4X & 2404.10X]	{n/a}
☐	May not drain to public way or nuisance location _____ [2427.9]	{309.1}
☐	Drainpipe min ¾ in. with ⅛ in./ft. slope or AMI _____[manu]	{309.1}
☐	May drain to indirect waste _____[manu]	{309.1}

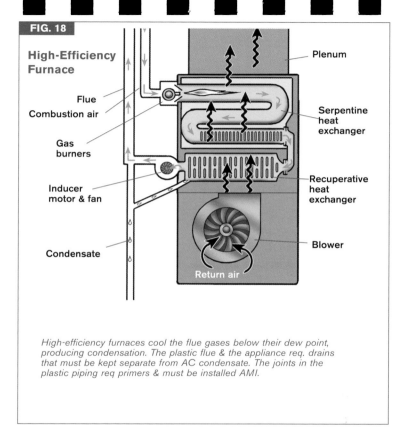

FIG. 18

High-Efficiency Furnace

Flue
Combustion air
Gas burners
Inducer motor & fan
Condensate
Return air
Plenum
Serpentine heat exchanger
Recuperative heat exchanger
Blower

High-efficiency furnaces cool the flue gases below their dew point, producing condensation. The plastic flue & the appliance req. drains that must be kept separate from AC condensate. The joints in the plastic piping req primers & must be installed AMI.

BOILERS & HYDRONICS

Modern high-efficiency WHs are often used with indirect WHs & for hydronic heating systems. Heat can be distributed through radiators, baseboard convectors, radiant slab-encased tubing, or duct heaters.

Steam & Hot Water Boilers

	09 IRC	09 UMC
☐ Install AMI	[2001.1]	{303.1}
☐ Installer to supply control diagram & operating manual	[2001.1]	{1020.0}
☐ Hot water boilers req pressure & temperature gauges F20	[2002.2]	{1004.3}
☐ Steam boilers req sight-glass & pressure gauge	[2002.3]	{1004.3}
☐ Pressure regulator req'd on water feed F20	[manu]	{manu}
☐ Shutoff valves req'd in supply & return piping EXC F20	[2001.3]	{1011.0}
• Single low-pressure steam boiler	[2001.3X]	{1011.0}
☐ Low-water cutoff control req'd	[2002.5]	{1011.0}
☐ Hydronic boilers req expansion tanks F19	[2003.1]	{1005.1}
☐ Tank support designed for twice waterlogged weight	[2003.1]	{1005.1}
☐ Tank capacity based on system volume T4	[2002.3]	{1005.4}
☐ PRV req'd F20	[2002.4]	{1011.0}
☐ PRV drain piped to within 18 in. of floor or receptor	[2002.4]	{1006.0}

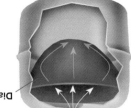

FIG. 19

Expansion Tank

Expanded water

Diaphragm

As water temperature increases & pressure rises, expanded water pushes against the diaphragm & compresses the air, preventing excessive pressure in the piping.

The tank must be sized per T4 for the total volume in the system, including the water in the boiler. The tank support must be designed for twice the waterlogged weight of the tank.

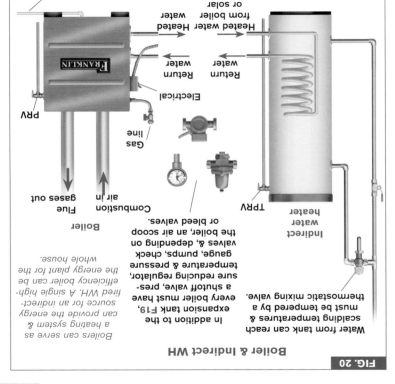

FIG. 20

Boiler & Indirect WH

Water from tank can reach scalding temperatures & must be tempered by a thermostatic mixing valve.

In addition to the expansion tank F19, every boiler must have a shutoff valve, pressure reducing regulator, temperature & pressure gauge, pumps, check valves &, depending on the boiler, an air scoop or bleed valves.

Boilers can serve as a heating system & can provide the energy source for an indirect-fired WH. A single high-efficiency boiler can be the energy plant for the whole house.

FRANKLIN

PRV

Boiler

Flue gases out

Combustion air in

Gas line

Electrical

Indirect water heater

TPRV

Return water

Return water

Heated water

Heated water from boiler or solar

Condensation is produced in the boiler & exhaust & must be discharged AMI.

Hydronic Piping: General

	09 IRC	09 UMC
☐ Provide means of system drain down _____	[2101.2]	{1201.2.7.8.8}
☐ Maintain backflow protection to potable water _____	[2101.3]	{UPC 603.0}
☐ No contact with material causing corrosion or damage	[2101.5]	{1201.2.7.8.5}
☐ Drilling, notching & protection per building code_____	[2101.6]	{1201.2.7.4}
☐ Provide for expansion & contraction _____	[2101.8]	{1201.2.7.5}
☐ Enter tee fittings on supply side NOT branch opening	[2101.7]	{n/a}

Exposed Piping

	09 IRC	09 UMC
☐ Support piping to avoid strain _____	[2101.9]	{1201.2.6}
☐ Pressure test min 100psi water for 15 {30} minutes	[2101.10]	{1201.2.8.3}
☐ Wrap/sleeve pipes through concrete walls or floors __	[2603.3]	{1201.2.7.8.1}

Embedded Piping (Radiant Heating)

	09 IRC	09 UMC
☐ Materials–steel pipe, Cu tubing {type L only in UMC}, PEX, PEX-AL-PEX, polybutylene CPVC, or polypropylene___	[2103.1]	{1204.1}
☐ Plastic pipe rated min 100psi at 180°F **F21** _____	[2103.1]	{1204.1}
☐ Pressure test 100psi for 30 minutes _____	[2103.4]	{1207.0}
☐ Maintain operating pressure on pipe when placing concrete _	[n/a]	{1203.2}

TABLE 4	MIN. EXPANSION TANK CAPACITY	[T2003.2]
System Volume (gal.)	**Pressurized Tank**	**Open Tank**
10	1.0	1.5
20	1.5	3.0
30	2.5	4.5
40	3.0	6.0
50	4.0	7.5
60	5.0	9.0
70	6.0	10.5
80	6.5	12.0
90	7.5	13.5
100	8.0	15.0

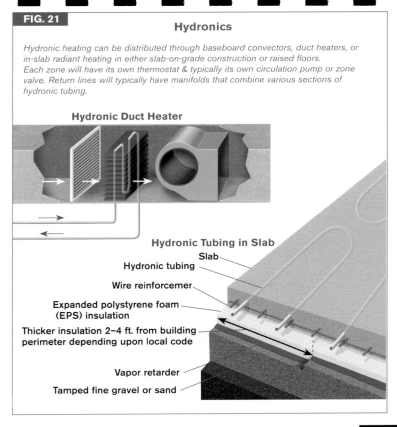

FIG. 21

Hydronics

Hydronic heating can be distributed through baseboard convectors, duct heaters, or in-slab radiant heating in either slab-on-grade construction or raised floors.
Each zone will have its own thermostat & typically its own circulation pump or zone valve. Return lines will typically have manifolds that combine various sections of hydronic tubing.

Hydronic Duct Heater

Hydronic Tubing in Slab
- Slab
- Hydronic tubing
- Wire reinforcemer
- Expanded polystyrene foam (EPS) insulation
- Thicker insulation 2–4 ft. from building perimeter depending upon local code
- Vapor retarder
- Tamped fine gravel or sand

MECHANICAL

BOILERS & HYDRONICS

GAS FLOOR FURNACES

In addition to these specific rules for gas-burning floor furnaces, the access & passageway illumination requirements on p.129 also apply. Grills for these furnaces can be hot & care must be taken to protect young children from them.

Underfloor Area

	IRC 09	UMC 09
☐ Must be L&L & installed AMI	[2437.1]	{912.1A}
☐ Unlisted furnaces only in noncombustible floors	[Ø]	{912.1B}
☐ Appliance must fit through access opening	[1305.1.4]	{304.0}
☐ IRC access opening min 22 × 30 in. (UMC trap door min 24 in. & wall opening 18 × 24 in.)	[1305.1.4]	{912.8}
☐ May not be in concrete slab on grade	[2437.2]	{912.1A}
• May not project into habitable space below	[2437.586]	{912.11}

Excavation Clearances

	IRC 09	UMC 09
☐ 6 in. to ground (2 in. if factory sealed) F22	[2437.4]	{912.7}
☐ 12 in. side clearance; 18 in. on control side F22	[2437.4]	{912.7}

FIG. 22

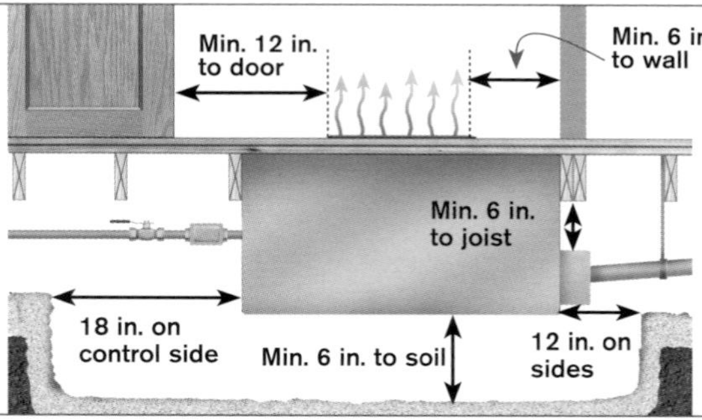

Floor Furnace

- 18 in. on control side
- 12 in. on sides
- Min. 6 in. to soil
- Min. 6 in. to joist
- Min. 12 in. to wall
- Min. 6 in. to door

Clearances

	IRC 09	UMC 09
☐ From sidewall—install AMI (typical 6 in. min) F23	[2436.3]	{928.1&2}
☐ From door swing [IRC: 12 in.] {UMC: AMI} F23	[2436.4]	{928.1&2}
☐ Do not rely on doorstops to maintain clearance	[2436.4]	{928.2}
☐ Clearance below structural projections AMI (typical 18 in.)	[2436.3]	{928.1&2}

Furnace Installation

	IRC 09	UMC 09
☐ OK in bedroom or bath that is not confined space	[2406.2]	{902.0B}
☐ Fan assist only if L&L & AMI	[2436.1]	{304.1}
☐ No ducts attached to wall furnaces	[2436.5]	{928.1A}
☐ Panels, grills & access doors not attached to walls	[2436.6]	{928.1E}
☐ Header plate at top of furnace AMI	[2436.1]	{928.1C}
☐ Stud bay depth AMI	[2436.1]	{928.1A}
☐ Unlisted furnaces not OK in combustible construction	[2404.3]	{928.1B}

Gas-fired wall furnaces that are not direct vent can be located only in rooms that are large enough to meet the combustion air requirements of the appliance. Because they need indoor air for combustion, they are typically found only in older buildings with high air infiltration rates. If a building is upgraded in terms of energy compliance, it might not be possible to use wall furnaces as the heat source. Vent installation on wall furnaces is especially important, in that clearances inside the wall are less than the normal minimums for B vents. When a wall furnace is installed in an existing building, one side of the wall above the furnace should be opened for inspection of vent clearances.

GAS WALL FURNACES

Above Floor

	IRC 09	UMC 09
☐ Not in doorway, landing, passageway, or exit way	[2437.2]	{912.4A}
☐ Flat-register type min 6 in. from wall F22	[2437.2]	{912.4B}
☐ Min 12 in. from door swing or draperies F22	[2437.2]	{912.4C}
☐ Two adjoining sides must have 18 in. clearance	[2437.2]	{912.4B}
☐ Wall-register type min 6 in. to inner corner	[2437.2]	{912.4B}
☐ Thermostat must be in same room	[2437.2]	{912.1C}

FIG. 23

Wall Furnace Clearances

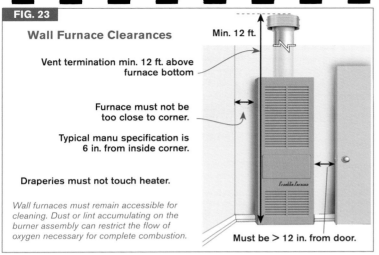

Min. 12 ft.

Vent termination min. 12 ft. above furnace bottom

Furnace must not be too close to corner.

Typical manu specification is 6 in. from inside corner.

Draperies must not touch heater.

Wall furnaces must remain accessible for cleaning. Dust or lint accumulating on the burner assembly can restrict the flow of oxygen necessary for complete combustion.

Must be > 12 in. from door.

Vent Installation

	09 IRC	09 UMC
☐ Cut top & floor plates flush to stud **F24**	[2427.6.1]	{928.1C}
☐ Furnace stud space vented by ceiling plate spacers AMI **F24**	[2427.6.1]	{928.1C}
☐ Subsequent ceiling stud plates firestopped AMI **F24**	[2427.6.1]	{928.1C}
☐ Single-story systems OK only in single story or top of multistory	[2427.6.1]	{928.1C}
☐ Multistory systems OK in single story or multistory	[2427.6.1]	{928.1C}
☐ Sleeve around vent in insulated assembly **F25**	[2426.4]	{n/a}
☐ Sleeve min 2 in. above insulation in insulated attic **F25**	[2426.4]	{n/a}
☐ Vent min height 12 ft. above bottom of furnace **F23**	[2427.6.4]	{802.6.2.2}

FIG. 24

Wall Furnace Flue

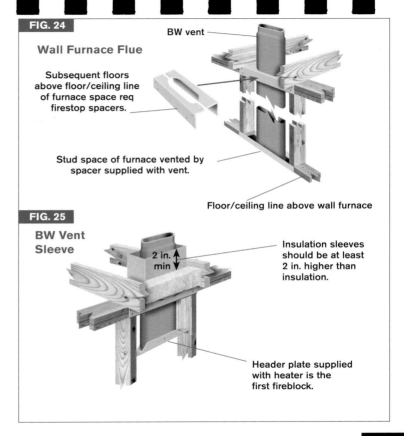

BW vent

Subsequent floors above floor/ceiling line of furnace space req firestop spacers.

Stud space of furnace vented by spacer supplied with vent.

Floor/ceiling line above wall furnace

FIG. 25

BW Vent Sleeve

2 in. min

Insulation sleeves should be at least 2 in. higher than insulation.

Header plate supplied with heater is the first fireblock.

GAS WALL FURNACES

ROOM HEATERS

Room heaters that are not direct vent must be supplied with sufficient combustion air from the interior & are typically found only in older buildings with high air-infiltration rates. Because these must connect to a venting system, they should be secured to prevent displacement of the vent. Though unvented heaters are recognized in the model codes, many jurisdictions prohibit their use. If they are in bedrooms or baths they should be provided with a nontamperable oxygen depletion sensor.

General

	09 IRC	09 UMC
☐ Secure to floor	[1307.2]	{303.4}
☐ Install AMI	[2446.1]	{303.1}
☐ Flame safeguard (pilot safety) req'd	[2446.1]	{305.0}

Unvented Heaters

	09 IRC	09 UMC
☐ Unvented heater may NOT be sole heat source	[2445.2]	{924.1}
☐ Unvented heater ≤ 6kBtu OK in bath	[2406.2]	{924.1.1X1}
☐ Unvented heater ≤ 10kBtu OK in bedroom	[2406.2]	{924.1.1X2}
☐ Rooms must meet unconfined space requirements	[2406.2]	{924.1}
☐ Max size 40kBtu	[2445.3]	{n/a}
☐ Max input ratings ≤ 20Btu/cu. ft. of room or space	[2445.5]	{n/a}
☐ Adjacent spaces with permanent large openings (doorway or archway) part of room volume	[2445.5]	{701.2}
☐ Unvented heater req's oxygen-depletion sensor	[2445.6]	{924.1.1X}

DIRECT-VENT APPLIANCES

Direct-vent appliances draw their source of combustion air from the same area where they vent combustion gases. This arrangement equalizes pressure on the inlet & outlet of the firebox, which is sealed & has no open flame on the building interior. Because these appliances do not use the interior air for combustion, they can be located in rooms that are considered confined spaces.

Direct-Vent Gas Wall Heaters F26

	09 IRC	09 UMC
☐ Install AMI	[2427.2.1, 2429.1]	{304.1}
☐ Indoor combustion air not req'd	[2407.1]	{928.1D}

FIG. 26

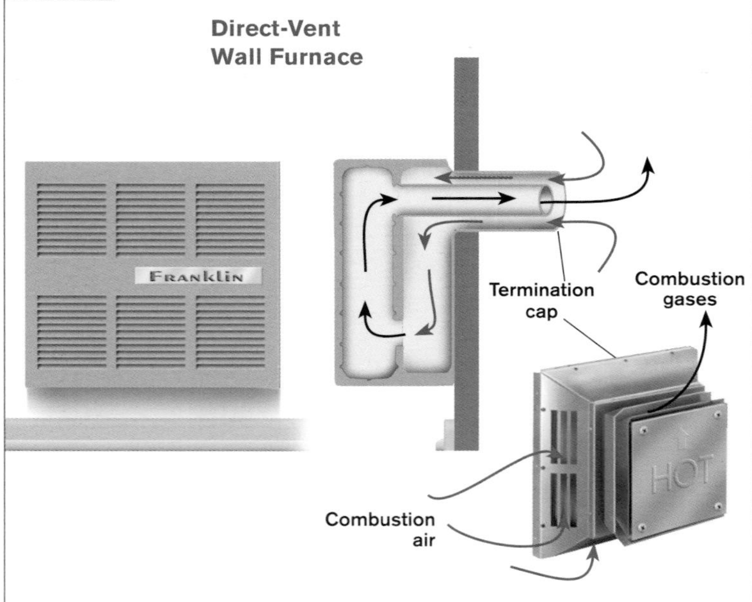

Direct-Vent Wall Furnace

FRANKLIN

Termination cap

Combustion gases

Combustion air

HOT

Direct-Vent Gas Fireplaces F27

	09 IRC	09 UMC
☐ OK in bedroom or bath	[2406.2]	{908.1X}
☐ Must be L&L & AMI	[2427.2.1]	{908.2D}
☐ Corrugated stainless steel tubing (CSST) or flex req's grommet through appliance wall	[2422.1.2.3X]	{1313.1}
☐ Vent termination clearances T5	[2427.8]	{802.8.3}

FIG. 27

Direct-Vent Gas Fireplace

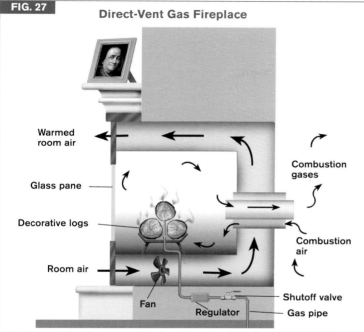

Warmed room air

Glass pane

Decorative logs

Room air

Fan *Regulator*

Combustion gases

Combustion air

Shutoff valve

Gas pipe

A direct-vent fireplace can vent horizontally out a sidewall or vertically to the roof. With a completely enclosed chamber, it draws in outside air for combustion & expels gases to the outside. The front glass enclosure allows radiant heat to pass into the room. It heats a room without robbing it of oxygen or of the heated air it provides & keeps it free of fumes & combustible materials, such as embers or ash.

Direct-Vent Appliance Vent Termination

	09 IRC	09 UMC
☐ OK to terminate through wall **F26,27** _____	[2427.8]	{802.8.3}
☐ Termination clearances **T5** _____	[2427.8]	{802.8.3}
☐ Bottom of vent terminal min 12 in. above grade _____	[2427.8]	{802.8.3}

TABLE 5	DIRECT VENT TERMINATION CLEARANCES FROM BUILDING OPENINGS [2427.8] {802.8.3}	
Appliance Rating		**Clearance[A] from Openings**
≤ 10kBtu		6 in.
> 10kBtu to 50kBtu		9 in.
> 50kBtu		12 in.

A. Measure stretched-string distance from edge of vent opening.

GAS APPLIANCES IN FIREPLACES

Vented Decorative Appliances in Fireplaces

	09 IRC	09 UMC
☐ Not allowed in bedroom if confined space _____	[2406.2]	{907.1}
☐ Must be L&L & installed AMI _____	[2432.1]	{907.2}
☐ Maintain open vent (block damper in open position) _____	[manu]	{manu}
☐ Fireplace screen req'd _____	[manu]	{907.3}
☐ Appliance with pilot or ignition system req's pilot safety ___	[2432.2]	{305.0}
☐ Shutoff inside firepit OK if AMI _____	[2420.5.1]	{1313.4}

Log Lighters

	09 IRC	09 UMC
☐ Log lighters must be AMI _____	[2433.1]	{manu}
☐ Req'd valve ≤ 6 ft. of fireplace & in same room _____	[2420.5]	{1313.4}
☐ Hard gas pipe inside firepit (no flex) _____	[2422.1.1]	{1313.1}

Unvented Gas Log Heaters

	09 IRC	09 UMC
☐ Not in factory-built fireplace unless L&L to UL127 & AMI [1004.4]		{n/a}

GAS APPLIANCE VENTING

Common forms of gas appliances, like WHs, have traditionally used a "gravity" vent system, where combustion gases are lighter than the surrounding air, so they rise by gravity to the outdoors. As appliances are becoming more efficient, other types of venting systems are being used, as shown in T6. The type of vent should match the appliance category & the manu's recommendations. The venting tables that are supplied by the manu & shipped with gas appliances must be used in vent systems that include induced draft appliances. Those tables are also in the codes.

TABLE 6	APPLIANCE VENTING CATEGORIES		
Category	**Condensation**	**Static Pressure**	**Typical vent**
I	No	Nonpositive	B Vent
II	Yes	Nonpositive	AMI
III	No	Positive	Stainless
IV	Yes	Positive	Plastic

General

	IRC 09	UMC 09
☐ Choose vent material based on appliance category T6	[2427.1]	{802.1}
☐ Category I induced draft is "gravity" vent appliance	[2427.1]	{802.1}
☐ Select type of venting system from T7	[2427.4]	{802.4.1}
☐ Properly support all vents	[2426.6]	{802.6.5}
☐ All vents L&L except plastic installed AMI or single wall	[2426.1]	{n/a}
☐ Plastic vents AMI & primer contrasting color	[2427.4.1.1]	{802.4.3}⁸
☐ Condensate drain also req'd for Category I or III if local experience shows need (recommended for some tankless WH)	[2427.8&9]	{802.9.2}
☐ Sheet-metal shield to 2 in. above attic insulation F29	[2426.4]	{n/a}
☐ Protect vents closer than 1½ in. from face of wall with steel shield plates extending min 4 in. beyond framing inside wall	[2426.7]	{n/a}
☐ No solid fuel & gas in same chimney flue	[2427.5.6.1]	{802.5.5.1}
☐ Size Category II, III & IV appliance vents	[2427.6.8.3]	{802.6.3.2}

Single-Wall Vents

	IRC 09	UMC 09
☐ Not allowed in dwellings.	[n/a]	{802.7.4.1}⁹
☐ Only for runs from appliance space directly to outside	[2427.4]	{n/a}
☐ May not originate in attic or pass through inside wall	[2427.6]	{n/a}
☐ Min 6 in. clear to combustible for single wall pipe	[2427.8]	{n/a}
☐ Termination min 2 ft. above roof	[2427.3]	{n/a}
☐ Termination min 2 ft. higher than building within 10 ft.	[2427.3]	{n/a}
☐ Not allowed outdoors in cold (freezing) climates	[2427.2]	{n/a}

TABLE 7	TYPE OF VENTING SYSTEM TO BE USED [T2427.4] [T8-1]		
Appliances	**Type of Vent**	**IRC**	**UMC**
Listed Category I Appliances listed for B vent	Type B gas vent	2427.6	802.6
	Chimney	2427.5	802.5
	Single-wall metal pipe	2427.7	Ø
	Listed chimney lining for gas	2427.5.2	802.5.13
	Special vent listed for appliance	2427.4.2	802.4.3
Listed vented wall furnaces	Type B-W gas vent	2427.6	802.6 928.0
Category II, III & IV appliances	AMI	2427.4.1 2427.4.2	802.4.2 802.4.3
Unlisted appliances	Chimney	2427.5	802.5
Decorative appliances in vented fireplaces	Chimney	2427.5	802.2
Direct-vent appliances	AMI	2427.1	802.5
Appliances with integral vent	AMI	2427.2	802.6

Vent Size Using Tables   09 IRC   09 UMC

☐ Tables can be used for all Category I appliances____ [2427.6.8.1] {802.6.3.1}
☐ Req'd to be used if appliance is induced draft____ [2427.10.3.1] {802.10.3.1}
☐ Connector not > 2 sizes larger than flue collar _____ [2428.3.17] {803.1.10}
☐ When vertical vent > than connector, use vertical diameter
    to determine table min & connector diameter for
    table max _____ [2428.2.8] {803.1.7}
☐ Use double-wall vent tables only for vents not exposed to
    outdoors below roofline (B vent in unvented chase insulated
    to R-8 or in unused masonry chimney flue not
    considered outdoors) _____ [2428.2.9] {803.1.8.1}
☐ Zero lateral values only if straight vertical vent connects directly to top outlet
    draft hood or flue collar_____ [2428.2.4] {803.1.3}
☐ No elbows if using "zero lateral length" part of table__ [2428.2.3] {803.1.2}
☐ Vent tables with lateral length allow for 2-90° elbows_ [2428.2.3] {803.1.2}
☐ Reduce table capacity 5% each elbow up to 45° & 10%
    each elbow > 45° up to 90°_____ [2428.2.3] {803.1.2}
☐ Reductions for elbows in common vents as above ___ [2428.3.6] {803.2.6}
☐ Reductions for common vent connectors as above___ [2428.3.7] {803.2.7}

Multiple Appliances Vented in Common   09 IRC   09 UMC

☐ Tables req'd to be used if induced draft included _ [2427.10.3.1] {802.10.3.1}
☐ Join multiple connectors as high as possible per available
    headroom & clearance **F28** _____[2427.10.3.4] {802.10.3.4}
☐ Connect smaller above larger EXC **F28** _____ [2427.10.4] {802.10.4.1}
   • OK if both at same level if max 45° from vertical _ [2427.10.4.1] {802.10.4.1}
☐ If both appliances have draft hoods, OK to size vent for
    100% of larger + 50% of smaller _____ [2427.10.3.4] {802.10.3.4}
☐ Reduce connector table capacity 5% each elbow up
    to 45° & 10% each elbow > 45° up to 90° **F28** _____[2428.3.7] {803.2.7}

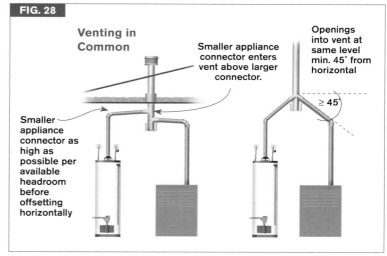

FIG. 28

Venting in Common

Smaller appliance connector enters vent above larger connector.

Smaller appliance connector as high as possible per available headroom before offsetting horizontally

Openings into vent at same level min. 45° from horizontal

≥ 45°

Forced Vents (Category IV)   09 IRC   09 UMC

☐ All mechanical draft systems L&L & installed AMI _____ [2427.3.3] {802.3.4.1}
☐ Forced draft system must be gas tight _____[2427.3.3] {802.3.4.3}
☐ No natural & forced-vent to common flue _____ [2427.3.3] {802.3.4.4}
☐ Terminate min 7 ft. above ground where adjacent to public
    walkways _____ [2427.3.3] {802.3.4.6}
☐ Terminate 3 ft. above forced air inlets within 10 ft._____ [2427.8] {802.8.1}
☐ Terminate min 4 ft. to side or below or 1 ft. above building
    openings, min 1 ft. above ground level EXC _____ [2427.8] {802.8.2}
   • Termination can be same as direct vent (**p.142**) if AMI_ [2427.8] {802.8.1&2}
☐ Collect & dispose of condensate from vent (see **p.130**)_ [2427.9] {802.9}

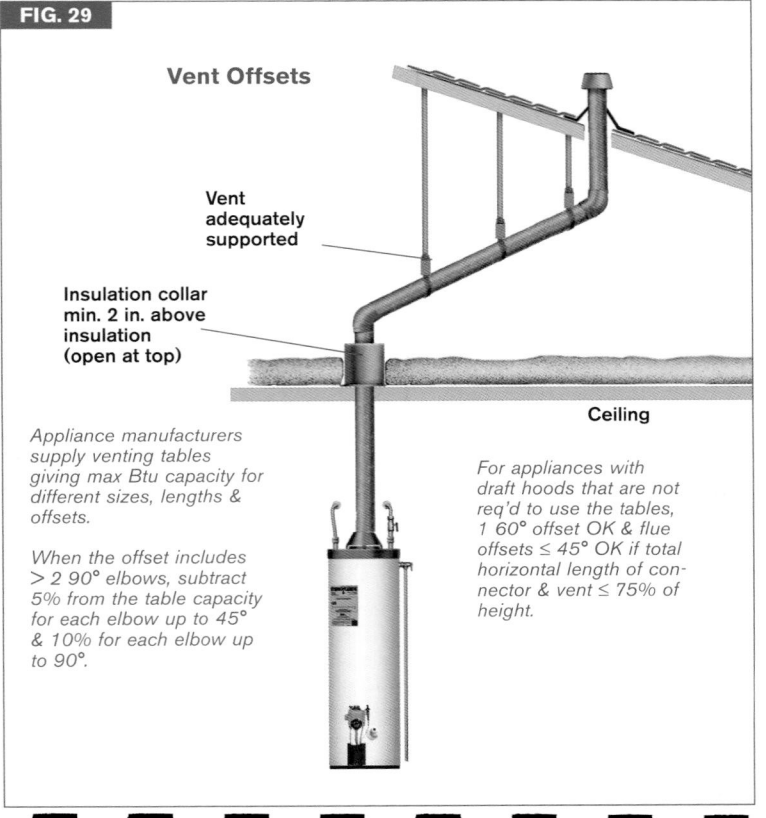

FIG. 29

Vent Offsets

Vent adequately supported

Insulation collar min. 2 in. above insulation (open at top)

Ceiling

Appliance manufacturers supply venting tables giving max Btu capacity for different sizes, lengths & offsets.

When the offset includes > 2 90° elbows, subtract 5% from the table capacity for each elbow up to 45° & 10% for each elbow up to 90°.

For appliances with draft hoods that are not req'd to use the tables, 1 60° offset OK & flue offsets ≤ 45° OK if total horizontal length of connector & vent ≤ 75% of height.

Connectors: General   09 IRC   09 UMC

☐ Must be as straight as practical_____[2427.10.6] {802.10.6}
☐ Min ¼ in./ft. slope toward appliance **F28** _____[2427.10.8] {802.10.8}
☐ No dips or sags_____[2427.10.8] {802.10.8}
☐ Must be as short as practical_____[2427.10.9] {802.10.9.1}
☐ Provide adequate support **F29** _____ [2427.10.10] {802.10.10}
☐ Entire connector must be accessible _____ [2427.10.12] {802.10.12}
☐ Attach to appliance with screws or AMI _____[2427.10.7] {802.10.7}
☐ Necessary size increases req'd to be made only at appliance
    outlet connection_____[2427.10.3.5] {802.10.3.5}
☐ Max horizontal length 18 in. per in. of connector diameter
    when using venting tables _____ [2428.3.2]   {803.2.2}

Single-Wall Connectors for Category I Appliances   09 IRC   09 UMC

☐ Not in attic, crawl space, or other unconditioned
    space EXC_____[2427.10.2.2] {802.10.2.2}
   • OK [in unconditioned basement or garage] within exterior walls
     if local 99% winter design temperature ≥ 5°F ____[2427.10.2.2X] {802.10.2.2X}
☐ Horizontal connector max 75% of vertical vent _____[2427.10.9] {802.10.9.2}
☐ Min 6 in. clearance to combustibles _____[2427.10.5]   {802.10.5}
☐ May not pass through interior wall, floor, or ceiling __ [2427.10.14]{802.10.14.1}

Type B Double-Wall Vent Connectors   09 IRC   09 UMC

☐ Min clearance to combustibles per L&L (typical 1 in.) [2427.10.5]   {802.10.5}
☐ Max horizontal length 100% of vertical vent _____[2427.10.9] {802.10.9.3}

Connectors to Chimneys   09 IRC   09 UMC

☐ Inspection req'd before connecting to existing chimney [2427.5.5]   {802.5.4}
☐ Must be lined with clay or metal EXC _____ [2427.5.5.1]   {802.5.4.2}
   • OK for replacement of like appliance if inspected_ [2427.5.5.1][10] {802.5.4.2}[10]
☐ Enter chimney above bottom (min 12 in. IRC) _____ [2425.9] {802.10.11}
☐ Chimney cross-sectional area not > 7× size of gas vent [2427.5.4]   {802.5.3}

Connectors to Chimneys (cont.)   09 IRC   09 UMC

☐ No connection to chimney serving solid fuel _____ [2427.10.13] {802.10.13}
☐ Secure connector to prevent blocking chimney ____ [2427.10.11] {802.10.11}
☐ No decorative shrouds unless L&L & AMI _____ [2427.5.3] {802.5.2.4}

Vents for appliances with draft hoods can be sized & installed in accordance with the pre-scriptive rules in the following section, or by using the tables that come with the appliance.

Vent Size (Appliances with Draft Hoods)   09 IRC   09 UMC

☐ Min size same as flue collar _____ [2427.6.8.1] {802.6.3.1}
☐ Max size 7× area of smallest flue collar _____ [2427.6.8.1] {802.6.3.1}
☐ If 2 appliances, 100% of larger + 50% of smaller ___ [2427.6.8.1] {802.6.3.1}
☐ Offsets 45° max except one of 60° OK **F29** _____ [2427.6.8.2] {802.6.1.1}

Gas Vent Terminations   09 IRC   09 UMC

☐ Gas vents must extend above roof EXC_____ [2427.6.3] {802.6.2}
  • Direct-vent appliances _____ [2427.6.3] {802.6.2}
  • Appliances with integral vents_____ [2427.6.3] {802.6.2}
  • Mechanical-draft appliances AMI (see **p.137**) _____ [2427.6.3] {802.6.2}
☐ Roof penetration req's flashing _____ [2427.6.5] {802.6.1}
☐ Must have listed cap_____ [2427.6.5] {802.6.1}
☐ Decorative shrouds only if L&L & AMI _____ [2427.6.3.1] {802.6.2.4}
☐ Chimneys min 3 ft. above roof **F30** _____ [2427.5.3] {802.5.2.1}
☐ Chimneys min 2 ft. higher than building within 10 ft. **F30** [2427.5.3] {802.5.2.1}
☐ Vent termination min 5 ft. vertical above flue collar ____ [2427.6.4] {802.6.2.1}
☐ B vents ≤ 12 in. diameter per **F31, T8** if > 8 ft. from wall _ [2427.6.4] {802.6.2}
☐ B vents min 2 ft. above vertical walls within 8 ft. **F31** ___ [2427.6.3] {802.6.2}
☐ B vents > 12 in. diameter min 2 ft. above roof & per **F30** [2427.6.4] {802.6.2}
☐ Type B or L min 5 ft. vertical above flue collar _____ [2427.6.3] {802.6.2.1}
☐ Wall furnace min 12 ft. from bottom of furnace **F23** ___ [2427.6.4] {802.6.2.2}
☐ Direct vent per **T5** _____ [2427.8] {802.8.3}

Gas Vent Terminations (cont.)   09 IRC   09 UMC

☐ "L" vent (oil vent used for gas or oil) termination:
  • Min 2 ft. above roof **F30**_____ [1804.2.4] {801.2}
  • Min 2 ft. above any portion of building within 10 ft. **F30** [1804.2.4] {801.2}

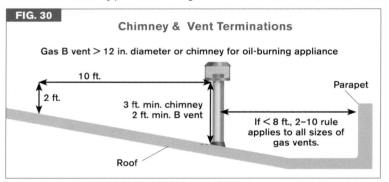

FIG. 30

Chimney & Vent Terminations

Gas B vent > 12 in. diameter or chimney for oil-burning appliance

10 ft.

2 ft.

3 ft. min. chimney
2 ft. min. B vent

Parapet

If < 8 ft., 2–10 rule applies to all sizes of gas vents.

Roof

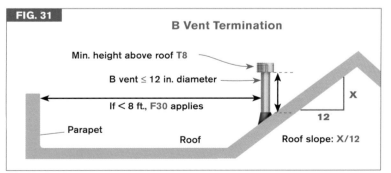

FIG. 31

B Vent Termination

Min. height above roof **T8**

B vent ≤ 12 in. diameter

If < 8 ft., **F30** applies

Parapet

Roof

X

12

Roof slope: X/12

TABLE 8	B VENT TERMINATION HEIGHT (F31) [F2427.6.3] {F8-2}		
Roof Slope	Min. Height (ft.)	Roof Slope	Min. Height (ft.)
Flat to 6/12	1	> 11/12 to 12/12	4
> 6/12 to 7/12	1¼	> 12/12 to 14/12	5
> 7/12 to 8/12	1½	> 14/12 to 16/12	6
> 8/12 to 9/12	2	> 16/12 to 18/12	7
> 9/12 to 10/12	2½	> 18/12 to 20/12	7½
> 10/12 to 11/12	3¼	> 20/12 to 21/12	8

OIL-FIRED APPLIANCES

The IRC provides specific rules for oil-burning appliances in Chapters 14 & 18. The UMC defers to NFPA 31 for these appliances. Jurisdictions using the UMC would use the NFPA 31 columns below.

General	09 IRC	NFPA 31
☐ Appliance clearances AMI EXC_____	[1306.1]	{4.3.2}
• New central furnaces in closets req clearances to **F32** _	[1305.1.1]	{n/a}
☐ Oil shutoff valve req'd_____	[2204.2]	{10.5.1}

Oil-Fired Floor Furnaces	09 IRC	NFPA 31
☐ Must be L&L for combustible construction_____	[1408.1]	{10.9.1}
☐ Install AMI_____	[1408.1]	{4.3.2}
☐ Floor register types min 6 in. from wall_____	[1408.3]	{10.9.4}
☐ Wall register type min 6 in. from inside corners_____	[1408.3]	{10.9.5}
☐ Min 12 in. from draperies or door in any position_____	[1408.3]	{10.9.4}
☐ Draperies OK above furnace if min 5 ft. clearance_____	[1408.3]	{n/a}
☐ Not OK to project into habitable space below_____	[1408.3]	{10.9.9}
☐ Must be supported independently of grill_____	[1408.5]	{10.9.3}
☐ Not OK to support from ground_____	[1408.5]	{10.9.7}
☐ Furnace must be accessible (not OK in slab)_____	[1408.3&4]	{10.9.7}

Oil-Fired Floor Furnaces (cont.)	09 IRC	NFPA 31
☐ Opening to underfloor area min 18 × 24 in. EXC_____	[1408.4]	{10.9.7.1}
• Floor trap door min 22 × 30 in._____	[1408.4]	{n/a}
☐ Min 6 in. clearance to ground_____	[1408.5]	{10.9.6}
☐ Min 6 in. clearance to sides below floor_____	[n/a]	{10.9.9.2}
☐ Provide adequate combustion air_____	[1701.1]	{10.9.8}
☐ Chimney connector clearance min 9 in. EXC_____	[1803.3.4]	{10.9.11}
• Lesser clearances with clearance reduction system **T9**_____	[1803.3.4]	{10.9.11}

Oil-Fired Recessed Wall Furnaces	09 IRC	NFPA 31
☐ Must be L&L for combustible construction_____	[1409.1]	{10.13.1}
☐ Install AMI_____	[1409.1]	{10.13.3}
☐ Locate so no fire hazard to walls, floors, or furnishings__	[1409.2]	{10.13.4}
☐ No doors within 12 in. of face of furnace_____	[1409.2]	{10.13.4}
☐ Doorstop cannot be used to maintain req'd clearance __	[1409.2]	{10.13.4}
☐ Min 3 ft. from wall opposite register_____	[1409.1]	{10.13.5}
☐ Panels, grills & access doors not attached to walls_____	[1409.4]	{10.13.6}
☐ Provide adequate combustion air_____	[1701.1]	{10.13.7}

Oil-Fired Room Heaters	09 IRC	NFPA 31
☐ Must be L&L & installed AMI_____	[1410.1]	{n/a}
☐ Noncombustible floor or assembly min 18 in. (12 in. NFPA 31) beyond all sides of appliance EXC_____	[1410.2]	{10.6.3.1}
• Appliances L&L for installation AMI without protection	[1410.2X]	{10.6.3}
☐ Clearances 24 in. front, 6 in. sides, rear & above_____	[manu]	{T10.6.1}

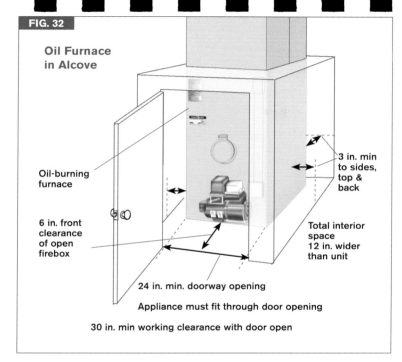

FIG. 32

Oil Furnace in Alcove

Oil-burning furnace

6 in. front clearance of open firebox

3 in. min to sides, top & back

Total interior space 12 in. wider than unit

24 in. min. doorway opening

Appliance must fit through door opening

30 in. min working clearance with door open

COMBUSTION AIR FOR OIL-FIRED APPLIANCES

The IRC no longer includes a separate set of rules for combustion air for oil-burning appliances & instead simply refers to NFPA 31.

General   NFPA 31

☐ Source from outside building if unusually tight construction _____ {5.2.1}
☐ Consider effect from exhaust fans (kitchen, bath, laundry) _____ {5.2.3}
☐ Screen req'd on outside openings, mesh openings ≥ ¼ in. _____ {5.6.2}
☐ Consider restrictive effect of louvers on openings: _____ {5.6.3}
  • Net free area 60–75% for metal louvers
  • Net free area 20–25% for wood louvers

Indoor Air Source   NFPA 31

☐ OK only for buildings of ordinary tightness_____ {5.3.1}
☐ Infiltration sufficient for unconfined space _____ {5.3.1}
☐ Unconfined space = ≥ 50 cu. ft./kBtu/hr. of all appliances
  in space **F16** _____ {3.3.60}
☐ Confined space req's openings to unconfined space_____ {5.4.1.3}
☐ Openings to unconfined space min 1 sq. in./kBtu/hr. **T3**_____ {5.4.1.2}
☐ Openings located near top & bottom of confined space **F15** _____ {5.4.1.1}

Outside Air Source   NFPA 31

☐ Openings located near top & bottom of confined space **F11** _____ {5.4.2.1}
☐ Openings to vented attic or crawlspace equivalent to outdoors **F13** __ {5.4.2.2}
☐ Direct exterior openings each sized at 1 sq. in./4kBtu/hr. **F11** _____ {5.4.2.3}
☐ Vertical ducts each sized at 1 sq. in./4kBtu/hr. **F8, 9** _____ {5.4.2.3}
☐ Horizontal ducts each sized at 1 sq. in./2kBtu/hr. **F12**_____ {5.4.2.3}

OIL-FIRED APPLIANCE CHIMNEYS & VENTS

Oil-fired appliances can be vented to listed L vents or to masonry or listed chimneys. IRC Chapter 18 deals with this subject. The UMC defers to NFPA 211 for oil-fired appliances, though NFPA 31 also contains similar rules. NFPA 211 does not address as many topics on oil-fired vents as NFPA 31 & for consistency with the rest of the codes in this section we are providing the NFPA 31 rules below.

General — 09 IRC — NFPA 31

	09 IRC	NFPA 31
☐ Appliances must be listed	[1302.1]	{13.2}
☐ Fuel-burning appliances req venting to outdoors	[1801.1]	{6.2.1}
☐ Vent system AMI of connected appliance	[1801.2]	{6.3.1}
☐ Draft regulator req'd if connected to chimney F33 EXC.	[1802.3]	{6.4.1}
• Arrangements that prevent excessive chimney draft	[n/a]	{6.4.1}
• Appliances L&L for use without draft regulator	[n/a]	{6.4.1}
☐ No manually operated dampers	[1802.2.1]	{6.4.2}
☐ Automatic dampers req burner interlock	[1802.2.2]	{6.4.3}
☐ Unused openings not OK in vent system	[1801.10]	{n/a}

Chimneys & Type L Vents — 09 IRC — NFPA 31

	09 IRC	NFPA 31
☐ Chimneys min size 2 flue collar, max size 3x	[1805.3.1]	{manu}
☐ Inspect chimney before installing replacement appliance	[n/a]	{6.6.7}
☐ Installer to verify that chimney is proper size	[n/a]	{6.6.7}¹¹
☐ If deterioration seen, inspect per NFPA 211 (p.155)	[n/a]	{6.6.7.2}
☐ Type L vents must be L&L & installed AMI	[1804.3]	{6.7.1.2}
☐ Type L vent termination min 2 ft. above roof F30	[1804.2.4]	{6.7.1.4}
☐ Chimney termination min 3 ft. above roof F30	[1805.1]	{6.6.6}
☐ Vent or chimney termination 2 ft. above any portion of building within 10 ft. F30	[1804.2.4 & 1805.1]	{6.7.1.4}
☐ Masonry chimneys req cleanout	[1801.3.3]	{6.6.1}
☐ Masonry chimneys req liner	[1805.1]	{6.6.8}

Chimney Connectors — 09 IRC — NFPA 31

	09 IRC	NFPA 31
☐ Connectors as short & straight as practical	[1803.3]	{6.5.1}
☐ Min rise ¼ in./ft. F33	[1803.3]	{6.5.10}
☐ Secure support req'd	[1803.3]	{6.5.13}
☐ Joints screwed (min 3 screws NFPA 31)	[1803.3]	{6.5.14}
☐ Diameter min size of flue collar of appliance	[1803.3.3]	{6.5.7}
☐ Horizontal distance max 75% of vertical	[1803.3.2]	{6.5.1.2}
☐ Horizontal distance max 10 ft. without draft fan	[n/a]	{6.5.1.1}
☐ Draft fans req's burner interlock	[n/a]	{6.3.2}

FIG. 33

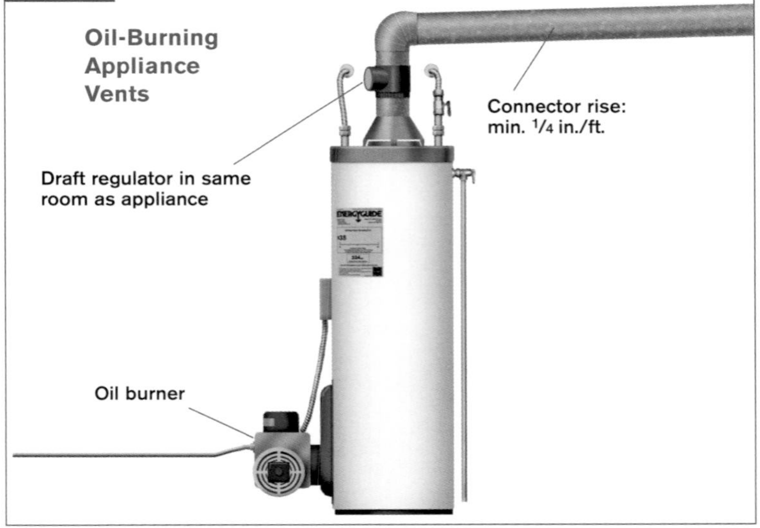

Oil-Burning Appliance Vents

Connector rise: min. ¼ in./ft.

Draft regulator in same room as appliance

Oil burner

OIL TANKS & PIPING

While the IRC & NFPA 31 provide rules for buried tanks, many jurisdictions do not permit them, as the potential for groundwater contamination is present should they develop a leak. Storage tanks may also be under the jurisdiction of the local fire protection district. There are special considerations with tanks in flood-prone areas or those with high seismic risk.

Tanks: General — 09 IRC — NFPA 31

- ☐ Tanks must be L&L _____ [2201.1] {7.2.1}[12]
- ☐ Install above design flood elevation or per NFPA 30__ [2201.6] {7.2.8.1}
- ☐ Restrain against earthquake movement per local codes[1307.2] {7.3.8.2}
- ☐ Tanks & supports req solid concrete foundations _____[n/a] {7.3.1}[13]
- ☐ Design foundation to minimize settling & corrosion___ [2201.2] {7.3.2}
- ☐ Max 660 gal above ground or inside building EXC ___ [2201.2] {n/a}
  - • Systems compliant with NFPA 31 _____[2201.2X] {7.2.7.2-5}

Outside Tanks — 09 IRC — NFPA 31

- ☐ Outside tank supports firmly anchored to foundation _____[n/a] {7.3.3}
- ☐ Tanks ≤ 275 gal (660 gal IRC) min 5 ft. from PL___ [2201.2.2] {7.8.2}
- ☐ Tanks > 275 gal & ≤ 660 gal min 10 ft. from PL _____[n/a] {7.8.2}
- ☐ Tanks > 660 gal per NFPA 30_____[2201.2X] {7.8.3}
- ☐ External tanks req corrosion-resistant coating _____ [2201.2.2] {7.8.4}

Inside Tanks — 09 IRC — NFPA 31

- ☐ Inside tanks > 60 gal only on lowest floor EXC_____[2201.2X] {7.5.4}
  - • Spill containment & no floor or open space below _[2201.2X] {7.5.5}
- ☐ Tanks > 10 gal min 5 ft. from any fire or open flame [2201.2.1] {7.5.7}

Abandoned Tanks — 09 IRC — NFPA 31

- ☐ Temporarily unused tanks emptied, cleaned & fill pipe filled with concrete & all other piping capped [2201.7] {7.12}
- ☐ Remove per International Fire Code or NFPA 30 [2201.7] {7.14}
- ☐ Also remove exterior piping of abandoned tanks [2201.7] {7.13.1}

Fill & Vent Piping — 09 IRC — NFPA 31

- ☐ Each tank or tank system req's separate fill & vent **F34** _ [2203.4] {7.5.9}
- ☐ Fill & vent pipes must terminate outside building **F34**_[2203.3&5] {8.4.7}
- ☐ Tank fill & vent piping min schedule 40 steel or brass ___ [2202.1] {8.2.1.1}
- ☐ No cast-iron fittings_____ [2202.2] {8.3.4}
- ☐ Fill pipe min 1¼ in. & pitched toward tank _____ [manu] {8.5.1}
- ☐ Vent pipe min 1¼ in. & pitched toward tank _____ [2203.4] {8.6.1}
- ☐ Fill piping min 2 ft. from building openings_____ [2203.3] {8.5.2}
- ☐ Each tank req's fill gauge (only inside tanks IRC) _____ [2201.5] {7.5.10}
- ☐ Fill gauge may be visual or audible _____[n/a] {8.10.3}
- ☐ No glass gauges or gauges subject to breakage _____ [2201.5] {8.10.4}
- ☐ Vent termination min 2 ft. from building openings _____ [2203.5] {8.6.2}
- ☐ Vent terminal screened & with weatherproof cap **F34** __ [2203.5] {8.6.3&4}
- ☐ Vent terminal above snow level & visible from fill _____ [2203.5] {8.6.2.1&2}

Piping & Tubing to Appliances — 09 IRC — NFPA 31

- ☐ Fuel supply schedule 40 steel or brass or seamless Cu, brass, or steel tubing _____ [2202.1] {8.2.2.1}
- ☐ Tubing req's corrosion-resistant coating or protective conduit to within 12 in. of tank or appliance _____[n/a] {8.2.2.2.1}
- ☐ Min tubing size ⅜ in. outside diameter Type L Cu_____ [2203.2] {8.7.1}
- ☐ Manual shutoff req'd at tank outlet _____ [2204.2] {8.7.1}
- ☐ Shutoff req'd at building entrance if tank outdoors _____[n/a] {9.2.11.1}
- ☐ Cross-connection of tanks to 660 gal total OK **F34**____ [2203.6] {8.9.1}
- ☐ Fusible link safety shutoffs req'd within 6 in. of filter on tank side & within 12 in. of inlet connection to burner **F34**_____[n/a] {8.10.6}[14]

PROPANE (LP GAS)

The IRC & UMC defer to NFPA 58, the *Liquefied Petroleum Gas Code*, published by NFPA. LP gas liquefies under moderate pressure & vaporizes upon release of the pressure. Horizontal storage tanks are manufactured to standards from the American Society of Mechanical Engineers (ASME) & portable cylinders are manufactured to U.S. Department of Transportation (DOT) standards. As a liquid or gas, propane is heavier than air & when gas leakage occurs in a pit or basement, an invisible pool of combustible material can accumulate until it rises to the level of an ignition source, such as the pilot or igniter on an appliance. One method of protection against this hazard is an interlock that would shut off the gas flow. Another method is to install a drain from the area containing the equipment & piping, as in F36.

Horizontal ASME Tanks

NFPA 58

- ☐ Tank clearances F35 .... [6.3.3&8&9 & Annex I]
- ☐ Tank not allowed indoors .... [6.2.1]
- ☐ Protect tanks from damage by vehicles .... [6.1.2]
- ☐ Masonry or concrete foundation req'd under tanks .... [6.6.3.1]
- ☐ Supports must be corrosion resistant & noncombustible .... [6.6.3.5]
- ☐ Secure tank against flotation in flood hazard areas .... [6.6.1.6]
- ☐ Secure tanks in seismic areas as approved by local AHJ .... [5.2.4.3D]
- ☐ Min 10 ft. from easily ignitable material (weeds, firewood, dry grass) .... [6.4.5.2]
- ☐ Replace containers with excessive dents or corrosion .... [5.2.1.4]

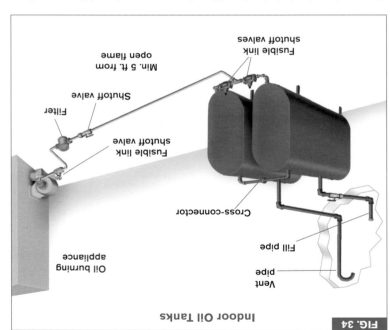

FIG. 34

Indoor Oil Tanks

- Oil burning appliance
- Filter
- Shutoff valve
- Fusible link shutoff valve
- Fusible link shutoff valves
- Min. 5 ft. from open flame
- Cross-connector
- Fill pipe
- Vent pipe

Compared to other fuels, heating oil has the greatest number of Btus per gallon (approximately 139K). It can be safely stored indoors as it is less readily ignitable than gasoline or propane. It is a commonly used fuel in the Northeast & relatively rare in western states or areas where natural gas is readily available. It displaced coal as the primary fuel source of the country. Heating oil typically has a red dye added to distinguish it from diesel oil.

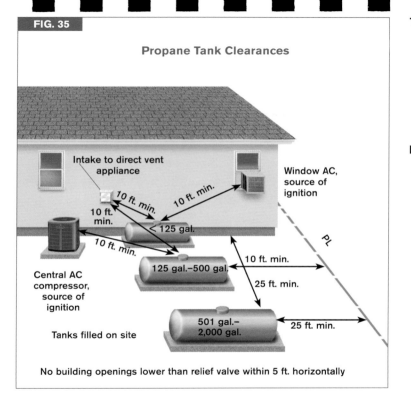

FIG. 35

Propane Tank Clearances

Intake to direct vent appliance

Window AC, source of ignition

10 ft. min.

10 ft. min.

10 ft. min.

< 125 gal.

PL

Central AC compressor, source of ignition

10 ft. min.

125 gal.–500 gal.

10 ft. min.

25 ft. min.

Tanks filled on site

501 gal.– 2,000 gal.

25 ft. min.

No building openings lower than relief valve within 5 ft. horizontally

Tank Valves & Regulators   NFPA 58

- ☐ Vapor & liquid shutoff valve req'd _____ [5.7.4.1]
- ☐ Shutoff valve req'd to be readily accessible _____ [5.7.8.1G]
- ☐ PRV req'd _____ [5.7.4.1 & 5.7.2.5]
- ☐ Rain cap req'd over PRV _____ [6.7.2.4]
- ☐ No shutoff valve between tank & PRV _____ [6.7.2.8]
- ☐ No shutoff valves on outlet of PRV _____ [6.7.2.10]
- ☐ First-stage or high-pressure regulator req'd outdoors _____ [6.8.1.4]
- ☐ Regulator must be designed to resist elements (freezing) _____ [6.8.1.5]

Piping & Tubing Systems   NFPA 58

- ☐ Piping material black, galvanized, brass, PE, or polyamide AMI _____ [5.9.3.1]
- ☐ Tubing brass, Cu L or K, CSST PE, or polyamide AMI _____ [5.9.3.2]
- ☐ Plastics OK only underground & outdoors _____ [6.9.4.1]
- ☐ Min 14AWG tracer wire req'd to be run with plastic pipe _____ [6.9.4.6]
- ☐ No cast-iron pipe fittings _____ [5.9.4.1}
- ☐ Buried metal pipe min 12 in. cover (18 in. if damage likely) EXC_____ [6.8.3.12]
  - • Conduit or shielding OK when 12 in. cover not possible _____ [6.9.3.12]
- ☐ Underground metal piping req's corrosion protection_____ [6.9.3.14]
- ☐ LP gas piping not OK as grounding electrode_____ [6.9.3.15]
- ☐ Grounding & bonding for static electricity protection not req'd _____ [6.22.1.3]
- ☐ Dielectric fitting req'd between above- & below-ground pipe_____ [6.9.3.16][15]
- ☐ Dielectric fitting to be above ground & outdoors _____ [6.9.3.16][15]

FIG. 36

Propane Appliance in Basement

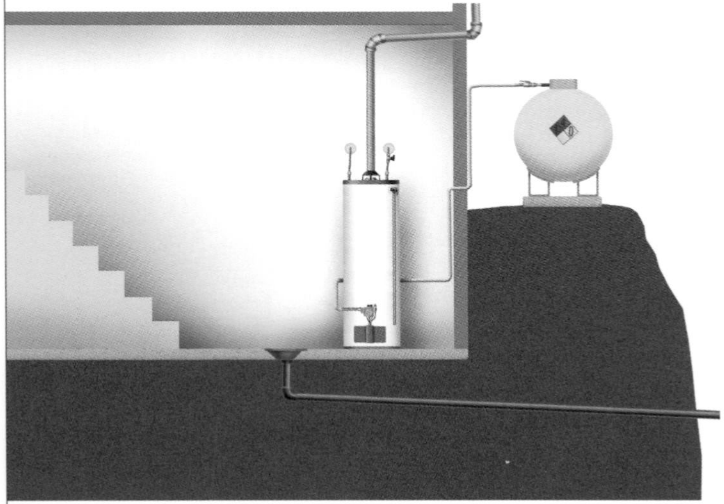

To prevent accumulation of an explosive concentration of propane from a leak, below-grade pits or basements should have a detector interlocked to the fuel line or a gravity drain to daylight.

FREESTANDING FIREPLACE STOVES (SOLID FUEL)

The IRC (section 1414.1) req's fireplace stoves to be listed, labeled & tested in accord with UL737, which in turn references the current edition of NFPA 211. NFPA 211 also recognizes unlisted appliances & provides specific rules to maintain safe clearances & protection of adjacent combustible material.

Fireplace Stoves & Solid-Fuel Room Heaters   NFPA 211

☐ Install listed equipment per L&L & AMI _____ [12.1]
☐ Install unlisted equipment AMI & per NFPA 211 _____ [12.1]
☐ Fire screen req'd per L&L_____ [12.1]
☐ Not in alcove or enclosed space < 512 cu. ft. unless so listed _____ [12.2.2]
☐ Not OK in garage _____ [12.2.4]
☐ Noncombustible floor material 18 in. beyond stove on all sides EXC _[12.5.1.4]
  • L&L floor protection assemblies OK AMI_____[12.5.1.5]
☐ List appliance floor protection AMI _____[12.5.1.1]
☐ Unlisted appliance floor protection:_____[12.5.1.2]
  • If legs provide ≥ 6 in. of ventilated clearance under stove, 2 in. thick
   masonry + metal **F37**_____[12.5.2.1]
  • If legs provide 2 in. to 6 in. of ventilated clearance, 4 in. thick hollow
   masonry + metal **F37**_____[12.5.2.2]
☐ If legs provide < 2 in. clear, floor req'd to be noncombustible _____[12.5.2.3]
☐ Fuel storage (firewood) min 36 in. from appliance_____ [T12.6.1 note a]
☐ 36 in. side, top & front clearance from appliance to combustibles EXC [12.6.1]
  • Listed appliance clearance to combustibles AMI _____[12.6.1.1]
  • Lesser clearances OK per **T9, F37,38** _____[12.6.2.1]

Connectors   NFPA 211

☐ Must be accessible for inspection, cleaning & replacement_____ [9.7.11]
☐ Single wall min 18 in. clear to combustibles EXC **F37** _____[T9.5.1.1]
  • Lesser clearance with approved clearance-reduction system **T9** __ [9.5.1.2.1]

Connectors (cont.) NFPA 211

☐ Not to pass through wall EXC _____ [9.7.4]
 • Listed pass-through system _____ [9.7.4]
☐ System designed per NFPA 211 figure 9.7.5_____ [9.7.5]
☐ Maintain min ¼ in./ft. rise from appliance collar to chimney _____ [9.7.7]

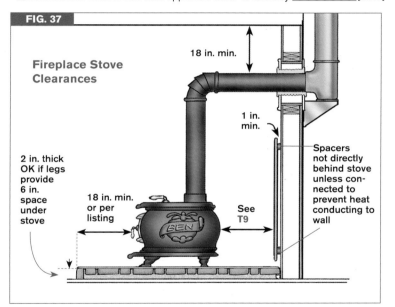

FIG. 37

Fireplace Stove Clearances

18 in. min.

1 in. min.

2 in. thick OK if legs provide 6 in. space under stove

18 in. min. or per listing

See T9

Spacers not directly behind stove unless connected to prevent heat conducting to wall

Connection to Masonry Fireplace (Stoves & Fireplace Inserts) NFPA 211

☐ Connector must extend to flue liner—not just to firebox_____ [12.4.5.1]
☐ If connector enters direct through chimney wall above smoke chamber, noncombustible seal req'd below entry _____ [12.4.5.1]
☐ No dilution of combustion products in flue with habitable space air __[12.4.5.1]
☐ Flue not less than size of appliance collar _____ [12.4.5.1]
☐ Flue diameter max 2× appliance collar if chimney walls exposed to exterior below roof_____ [12.4.5.1]
☐ Flue diameter max 3× appliance collar if no part of chimney walls exposed to exterior below roof_____ [12.4.5.1]
☐ Installation must allow for chimney inspection & cleaning_____ [12.4.5.1]

RECOMMENDED INSPECTIONS OF EXISTING CHIMNEYS NFPA 211 14.3

Level 1: All readily accessible areas of chimney, structure & flue. To be performed annually, during routine cleaning & when replacing appliances with a similar appliance.

Level 2: Level 1 + video scan of flue. Verify clearances & suitability of flues. To be performed upon resale of property; upon addition or removal of appliances, adding or replacing with dissimilar appliance, or after operating malfunction.

Level 3: Level 1 & 2 + removal of components as necessary to gain access. To be performed when Level 1 or Level 2 cannot identify conditions deemed critical to renewed or continued use; fire or damage investigations.

CLEARANCE REDUCTION SYSTEMS

Clearance reduction systems are used with solid-fuel, oil-burning & gas-burning appliances. They provide a practical means of installing appliances in spaces where they otherwise might not fit or would take too much space in a room. They may not be used with appliances in closets (alcoves); those appliances req clearances in accordance with the nameplate label. Tables 9.5.1.2 & 12.6.2.1 in NFPA 211 have the same values as **T9** for fireplace stoves, which otherwise req 36 in. clearance. The UMC uses **T9** as Table 5.3 & in NFPA 54 it is table 10.2.3(b).

General   09 IRC

☐ Clearance reductions allowed per **T9** _____ [1306.2, 1803.3.4, 2409.2]
☐ Gas appliance & vent connector reductions per **F39, T9** _____ [2409.2]
☐ Solid fuel appliances not allowed to be reduced to < 12 in. EXC ___ [1306.2.1]
  • Appliances listed for < 12 in. & installed AMI _____ [1306.2.1]
☐ No spacers directly behind appliance or connector **F37,38** _____ [F1306.2]
☐ Spaces noncombustible (stacked washers, conduit, etc.) _____ [F1306.2]
☐ Ventilated air space min 1 in. & open at edges **F37, 38, T9** _____ [1306.2]
☐ Air space in corner open top & bottom **F38, T9** _____ [1306.2]
☐ Air space on flat wall open top & bottom or side & top **F37, T9** _____ [1306.2]

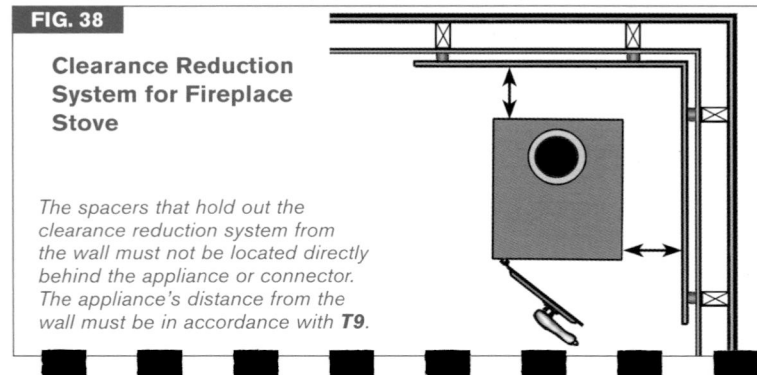

FIG. 38

Clearance Reduction System for Fireplace Stove

The spacers that hold out the clearance reduction system from the wall must not be located directly behind the appliance or connector. The appliance's distance from the wall must be in accordance with T9.

TABLE 9	CLEARANCE REDUCTION (in.) [T1306.2] {T5-3}									
Unprotected	36	18	12	9	6	36	18	12	9	6
Method[A]	Protected Wall Clearance					Protected Ceiling Clearance				
3½ in. masonry	24	12	9	6	5	n/a	n/a	n/a	n/a	n/a
3½ in. masonry with air space[B]	12	6	6	6	6	n/a	n/a	n/a	n/a	n/a
½ in. insulation board over fiber or mineral wool batts[C]	18	9	6	5	3	24	12	9	6	4
½ in. insulation board with air space[B]	12	6	4	3	3	18	9	6	5	3
24 gage Zi steel with air space[B]	12	6	4	3	2	18	9	6	5	3
1 in. insulating batts between 2 sheets 0.024 Zi steel with air space	12	6	4	3	3	18	9	6	5	3
24 gage Zi steel over reinforced batts with air space[B]	12	6	4	3	3	18	9	6	5	3

A. Clearances are measured in closest stretched-string distance. In no case can a solid-fuel burning appliance clearance be reduced to < 12 in.

B. Air spaces must be a min. of 1 in. & ventilated by being open at bottom & top edges or sides & top edges. Spacers must be noncombustible & cannot be mounted directly opposite the appliance or connector.

C. Insulation fiber or mineral wool must have thermal conductivity ≤ 1.0 Btu/sq. ft. Mineral wool blankets or board min. density 8 lbs./cu. ft. & min. melting point 1500°F.

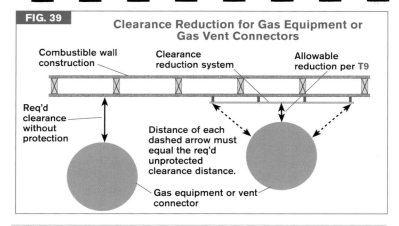

FIG. 39

Clearance Reduction for Gas Equipment or Gas Vent Connectors

Combustible wall construction

Clearance reduction system

Allowable reduction per **T9**

Req'd clearance without protection

Distance of each dashed arrow must equal the req'd unprotected clearance distance.

Gas equipment or vent connector

ELECTRIC HEAT

Electric resistance heating can be in the form of central forced-air furnaces, base-board heaters, radiant ceiling panels, duct heaters & even exotic systems such as electric heat in ceramic tile bath floors. The wiring for electric heating must be sized to 125% of the load to ensure that it does not also become a heater.

General	09 IRC	11 NEC
☐ Circuits considered continuous load	[3702.10]	{424.3B}
☐ Circuits for continuous loads sized to 125% of load	[3701.2]	{210.20A}
☐ All electric heating equipment must be L&L	[3403.3]	{424.6}
☐ Factory-applied nameplates must include:	[1303.1]	{424.6}

• Label with manu name, model & serial no.
• Operating & maintenance instructions or publication no. of manual
• Rating in volts, amperes, or watts, no. of phases if > 1
• Req'd clearances from combustibles

Central Electric Heat	09 IRC	11 NEC
☐ Disconnect in sight of equipment unless breaker capable of being locked in OFF position	[4101.5]	{424.19}
☐ Locking means must remain without lock installed	[T4101.5]	{424.19}

Baseboard Heaters	09 IRC	11 NEC
☐ Must be L&L & installed AMI	[3403.3]	{424.6}
☐ Branch circuits for 2 or more units can be 15, 20, 25, or 30 amps	[3702.10]	{210.3}
☐ No receptacles above heaters: integral receptacles with heaters can substitute for req'd receptacles in rooms **F40**	[1405.1]	{424.9}

FIG. 40

Electric Baseboard Heaters

Listing instructions prohibit installation of baseboard heaters under receptacles

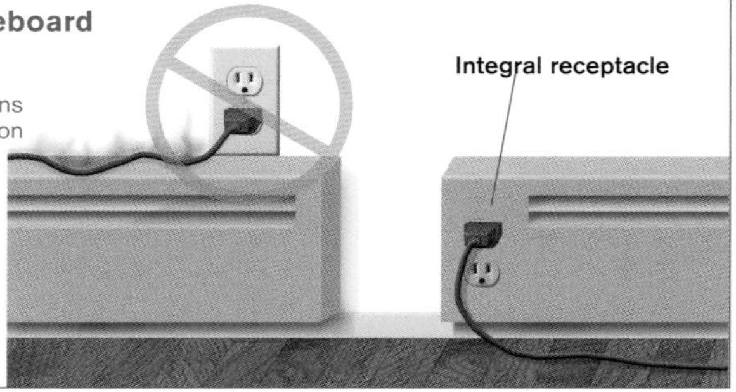

Integral receptacle

Heating Cables in Concrete or Masonry Floors	11 NEC
☐ Min 1 in. spacing between cables	{424.44B}
☐ Leads protected where leaving floor	{424.44E&F}
☐ GFCI protection req'd for cables in bathroom floors	{424.44B}
☐ Secure in place while concrete or other finish applied	{424.44C}
☐ Inspection & approval req'd before covering	{424.44G}

CLOTHES DRYER EXHAUST

Electric Clothes Dryer Exhaust

	09 IRC	09 UMC
☐ L&L ductless (condensing) dryers OK per L&L	[1502.2X]	{n/a}
☐ Closet installation req'd make-up air opening min 100 sq. in.	[n/a]	{504.3.2}
☐ Flexible transition ducts L&L & single piece	[1502.4.3]	{504.3.2.1X}
☐ Connectors not concealed & max 8 ft. {6 ft. UMC} F41	[1502.4.3]	{504.3.2.1X}
☐ Duct smooth metal, no screws in air flow F41	[1502.4.1]	{504.3.2.1}
☐ Support & secure at max 4 ft. intervals[16]	[1502.4.2]	{n/a}
☐ Duct min 4 in. diameter[17]	[1502.4.1]	{504.3.2}
☐ IRC: Max length AMI or 25 ft. minus bends per T10[18]	[1502.4.4]	{n/a}
☐ UMC: Max length 14 ft. minus 2 ft. each 90° turn > 2	[n/a]	{504.3.2.2}
☐ No mixing with or passage through other systems	[1502]	{504.3.1}
☐ End outside in backdraft damper & no screens F42	[1502.3]	{504.3.1}
☐ Min 3 ft. from other building openings	[1502.3]	{n/a}
☐ Shield plates < 1¼ in. from framing surface F41[19]	[1502.5]	{n/a}
☐ Length of concealed duct on tag on 5 ft. of connection[20]	[1502.4.5]	{n/a}

Gas Clothes Dryer Exhaust

	09 IRC	09 UMC
☐ Closet req's makeup air opening [min 100 sq. in. IRC]	[2439.4]	{905.3A}
☐ Flexible transition ducts (connectors) L&L & single piece	[2439.5.4]	{905.4C}
☐ Connectors not concealed [& max 8 ft. in IRC] F41	[2439.5.4]	{905.4C}
☐ Duct smooth metal, no screws in air flow F41	[2439.5.1]	{905.4B}
☐ Support intervals max 4 ft. spacing[16]	[2439.5.2]	{n/a}
☐ Duct min 4 in. diameter	[2439.5.1]	{n/a}
☐ Max 35 ft. minus bends per T10 or AMI F41[18]	[2439.5.5.1]	{n/a}
☐ No mixing with or passage through other systems	[2439.1]	{905.4A}
☐ End outside in backdraft damper & no screens F42	[2439.3]	{n/a}
☐ Shield plates < 1¼ in. from framing surface F41[19]	[2439.5.3]	{n/a}
☐ Length of concealed duct on tag on 5 ft. of connection[20]	[2439.5.6]	{n/a}

TABLE 10	DRYER FITTING EQUIVALENT LENGTH [T1502.4.4.1 & T2439.5.5.1]	
Fitting Radius	**Equivalent Length**	
	45° Elbow	**90° Elbow**
4 in. mitered	2 ft. 6 in.	5 ft.
6 in. smooth	1 ft.	1 ft. 9 in.
8 in. smooth	1 ft.	1 ft. 7 in.
10 in. smooth	9 in.	1 ft. 6 in.

Electric Radiant Heat Systems

	09 IRC	11 NEC
☐ Install AMI	[1406.1]	{424.93A1}
☐ Install panels parallel to framing	[1406.3]	{424.93B2}
☐ Fasteners > ¼ in. from heating element	[1406.3]	{424.93B3}
☐ Min 8 in. distance from surface-mounted fixture boxes	[n/a]	{424.93A3}
☐ Min 2 in. distance from recessed fixtures & trim	[n/a]	{424.93A3}
☐ No field modification of panels unless so listed	[1406.3]	{424.93B4}
☐ Wiring above heated ceiling min 2 in. clearance	[n/a]	{424.94}
☐ Wiring above heated ceiling considered as 50°C ambient unless over 2 in. thermal insulation	[n/a]	{424.94}

Electric Duct Heaters

	09 IRC	11 NEC
☐ Install AMI	[1407.1]	{424.66}
☐ If used in system with AC, must be L&L for same	[1407.1]	{424.62}
☐ If < 4 ft. from heat pump/air-conditioning, both must be listed for such clearances	[1407.3]	{424.61}
☐ Install with manu recommended clearance from Class 1 ducts unless L&L for direct connection	[1407.2]	{424.66}
☐ Lockable breaker req'd or disconnect within sight	[4101.5]	{424.65}
☐ Each unit req'd integral limit controls & manual reset	[1407.1]	{424.64}
☐ Must be accessible for servicing	[1407.4]	{424.66}
☐ Interlock req'd to prevent heat if fan not operating	[1407.5]	{424.63}

FIG. 41

Dryer Exhaust

If duct length based on manu instructions, copy must be provided to AHJ & duct must be inspected.

The Consumer Product Safety Commission (CPSC) estimates that up to 16,000 home fires a year originate at clothes dryers. Common causes of these fires are lint buildup from improperly installed or maintained exhaust ducts. Screws should not penetrate to the interior of the duct as they accumulate lint, which leads to blockage.

NOTICE
Concealed duct length 39 ft.

UMC length 14 ft., up to 2 90° bends, deduct 2 ft. for each additional 90°

IRC length 25 ft. for electric, 35 ft. for gas or AMI

FIG. 42

Backdraft Damper

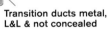

Deduct for bends T10

Transition ducts metal, L&L & not concealed

End outside & no screens

Dryers with specific manu instructions are allowed longer lengths than otherwise permitted by code.

VENTILATION & EXHAUST SYSTEMS

Building codes tell us when we must provide ventilation for interior spaces & mechanical codes tell us how to provide it. These topics overlap with energy codes & green building standards. The standard recognized in many energy codes is *ASHRAE 62.2, Ventilation & Acceptable Indoor Air Quality in Low-Rise Residential Buildings.* Check with your local jurisdiction to determine if these standards apply in your area. Whole-house ventilation is intended to dilute the contaminants from materials found in furnishings, furniture & building products. Localized exhaust removes contaminants from specific sources, such as kitchens & baths.

Whole Building Ventilation · ASHRAE 62.2

☐ Mechanical exhaust, supply, or combination system req'd _____ [4.1]
☐ Min ventilation rate must comply with **T11** _____ [4.1]
☐ Local exhaust fans can supply whole-house continuous ventilation_____ [4.2]
☐ Central-fan-integrated supply timer must operate min 10% of time_____[4.4X]
☐ OK to override with "fan on" control at thermostat _____ [4.4]
☐ Whole building or continuous ventilation fans max 1.0 sone EXC _____ [7.2.1]
  • Heating, venting & air-conditioning (HVAC) air handlers _____ [7.2X]
  • Remote-mounted fans with min 4 ft. ductwork between grill & fan _____ [7.2X]

TABLE 11	MIN. VENTILATION RATES (CFM) (ASHRAE 62.2 T4.1A)				
Floor Area (sq. ft.)	**Number of Bedrooms**				
	0–1	2–3	4–5	6–7	> 7
< 1,500	30	45	60	75	90
1,501–3,000	45	60	75	90	105
3,001–4,500	60	75	90	105	120
4,501–6,000	75	90	105	120	135
6,001–7,500	90	105	120	135	150
> 7,500	105	120	135	150	165

The requirement can be met from multiple sources, such as 2 or more continuously operating exhaust fans adding up to at least the req'd amount in the table.

Local Exhaust ASHRAE 62.2

☐ Mechanical exhaust req'd each kitchen_____ [5.1]
☐ Kitchen exhaust min 100cfm intermittent or 5ACH continuous ___[T5.1 & T5.2]
☐ Vented range hood req'd if 100 cfm is < 5 kitchen ACH _____ [T5.1]
☐ Bathroom 50cfm intermittent or 20cfm continuous_____[T5.1 & T5.2]
☐ Controls may be humidistat, timer, or occupancy sensor provided
   that occupant has ability to override control_____ [5.2.1]

Additional Air Quality Requirements ASHRAE 62.2

☐ Clothes dryers req exterior exhaust except condensing dryers _____ [6.3]
☐ Air inlets min 10 ft. from contaminants such as plumbing vents_____ [6.8]
☐ Exhaust ventilation may not deplete combustion air to appliances within
   pressure boundary (sum of 2 largest exhaust max 15cfm/100 sq. ft.)_____ [6.4]
☐ Door from attached garage to house weatherstripped _____ [6.5.1]
☐ Duct leakage outside pressure boundary max 6% _____ [6.5.2]
☐ Central furnace or AC system filter min efficiency reporting value
   MERV 6 _____ [6.7]
☐ Habitable spaces req ventilation openings ≥ 4% of floor area _____ [6.6.1]
☐ Utility rooms req ventilation openings ≥ 4% of floor area EXC_____ [6.6.2]
  • Utility rooms with dryer exhaust duct _____[6.6X]

Bathroom Exhaust & Ventilation 09 IRC ASHRAE

☐ Mechanical ventilation 50cfm intermittent or 20cfm continuous
   direct to exterior OK EXC **F43** _____ [303.×] {5.1}
  • Natural ventilation openings min 1.5 sq. ft. OK _____ [303.3] {Ø}
☐ Air may not be exhausted into attic _____ [1501.1] {n/a}
☐ Toilet rooms req vent openings ≥ 4% of floor area EXC _____[n/a] {6.6.2}
  • Toilet compartments within bathrooms_____[n/a] {6.6X}
☐ Air exhaust & intake openings req screens_____ [303.5] {n/a}

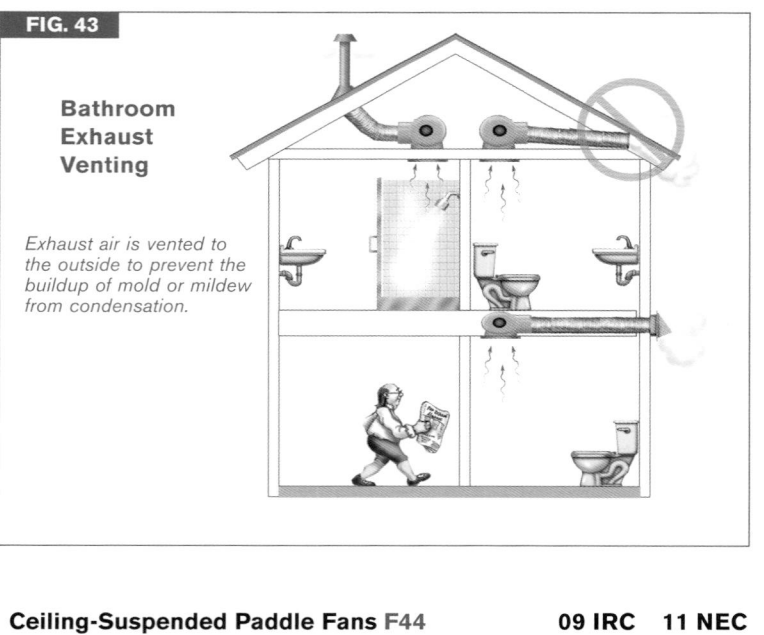

FIG. 43

Bathroom Exhaust Venting

Exhaust air is vented to the outside to prevent the buildup of mold or mildew from condensation.

Ceiling-Suspended Paddle Fans F44 09 IRC 11 NEC

☐ Listed box for fan support (no standard boxes) _____ [3905.9] {314.27C}
☐ Listed fan boxes without weight marking OK up to 35 lb. [3905.9] {314.27C}
☐ Fan > 35 lb. & < 70 lb., fan box L&L for suitable weight _ [3905.9] {314.27C}
☐ Independent support for fans > 70 lb. _____ [3905.9] {314.27C}

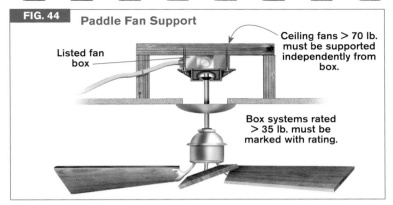

FIG. 44   **Paddle Fan Support**

Listed fan box

Ceiling fans > 70 lb. must be supported independently from box.

Box systems rated > 35 lb. must be marked with rating.

KITCHENS

Freestanding Ranges	09 IRC	09 UMC
☐ Must be listed as household type–not commercial _____	[2447.3]	{n/a}
☐ Vertical clearance to combustibles min 30 in. EXC_____	1901.1	{916.1B}
• Lesser clearances AMI_____	[1901.1]	{916.1B}
• 24 in. OK with metal hood or metal over millboard ____	[n/a]	{916.1B}
☐ Side clearance to combustibles AMI EXC _____	[1901.2]	{916.1A}
• 6 in. min sides & rear for unlisted appliances_____	[∅]	{916.1A}

Built-in Ranges	09 IRC	09 UMC
☐ Install AMI _____	[1901.2 & 2447.1]	{916.2A&C}
☐ Vertical clearance to combustibles min 30 in. EXC **F45**__	1901.1	{916.2B}
• Lesser clearances AMI **F45**_____	[1901.1]	{916.2B}
• 24 in. OK with metal hood or metal over millboard ____	[n/a]	{916.2B}
☐ Must be level _____	[n/a]	{916.2D}

Hood for Open-Top Broilers	09 IRC	09 UMC
☐ Hood req'd & must extend as wide as broiler unit _____	[1505.1]	{920.3}
☐ Min ¼ in. clearance to combustibles _____	[1505.1]	{920.3}
☐ Min 24 in. from cooking surface to combustible materials	[1505.1]	{920.3}
☐ Must be ducted to outdoors & have backdraft damper __	[1505.1]	{504.1}

Range Hoods	09 IRC	09 UMC
☐ Must go outdoors (min 3 ft. from openings UMC) EXC _	[1503.1]	{504.5}
• Ductless (recirculating) range hoods OK_____	[1503.1X]	{303.1}
☐ Exterior openings screened with ¼ in. to ½ in. mesh ___	[1503.1]	{n/a}
☐ Min 100 cfm intermittent or 25 cfm continuous_____	[1503.3]	{n/a}
☐ PVC OK for downdraft duct under slab _____	1503.2X]	{504.2X}

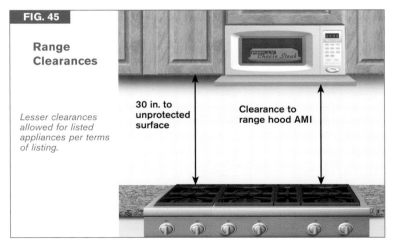

FIG. 45

Range Clearances

Lesser clearances allowed for listed appliances per terms of listing.

30 in. to unprotected surface

Clearance to range hood AMI

GLOSSARY

GLOSSARY

A

ABS (acrylonitrile-butadiene-styrene): A plastic pipe that is usually black and used for DWV. It can also be used for building water supply and for vent piping of high-efficiency condensing appliances.

Accessible: Capable of being exposed without damage to the building or component structure or finishes, and that may req removal of access doors or fasteners req tools.

AHJ (Authority Having Jurisdiction): An organization responsible for enforcing the code, typically the building department & its authorized representatives.

Airbreak: A physical separation in which a discharge pipe from a fixture, appliance, or device drains indirectly into a receptor & enters below the flood level rim of a receptor, such as a clothes washer standpipe.

Air conditioning: The process of heating, cooling, humidifying, dehumidifying, filtering, or otherwise treating air in a building. Most nontrade persons relate this term to cooling only.

Air handler: A blower or fan enclosed in a metal box used for the purpose of distributing supply air to a room, space, or area.

Alcove: A room or space such as a closet that is not large in relationship to the appliances within it. It would be less than 16 times the volume of a boiler or 12 times the volume of a furnace. For measurement purposes, only the portion of the room up to 8 ft. above the floor counts in determining volume. Also see "Room large in comparison to size of equipment."

Approved: Accepted by the Authority Having Jurisdiction (AHJ), UL & other testing laboratories do not approve materials; they test products & determine their conformity to published standards. Only the AHJ can approve them.

B

Backflow: A flow of water or other liquids, mixtures, or substances into the distributing pipes of a potable supply of water from any source other than its intended source.

Bathroom: In ASHRAE, a bathroom is a room containing a tub, shower, spa, or other source of moisture. A half bath contains only a water closet & lavatory & is not considered a bathroom. In the NEC, a bathroom is a room containing a basin & another plumbing fixture.

Btu (British thermal unit): The quantity of heat necessary to raise the temperature of 1 lb. of water 1°F.

Building thermal envelope: The basement walls, exterior walls, floor, roof & any other building element that encloses conditioned spaces.

C

Central-fan-integrated supply system: A method of supplying whole-house ventilation through a makeup air duct connected to the supply ducts of a forced-air system & a timer on the furnace fan control.

Check valve: A device used to prevent the flow of liquids in a direction not intended in the design of the system. Check valves are not backflow preventers. They are often used in solar systems.

Chimney: A primarily vertical structure containing one or more flues, for the purpose of carrying gaseous products of combustion & air from an appliance to the outside atmosphere. Factory-built chimneys must be listed & labeled. Masonry chimneys are field constructed of solid masonry units, bricks, stones, or concrete.

Chimney connector: A pipe connecting a fuel-burning appliance to a chimney flue.

Closet: *See "Alcove."* **F32**

Combustible material: Any material not defined as noncombustible. The extent of combustibility of surface materials is measured in flame spread index & smoke-developed index. Many HVAC components req specified clearances from combustible material, including the paper facing of gypsum board.

Combustion air: Air req'd for combustion of a fuel. It includes air that is burned with the fuel, air for dilution of the flue gases and that is introduced into draft hoods, and ventilation air that cools appliances.

Common vent: A pipe venting two trap arms on the same branch, either back to back or one above another.

Concealed: Not exposed to view without removal of building surfaces or finishes.

Confined space: A room or space having a volume less than 50 cu. ft. for each 1,000Btu input rating of all fuel-burning appliances in the room or space.

CPVC (chlorinated polyvinyl chloride): Plastic pipe designed for hot & cold water. Water distribution pipe is typically cream-colored, and orange CPVC is used for automatic fire sprinkler piping.

D

Decorative appliance for installation in fireplaces: An assembly with artificial logs & with gas burners to simulate a solid-fuel fire, installed inside a fireplace otherwise capable of burning solid fuel. They can be either manually or automatically operated. If automatic, they must include a flame safeguard device.

Decorative shroud: A partial enclosure for aesthetic purposes that surrounds or conceals the termination of a chimney or vent. Decorative shrouds must be specifically listed for the chimney or vent assembly & are often installed incorrectly.

Dilution air: Air that combines with flue gases at the draft hood of an appliance. *See "Combustion air."*

Direct-vent appliances: Appliances that are constructed & installed so that all air for combustion is derived from the outside atmosphere & all flue gases are discharged directly to the outside atmosphere, usually by a coaxial flue pipe inside the combustion air pipe.

Draft: The flow of gases or air through a chimney or flue caused by pressure differences. An induced draft appliance has a fan to overcome the resistance of the combustion chamber while still delivering flue gas to the vent at nonpositive pressure relative to the atmosphere. A forced draft appliance delivers flue gas under positive pressure. Natural draft is caused by the height of the chimney & the difference in temperature of hot gases & outside atmosphere.

Draft hood: A nonadjustable device integral to an appliance or made part of the appliance connector. It provides for the escape of flue gases from the appliance in the event of insufficient draft, backdraft, or stoppage. A draft diverter (typical on WH) prevents backdraft from entering the appliance & neutralizes the stack effect on the operation of the appliance.

Draft regulator: A device that functions to maintain a desired draft in the appliance by automatically reducing the draft to the desired value. These are usually adjustable, such as the barometric damper on an oil-burning appliance flue. A double-acting barometric draft regulator is free to move in either direction & protect against both excessive draft (that could allow the flame to lift) & backdraft.

Duct: A continuous passageway for the transmission of air (usually forced) made of factory-built components.

E

Energy-recovery ventilator (ERV): Same as heat-recovery ventilator, with a heat exchanger core that removes humidity. ERVs req a drain to remove water that condenses in them.

Forced draft: A vent system using a fan or other mechanical means to expel flue gases under positive static vent pressure.

Furnace: A device that is completely self-contained & designed to supply heated air to spaces remote from or adjacent to the furnace location. A central furnace uses ducts to supply to spaces.

G

Gas connector: Tubing or piping that connects the gas supply piping to the appliance.

H

Habitable room: A room used for living, sleeping, eating, or cooking. Bathrooms, closets, halls, storage spaces & laundry rooms are not considered habitable rooms.

Hangers: *See "Supports."*

Hearth: The floor area with the fire chamber of a fireplace or fireplace stove.

Hearth extension: The surfacing applied to the floor area in front of & to the sides of the hearth opening of a fireplace or fireplace stove.

Heat pump: A system that uses the change of state of a refrigerant to extract heat from one substance & transfer it to another area of the same or a different substance. Heat pumps can provide both heating & cooling.

Heat-recovery ventilator (HRV): A combination ventilation system that replaces indoor air with outdoor air that passes through a heat exchanger. The heat exchanger tempers the outdoor air to minimize energy losses. HRVs that also remove humidity from the indoor air are referred to as energy-recovery ventilators.

Horizontal: Any pipe or vent that is less than 45° from horizontal.

H.S.P.F. (Heating-seasonal-performance factor): The measure of a system's efficiency in heating mode. The higher the number, the more efficient the system.

Evaporative cooler: A device used for reducing the sensible heat of air for cooling by the process of evaporation of water into an airstream. Also known as a "swamp cooler". Evaporative coolers are used in hot, dry climates & for makeup air in commercial kitchens.

F

Factory-built fireplace: A fireplace composed of listed factory-built components assembled in accordance with the terms of the listing to form the completed fireplace. The appliance must be suitable for solid fuel & be equipped with a listed & properly installed chimney.

Fan-assisted appliance: An appliance equipped with an integral mechanical means to either draw or force products of combustion through the combustion chamber &/or heat exchanger.

Fireblock: Building materials installed to resist the free passage of flame to other areas through small concealed spaces of the building.

Fireplace stove: A freestanding solid-fuel burning device designed to be operated with the firebox door either open or closed.

Firestop: Until the early 1990s, this term was used for what today is called fireblocking. A penetration firestop assembly is a group of materials installed to resist the passage of flame through an assembly, typically around a duct, vent, or chimney passing through a rated ceiling, floor, or wall.

Flame safeguard: A device that will automatically shut off the fuel supply to a main burner or group of burners when the means of ignition of those burners becomes disabled & when flame failure occurs.

Flue: A passageway intended to carry hot gases through a chimney. The term is also used as a substitute for "vent"

Flue collar: The outlet of an appliance designed for the attachment of a draft hood, vent connector, or venting system.

Flue gases: Products of combustion plus excess air in an appliance flues or heat exchangers.

In sight: *See "Within sight."*

Indirect-fired WH: A water heater with a storage tank equipped with a heat exchanger used to transfer heat from an external source to heat potable water. The storage tank could derive its heat source from an external source, such as solar or a boiler, or an internal source.

Induced draft appliance: An appliance that utilizes a fan to overcome resistance of a heat exchanger & to assist in the delivery of flue gases to the appliance outlet (flue collar). Induced draft appliances typically deliver the flue gases to the flue collar at non-positive pressure due to the temperature of those gases relative to outside atmosphere. *See "Vented gas appliance categories."*

Induced draft burner: A burner that depends upon a draft that is induced by a fan that is integral to the appliance & is downstream from the burner.

Joint: Connection between two pipes:

Brazed joint: Joint obtained by joining metal parts with alloys that melt at temperatures > 840°F (449°C), but lower than the melting temperature of the parts to be joined.

Expansion joint: Loop, return bend, or return offset that accommodates pipe expansion & contraction.

Flexible joint: Joint that allows movement of one pipe without deflecting the other pipe.

Mechanical joint: Joint that uses compression to seal the joint.

Slip joint: Joint that incorporates a washer or special packing material to create a seal.

Soldered joint: Joint obtained by joining of metal parts with metallic mixtures or alloys that melt at a temperature < 800°F (427°C) & > 300°F (149°C).

Welded joint or seam: Joint or seam obtained by the joining of metal parts in the plastic molten state.

Label: A marking applied on a product that identifies the manu, the function or designation of the product, and the agency that has evaluated a sample of that product.

Labeled: Equipment, materials, or products affixed with a label or other identifying mark to attest that the product complies with identified standards or has been found suitable for a specific purpose. *See "Listed."*

Liquefied petroleum (LP) gas: LP or propane gas is composed primarily of propane, propylene, butanes, or butylenes or mixtures thereof that are gaseous under normal atmospheric conditions but capable of being liquefied under moderate pressure at normal temperatures. LP gas is typically stored in tanks on site. Unlike natural gas (CH_4), LP gas (C_3H_8) is heavier than air.

Listed: Equipment or materials on a list published by an approved organization that is concerned with product evaluation and that maintains periodic inspection of production of listed equipment or materials. The listing will state that the product meets specified standards or has been found suitable for a specific purpose.

Log lighter, gas-fired: A manually operated solid-fuel ignition device for installation in a vented solid-fuel burning fireplace. These devices are intended to help initiate a fire in a fireplace, as compared to a decorative appliance for installation in a fireplace.

Low-pressure hot-water heating boiler: A boiler furnishing hot water at pressures not exceeding 160 psi or temperatures not exceeding 250°F.

Low-pressure steam-heating boiler: A steam boiler that operates at pressures not exceeding 15 psi.

Luminaire: A complete lighting fixture including the lamp(s), mounting assembly & cover.

MECHANICAL

M

Makeup air: Air provided to replace air being exhausted.

N

National Recognized Testing Laboratory (NRTL): A testing facility recognized by OSHA as qualified to provide testing & certification of products & services. Examples of NRTLs are CSA, NDF & UL.

Natural-draft burner: A burner in which proper combustion depends on establishing a draft of flue gases that will rise by the pressure difference between the flue gases & outside atmosphere.

Noncombustible material: Material that passes a test procedure as set forth in ASTM E136 for defining noncombustibility of materials. This includes materials that will not ignite & burn when subjected to fire, or material having a structural base of noncombustible material with a surfacing material not $> 1/8$ in. thick & a flame-spread index not higher than 50. This does not apply to surface-finish materials, the entire material of which must be noncombustible from the standpoint of clearances to heating appliances.

O

Offset: A combination of elbows or bends in a line of piping that brings a section of pipe in or out of line, but into a line parallel with the other section.

Ordinary tightness: Buildings of ordinary tightness are those that do not meet the standards for "unusually tight construction."

P

PEX tubing: Water-supply or hydronic heat tubing made of cross-linked polyethylene. PEX-AL-PEX has a layer of aluminum sandwiched between layers of PEX.

Plenum: A chamber, other than the occupied space being conditioned, that forms part of the air circulation system.

Power vent: See "Forced draft."

Pressure boundary: The boundary separating indoor from outdoor air. A ventilated crawl space or attic would be outside the pressure boundary.

Pressure-relief valve (PRV): A device designed to protect against high pressure & to function as a relief mechanism.

R

Readily accessible: Access that does not req removing a panel or door. For electrical equipment, this also means not having to resort to use of a ladder.

Room heater, circulating: A room heater with an outer jacket surrounding the heat exchanger & with openings at the top & bottom designed to circulate air between the heat exchanger & outer jacket.

Room heater (liquid or gas fuel): A room heater installed in the space to be heated & not connected to duct.

Room heater, radiant: A room heater designed to transfer heat primarily by direct radiation.

Room heater (solid fuel): A solid-fuel burning appliance designed to be operated with the fire chamber door closed. See "Fireplace stove."

Room large in comparison to size of equipment: A room having at least 12 times the volume of a furnace or other air-handling device, or 16 times the volume of a boiler. When the ceiling is greater than 8 ft., the volume is calculated based on an 8 ft. height. See "Alcove."

S

Slope: Fall or pitch along a line of a pipe or vent.

Supports: Devices used to support or secure pipes, fixtures, or equipment.

T

Ton (cooling): The amount of heat energy req'd to melt 1 ton of ice (288,000 Btus). Air-conditioners & heat pumps are typically sized in terms of tonnage, based on melting 1 ton of ice in 1 day. Therefore 1 ton of AC = 288,000Btus/24hr. = 12,000 Btus. The tonnage of a unit is usually encoded in the model number as a multiplier of 12, i.e., the number 36 would equal a 3-ton unit.

U

Unconfined space: A room or space having at least 50 cu. ft. for each 1,000 Btu of the fuel-burning appliances contained in the room or space.

Unlisted: An appliance not shown to comply with nationally recognized standards by an approved testing agency. An unlisted appliance might still have nameplate instructions. The IRC does not accept unlisted appliances. The UMC leaves their acceptance to the AHJ.

Unusually tight construction: Construction with walls & ceilings having a vapor retarder of 1 perm or less with sealed or gasketed openings, weatherstripping on openable windows & doors, and caulking or sealant at joints. Buildings of unusually tight construction are req'd by many energy codes & have a targeted air infiltration rate < 0.35ACH.

V

Vent (fuel-burning appliances): A passageway for conveying flue gases from an appliance to the outside atmosphere.

Vent, Type B: A vent listed & labeled for use with appliances with draft hoods & other Category I appliances

Vent, Type BW: A vent listed & labeled for use with wall furnaces.

Vent, Type L: A vent listed & labeled for appliances requiring either type L (oil-fired appliance) vents or Type B vents.

Vent connector: A device that connects an appliance to a vent.

Vented decorative gas appliance: A vented appliance that does not provide heat & whose only function is the aesthetic effect of the gas flames.

Vented gas appliance categories:

Category I: An appliance that operates with nonpositive vent static pressure & with a gas vent temperature that avoids excessive condensate production in the vent.

Category II: An appliance that operates with nonpositive vent static pressure & a vent gas temperature that is capable of causing excessive condensate production in the vent.

Category III: An appliance that operates with a positive vent static pressure & with a vent gas temperature that avoids excessive condensate production in the vent.

Category IV: An appliance that operates with a positive vent static pressure & with a vent gas temperature that is capable of causing excessive condensate production in the vent.

Vertical: Any pipe or vent that is 45° or more from horizontal.

W

Within sight: Visible, unobstructed & not more than 50 ft. away.

Wood stove: See "Fireplace stove" or "Room heater (solid fuel)."

TABLE 12   SIGNIFICANT CODE CHANGES – 2009 IRC, UPC & UMC; 2011 NFPA 31 & NFPA 58

No.	Page	Code	Description
1	129, 136	2006 UMC 931.3X	Rule allowing furnace to be placed at edge of attic access & serviced from opening removed in 2009 UMC.
2	130	IRC 1403.2	Raised pad for heat pumps still req'd in IRC 1403.2. A broader section, 1308.3, deleted in 2009 IRC. If req'd outdoor AC equipment to be raised at least 3 in. above finished grade.
3	130	IRC 1411.3.1	Option added for pan with interlocked detector & fitting to allow the pan to drain.
4	132	UMC 311.3&4	2006 UMC did not regulate these return air sources.
5	136	IRC 2408.4 & UMC 904.3.1	Gas appliances supported on grade req a slab or equivalent extending at least 3 in. above grade. In UMC, this section applies only for under-floor appliances.
6	136	UMC 904.3.1.2	UMC now req's 6 in. clearance for suspended under-floor appliances.
7	137	IRC 2427.8 & UMC 802.8.4	Condensate drain fittings from Category IV appliances to be in accordance with manu instructions. These include a requirement for primer.
8	144	IRC 2427.4.1.1 & UMC 802.4.3	New specification that primers used with plastic vent piping be a contrasting color. Appliance manu instructions also apply to fittings.
9	144	UMC 802.7.4.1	UMC no longer allows single-wall gas vents in dwellings. Single-wall connectors still allowed.
10	146	IRC 2427.5.1 & UMC 802.5.4.2	Prior code editions had no exception allowing unlined masonry chimneys for gas-burning appliances. They are now acceptable for replacement appliances of same size & characteristics provided chimney passes an NFPA 211 inspection.
11	150	NFPA 31 6.6.7	Requirement added that installer verify that chimney is proper size for connected appliances.
12	151	NFPA 31 7.2.1	Storage tanks must now be listed to one of standards recognized in NFPA 31.
13	151	NFPA 31 7.3.1	In 2005 NFPA 31, tanks could be supported on ground or on foundations of concrete, steel, masonry, or pilings.
14	151	NFPA 31 8.10.6	2011 edition specifies locations of thermally operated safety shutoff valves.
15	153	NFPA 58 6.9.3.16	New requirement for dielectric fitting to electrically isolate above ground & below-ground portions of propane piping. Note that propane piping is not req'd by NFPA 58 to be electrically bonded, NEC does req bonding of any interior piping system capable of becoming energized.

TABLE 45

SIGNIFICANT CODE CHANGES – 2009 IRC, UPC & UMC; 2011 NFPA 31 & NFPA 58 (cont.)

No.	Page	Code	Description
16	158	IRC 1502.4.2 & 2439.5.2	New requirement that dryer exhaust ducts be supported & secured at 4 ft. intervals.
17	158	IRC 1502.4.1	New requirement that electric dryer exhaust ducts be min 4 in. diameter.
18	158	IRC 1502.4.4 & 2439.5.5.1	New rules provide tables for equivalent length of fittings & allow for concealed duct runs of any length allowed by manu.
19	158	IRC 1502.5 & 2439.5.3	Protective shield plates req'd when duct is less than 1¼ in. from face of framing.
20	158	IRC 1502.4.5 & 2439.5.6	When duct concealed within framing, label or tag stating equivalent duct length req'd within 6 ft. of duct connection.

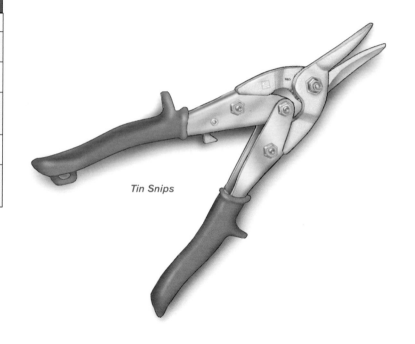

Tin Snips

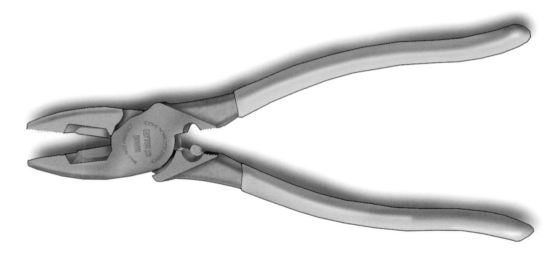

Side Cutters

Code ☑Check® Electrical 6th Edition

Based on the 2011 NEC® and the 2009 IRC®

BY REDWOOD KARDON & DOUGLAS HANSEN

Illustrations & Layout: Paddy Morrissey

Code Check Electrical 6th Edition is a field guide to common code issues in residential electrical installations. It is based on the **2011 National Electrical Code**–the most widely used electrical code in the United States–and the **2009 International Residential Code.** Before beginning any electrical project, check with your local building department. In addition to a model code, energy codes and special rules from utility companies could also apply.

Each code line in **Code Check Electrical** references the two codes named above. Many building jurisdictions use older versions of the codes. If you are in an area that still uses the 2008 NEC, look in the "**09 IRC**" column of code references to see if the item applies in your area and use the table on page 231 to see changes that were made in the **2008 NEC**, **2009 IRC**, and **2011 NEC**.

In places where the IRC does not reference a particular rule, the NEC rule might still apply, even where the IRC code is adopted. The IRC states that items not specifically mentioned in that code should comply with the NEC. This applies to issues such as old wiring, outside feeders, and photovoltaics, which are not covered in the IRC.

Thanks to Hamid Naderi, International Code Council, for his invaluable editorial input.

For information on electrical fundamentals and theory, visit:

http://www.codecheck.com/cc/OhmsLaw.html.

HOW TO USE CODE CHECK ELECTRICAL

Each text line ends with two code citations. The code numbers on the left, with straight brackets, refer to the 2009 IRC. The code numbers on the right, in braces, refer to the 2011 NEC. As in the following example from p. 179:

☐ Max 6 disconnects to shut off power_____ [3601.7]   {230.71}

This line states that there can be no more than 6 disconnects to shut off the power, and the rule is found in 3601.7 of the IRC and 230.71 of the NEC.

An "n/a" in a code line means the rule is not applicable to that particular code.

An "EXC" at the end of a line means that an exception–or exceptions–to the rule will follow in the next line, as on **p.187**:

☐ Size per service conductor size **T5** EXC _____ [3603.4]   {250.66}
• 6 AWG Cu largest size GEC needed if ending at rod [T3603.1]   {250.66A}

*This line states that the grounding electrode conductor must be sized per the size of the service conductors, in accordance with **Table 5**, except that the portion of the grounding electrode conductor that solely serves a ground rod need never be larger than 6 AWG.*

Significant code changes are highlighted by a code citation in a different color. The superscript after the code citation refers to the table on p.231. The following example is from **p.199:**

☐ GFCIs req'd to be in readily accessible locations   [n/a]   [210.8A][19] _____

GFCI devices must be located in an area where they remain readily accessible. The rule is not in the IRC. In the NEC it is a change in the 2011 code, summarized as change #19 in Table 23 on page 231.

Text lines ending in OR mean that an alternative rule follows in the next line, as on **p.202:**

☐ Separate 20A circuit for bath receptacles only OR   [3703.4]   [210.11C3] _____

• Dedicated 20A circuit to each bathroom   [3703.4X]   [210.11C3X] _____

A separate 20-amp circuit must be supplied for no other purpose than the bathroom receptacles. Alternatively, each bathroom can be supplied with its own 20-amp circuit, and then other outlets in that bathroom (such as lights) could be on the circuit.

ABBREVIATIONS

A = amp(s), amperage, amps, such as a 15A breaker
AC = alternating current
AC = air conditioning
AC = armored cable, a.k.a. "BX"
AFCI = arc-fault circuit interrupter
AHJ = Authority Having Jurisdiction
AI = aluminum
AMI = in accordance with manufacturer's instructions

AWG = American Wire Gauge
CATV = cable television
CO = carbon monoxide
cu. = cubic, as in cu. in.
Cu = copper
DC = direct current
EGC = equipment grounding conductor
EMT = electrical metallic tubing
ENT = electrical nonmetallic tubing, a.k.a. "Smurf tubing"

EV = electric vehicle
EXC = exception(s)
FMC = flexible metal conduit, a.k.a. "Greenfield"
ft. = foot, feet
GEC = grounding electrode conductor
GES = grounding electrode system
GFCI = ground-fault circuit interrupter
GFPE = ground-fault protection of equipment
hp = horsepower
IMC = intermediate metal conduit
in. = inch(es)
IRC = International Residential Code
kcmil = 1,000 circular mil units (conductor size)
L&L = listed & labeled, listing & labeling
lb. = pound(s)
LFMC = liquidtight flexible metal conduit, a.k.a. "Sealtight"
LFNMC = liquidtight flexible nonmetallic conduit
manu = manufacturer(s)
max = maximum
MC = metal-clad cable
min = minimum
NEC = National Electrical Code

ABBREVIATIONS (CONTINUED)

NFPA = National Fire Protection Association
NM = nonmetallic-sheathed (cable)
OCPD = overcurrent protection device
PV = photovoltaic
PVC = rigid polyvinyl chloride conduit (breaker or fuse)
req = require, requiring, requirement
req'd = required
req's = requires
RMC = rigid metal conduit
SCCR = short circuit current rating
SE = service entrance cable
SFD = single-family dwelling
sq. = square, as in sq. in.
temp = temperature
UF = underground feeder cable
USE = underground service entrance cable
TR = tamper-resistant
V = volt(s), such as a 120V circuit
VA = volt-ampere(s), units of apparent power
W = watt(s), units of true (useful) power
WR = weather-resistant

TABLE OF CONTENTS

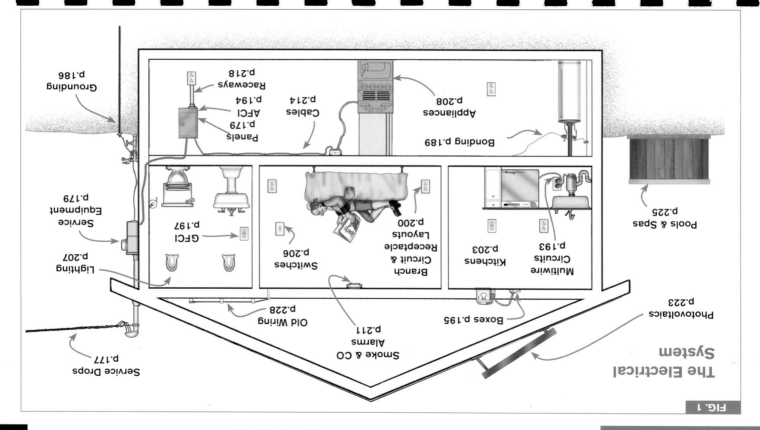

THE ELECTRICAL SYSTEM

FIG. 1

Accessible, as applied to wiring methods: Not permanently concealed or enclosed by building construction.

Accessible, as applied to equipment: Capable of being removed or exposed without damaging the building finish or structure. A piece of equipment can be considered accessible even if tools must be used or other equipment must be removed to gain access to it.

Accessible, readily: Capable of being reached quickly for operation or inspection without the necessity of using tools to remove covers, resorting to ladders, or removing other obstacles.

Alternating current (AC): Current that flows in one direction and then in the other in regular cycles, referred to as frequency or Hertz.

Apparent power: See "*Power*."

Approved: Acceptable to the authority having jurisdiction (AHJ). The AHJ will usually approve materials that are listed and labeled.

Arc-fault: An electric current propagated through air.

• **Arc-Fault Circuit Interrupter (AFCI):** A device intended to provide protection from the effects of arc faults by recognizing characteristics unique to arcing and by functioning to de-energize the circuit when an arc fault is detected.

• **AFCI, branch/feeder type:** "First generation" AFCI devices capable of interrupting parallel arcing faults. They do not meet the present code standard.

• **AFCI, combination type:** An AFCI meeting the standard for interrupting both series and parallel arcs.

Authority Having Jurisdiction (AHJ): The building official or persons authorized to act on his or her behalf.

Bonded, bonding: Connected to establish continuity and conductivity.

Branch circuit: The circuit conductors between the final overcurrent protection device (OCPD) (breaker or fuse) protecting the circuit and the outlet or outlets.

• **Branch circuit, general purpose:** Branch circuit that supplies 2 or more receptacles or outlets for lighting and appliances.

• **Branch circuit, individual:** Branch circuit supplying only 1 piece of equipment.

• **Branch circuit, multiwire, residential:** Branch circuit consisting of 2 hot conductors having 240V potential between them and a grounded neutral having 120V potential to each hot conductor **F17**.

• **Branch circuit, small appliance:** Branch circuit supplying portable household appliances in kitchens and related rooms and that has no permanently installed equipment connected to it (see **p.203** for exceptions).

Clothes closet: A non-habitable room or space intended primarily for storage of garments & apparel **F37**.

Controller: A device to directly open and close power to a load.

Derating: A reduction in the allowable ampacity of conductors because of ambient temperatures > 86°F, or more than 3 current-carrying conductors in the same raceway, or for cables without spacing between them.

Device: A piece of equipment that carries or controls electrical energy as its primary function, such as a switch, receptacle, or circuit breaker.

Equipment: A general term including materials, fittings, devices, appliances, luminaires (fixtures), apparatus, machinery, and the like used as a part of, or in connection with, an electrical installation.

Equipment grounding conductor (EGC): A wire or conductive path that limits voltage on metal surfaces and provides a path for fault currents **F16**.

Feeders: Conductors supplying panelboards other than service panels.

Flexibility after installation: Anticipated movement after initial installation, such as that caused by motor vibration or equipment repositioning.

Gooseneck: A curve at the top of a service entrance cable designed to prevent water from entering the open end of the cable.

Ground: The earth.

Grounded conductor: A current-carrying conductor that is intentionally connected to earth (a neutral).

Grounding electrode conductor (GEC): A conductor used to connect the service neutral or the equipment to a grounding electrode or to a point on the grounding electrode system **F6**.

GLOSSARY OF ELECTRICAL TERMS

Ground fault: An unintentional connection of a current-carrying conductor to equipment, earth, or conductors that are not normally intended to carry current.

• **Ground-Fault Circuit Interrupter (GFCI):** A device to protect against shock hazards by interrupting current when an imbalance of 6 milliamps or more is detected.

• **Ground-Fault Protected Equipment (GFPE):** A device to protect equipment from ground faults and allowing higher levels of leakage current than a GFCI.

Hertz: A measure of the frequency of AC. In North America, the standard frequency is 60 Hertz.

Individual branch circuit: See "Branch circuit, individual."

In sight: See "Within sight"

Interrupting rating: The highest current a breaker or fuse can interrupt without sustaining damage.

Lighting outlet: An outlet intended for the direct connection of a lampholder or a luminaire.

Load: The demand on an electrical circuit measured in amps or watts.

Location, damp: An area protected from the weather, yet subject to moderate degrees of moisture, such as a covered porch.

Location, dry: A location not normally subject to wetness.

Location, wet: All areas subject to direct saturation with water, and all conduits in wet outdoor locations or underground or in concrete or masonry in earth contact.

Luminaire (formerly lighting fixture): A complete lighting unit including parts to connect it to the power supply, and possibly parts to protect or distribute the light source. A lampholder, such as a porcelain socket, is not itself a luminaire.

Neutral conductor: The conductor connected to the neutral point of a system that is intended to carry current under normal conditions F17.

Open conductors: Individual conductors not contained in a raceway or cable sheathing, such as a typical service drop.

Outlet: The point on a wiring system at which current is taken to supply equipment. A receptacle or a box for a lighting fixture is an outlet; a switch is not an outlet.

Overcurrent: Any current in excess of the rating of equipment or conductor insulation. Overcurrents are produced by overloads, ground faults, or short circuits.

Overfusing: A fuse or breaker that has an overload rating greater than allowed for the conductor it is protecting.

Overload: Equipment drawing current in excess of the equipment or conductor rating and in such a manner that damage would occur if it continued for a sufficient length of time. Short circuits and ground faults are not overloads.

Panelboard: The "guts" of an electrical panel; the assembly of bus bars, terminal bars, etc., designed to be placed in a "cabinet." What is commonly called an electrical panel or load center is, by NEC terms, a panelboard mounted in a cabinet F16.

Power: There are 2 designations for AC electrical power. Apparent power (input) is expressed in V × A. True power (useful output) is expressed in watts.

Service: The conductors and equipment providing a connection to the utility F2.

Service drop: The overhead conductors supplied by the utility F2.

Service entrance conductors: The conductors from the service point to the service disconnect.

Service equipment: The equipment at which the power conductors entering the building can be switched off to disconnect the premises' wiring from the utility power source. A meter can be a part of or separate from the service equipment.

Service lateral: Underground service entrance conductors.

Service point: The connection or splice point at which the service drop and service entrance meet—it is the handoff between the utility and the customer.

Short circuit: A direct connection of current-carrying conductors without the interposition of a load, resulting in high levels of current.

Short Circuit Current Rating (SCCR): The amount of current that panelboards and switchboards must be able to carry during a short circuit condition without sustaining damage. See "Interrupting rating."

Snap switch: A typical wall switch, including 3-way and 4-way switches.

Uter: a concrete-encased grounding electrode, named after the developer of the system, Herbert Uter F6.

Unit switch: A switch that is an integral part of an appliance.

Within sight (also called "in sight"): Visible, unobstructed, and not more than 50 ft. away.

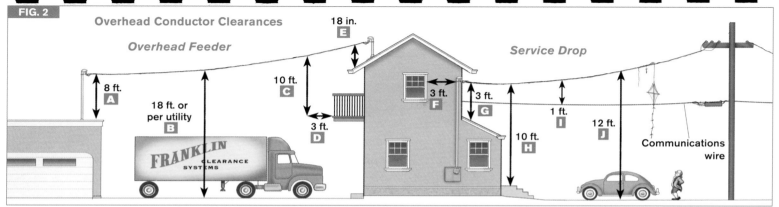

Figure 2: Overhead Conductor Clearances

Labels in figure:
- Overhead Feeder
- Service Drop
- 18 in. **E**
- 8 ft. **A**
- 18 ft. or per utility **B**
- 10 ft. **C**
- 3 ft. **D**
- 3 ft. **F**
- 3 ft. **G**
- 1 ft. **I**
- 10 ft. **H**
- 12 ft. **J**
- Communications wire
- FRANKLIN CLEARANCE SYSTEMS

OVERHEAD SERVICE DROP CLEARANCES

Service drop conductors typically have no outer jacket for physical protection and no overload protection at their source. They are protected by isolation and proper clearances. The codes specify minimum clearances, and the serving utility may have different rules that override the code. Check with your local jurisdiction to determine any variations from the standard clearances below.

Vertical above Roof F2   **09 IRC**   **11 NEC**

☐ <4-in-12 slope: min 8 ft. **A** EXC _____ [3604.2.1] {230.24A}
- 3 ft. OK if roof area guarded or isolated _____ [n/a] {230.24AX5}[1]

☐ ≥ 4-in-12 slope: min 3 ft. **G** EXC _____ [3604.2.1X2] {230.24AX2}
- 18 in. OK for ≤4 ft. over eaves **E** _____ [3604.2.1X3] {230.24AX3}

☐ Maintain req'd distance above roof for 3 ft. past edge EXC _____ [3604.2.1] {230.24A}
- Edge clearance above roof is not req'd when attached to side of building _____ [3604.2.1X4] {230.24AX4}

Vertical above Grade F2   **09 IRC**   **11 NEC**

☐ 10 ft. above final grade to lowest point of drip loop [3604.2.2] {230.24B1}
☐ Area accessible only to pedestrians: 10 ft. **H** ____ [3604.2.2] {230.24B1}
☐ General above grade & driveways: 12 ft. **J** _____ [3604.2.2] {230.24B2}
☐ Above roads or parking areas subject to truck traffic: 18 ft. **B** _____ [3604.2.2] {230.24B4}
☐ Any direction from swimming pool water: 22½ ft. ___ [4103.5] {680.8A}

Openings & Communication Wires F2   **09 IRC**   **11 NEC**

☐ Vertical above decks & balconies: 10 ft. **C** _____ [n/a] {230.9B}
☐ From side of area above decks & balconies: 3 ft. **D** _ [3604.1] {230.9A}
☐ Below or to sides of openable window: 3 ft. **F** _____ [3604.1] {230.9A}
☐ Communications wire ≥ 12 in. to parallel power wires **I** __ [n/a] {800.44A4}

The clearances from windows & doors apply to open conductors & not to conductors contained inside a raceway or a cable with an overall outer jacket. The codes do not have a requirement for min. clearance of open conductors above a window. Check to see if your local utility has a requirement.

SERVICE ENTRANCE CONDUCTORS

The connection between the service drop or lateral and the permanently installed building wiring is typically considered the "service point"—the handoff from the utility to the customer. From that point to the service equipment, the conductors are referred to as service entrance conductors. Though the utility does not have exclusive control of these conductors, they may still have jurisdiction over them, including the size of conduits and the placement of metering equipment.

General

	09 IRC	11 NEC
☐ Wire size for SFD per **T110**	[T603.1]	{T310.15B6}
☐ Min wire size for SFD 4 AWG Cu or 2 AWG Al **T110**	[T603.1]	{T310.15B6}
☐ Conductors & cables exposed to sunlight L&L as sunlight-resistant or covered with material L&L as sunlight-resistant	[3605.6]	{310.8D}
☐ Identify (white marking or tape) neutral at both ends	[3407.1]	{200.6B}
☐ Service heads/goosenecks above attachment point EXC	[3605.9.3]	{230.54C}
• Attachment within 24 in. OK when necessary	[3605.9.3X]	{230.54CX}
☐ No branch circuits or feeders in same raceway with service conductors	[3601.4]	{230.7}
☐ Form drip loop in conductors	[3605.9.5]	{230.54F}
☐ Individual open conductor insulating supports min 2 in. from building surfaces	[n/a]	{230.51C}

Service Entrance (SE) Cables

	09 IRC	11 NEC
☐ Protect SE cables subject to damage with metal conduit, PVC-80, EMT, or other approved means **F58,59,63**	[3605.5]	{230.50B}
☐ Secure SE cable every 30 in. & 12 in. from terminations	[3605.7]	{230.51A}
☐ Raintight service head or taped gooseneck req'd	[3605.9.2]	{230.54B}
☐ Seal SE cable to prevent water entry to box	[3605.9.6]	{230.54G}

COMMON UTILITY COMPLAINTS

Aside from code issues, utility company rules and standards must be followed. Most utilities publish their gas and electrical service requirements or post them online. The following items are not in the codes, and you should consult your local utility to comply with their rules on these issues.

Meter Base(s)

☐ Too close to gas meter
☐ Height incorrect
☐ Barrier post (bollard) needed to protect meter from vehicles on driveway
☐ Not readily accessible to meter readers

Service Entrance Conductors

☐ Insufficient conductor length at service head
☐ Insufficient clearance to communication lines
☐ Insufficient clearance above windows
☐ Height above standing surface (roof deck) too low
☐ Trees under service drop
☐ Customer performing own cutover from old service to new

Service Riser

	09 IRC	11 NEC
☐ Wiring method listed for electrical (no plumbing pipe)	[3605.2]	{230.43}
☐ Suitable for wet location if exposed to weather	[3605.8]	{230.53}
☐ Overhead raceway req's raintight service head	[3605.9.1]	{230.54A}
☐ Brace riser to utility or local specifications	[3604.5]	{230.28}
☐ Only power conductors on service risers—no CATV	[3604.5]	{230.28}
☐ Size raceway to max 40% fill **T17–T22**	[3904.6]	{T1&T4}
☐ Size raceway per utility _____	[utility]	{utility}

SERVICE PANELS

The term "service equipment" refers to the switches, circuit breakers, or fuses that disconnect power from the utility at the customer's end of the service conductors. A meter is not considered service equipment, though it is sometimes in the same enclosure as the service equipment. As with all electrical equipment that might req access for maintenance, examination, or repair, sufficient working space must be maintained around service equipment.

General

	09 IRC	11 NEC
☐ Enclosure L&L as suitable for service equipment __	[3601.6.1]	{230.66}
☐ Max 6 disconnects to shut off power_____	[3601.7]	{230.71}
☐ Service disconnects labeled as such _____	[3601.6.1]	{230.70B}
☐ In multiple-occupancy building, each occupant must have ready access to disconnect EXC_____	[3601.6.2]	{230.72C}
• OK for management to have only access to service disconnect supplying > 1 occupancy_____	[n/a]	{230.72CX}
☐ Max height of breaker 6 ft. 7 in. _____	[4001.6]	{240.24A}
☐ Provide working space **F3**_____	[3405.2]	{110.26}

WORKING SPACE

Working space around equipment is essential for worker safety. These requirements apply to any electrical equipment that might req examination, adjustment, servicing, or maintenance while energized. The spaces around electrical equipment should not be used for storage.

General **F3**

	09 IRC	11 NEC
☐ Front working clearance min 36 in. deep_____	[3405.2]	{110.26A1}
☐ Distance measured from face of enclosure or live parts	[3405.2]	{110.26A1}
☐ Work space extends to floor EXC _____	[3405.2]	{110.26A3}
• Related equipment may extend 6 in. beyond panel front	[3405.2]	{110.26A3}
☐ Clear width min 30 in. wide or width of equipment __	[3405.2]	{110.26A2}

General (cont.) **F3**

	09 IRC	11 NEC
☐ Panel need not be centered in space, hinged doors must be openable at least 90°_____	[3405.2]	{110.26A2}
☐ Working space not to be used for storage _____	[3405.4]	{110.26B}
☐ Illumination req'd for all indoor panels _____	[3405.6]	{110.26D}
☐ Min headroom for service & panels 6½ ft. _____	[3405.7]	{110.26A3}

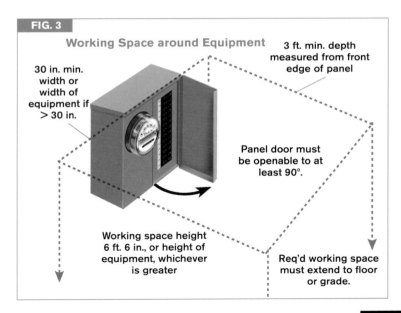

FIG. 3

Working Space around Equipment

3 ft. min. depth measured from front edge of panel

30 in. min. width or width of equipment if > 30 in.

Panel door must be openable to at least 90°.

Working space height 6 ft. 6 in., or height of equipment, whichever is greater

Req'd working space must extend to floor or grade.

SEPARATE BUILDINGS

Care must be taken to avoid objectionable currents on the grounding paths between buildings supplied by a common service. Install separate insulated neutral conductors, rather than using the grounding conductors as neutrals. The IRC does not address outside feeders and separate buildings except for the rules on grounding.

Outside Feeders   11 NEC

☐ Trees may not support overhead conductors _____ {225.26}
☐ Overhead feeder height rules same as services **F2** _____ {225.18&19}
☐ Provide proper cover for buried cable or conduit **F5, T1** _____ {300.5}
☐ Each building or structure req's GES EXC **F4** _____ {250.32A}
   • Building or structure with only 1 branch circuit with EGC _____ {250.32AX}
☐ Multiwire circuit considered 1 circuit for above rule _____ {250.32AX}
☐ Seal underground raceways where entering building _____ {225.27}[2]
☐ Max 1 feeder or branch circuit to each building _____ {225.30}
☐ Max 1 feeder or branch circuit back to original building _____ {225.30}[3]
☐ Disconnect req'd at each building **F4** _____ {225.31}
☐ Disconnect must be rated as service equipment EXC _____ {225.36}
   • Garages or outbuildings snap switches or 3-ways OK _____ {225.36X}
☐ EGC (4-wire feeder) req'd between buildings EXC _____ {250.32B}
   • Existing installations to separate buildings with no continuous metal
   paths, e.g., metal water pipe, etc., between 2 structures _____ {250.32BX}
☐ Do not bond neutral to EGC or enclosure in subpanel _____ {250.32B}

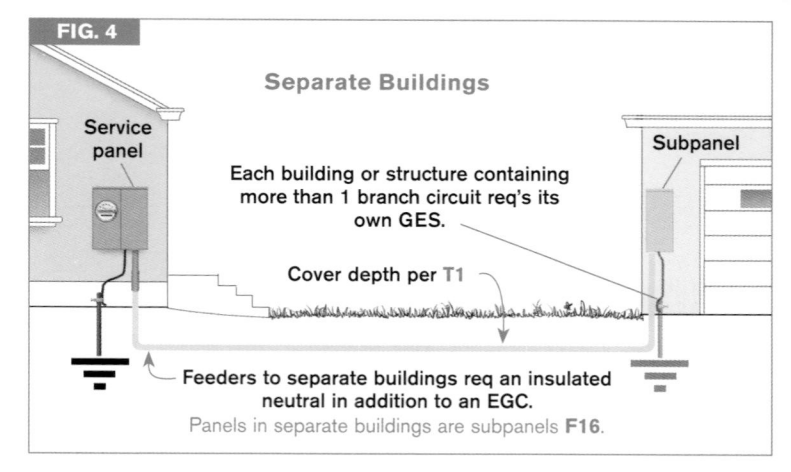

FIG. 4

Separate Buildings

Service panel

Subpanel

Each building or structure containing more than 1 branch circuit req's its own GES.

Cover depth per **T1**

Feeders to separate buildings req an insulated neutral in addition to an EGC.

Panels in separate buildings are subpanels **F16**.

MULTI-METER SERVICES

Services to 2-family and multi-family dwellings might come to a multi-meter panel, or to a "hot gutter" with splices ahead of any overcurrent protection. See p.189 for bonding requirements on such services

General   09 IRC   11 NEC

☐ Only 1 service per building _____ [3601.2]   {230.2}
☐ Provide each occupant access to service disconnect [3601.6.2] {230.72C}
☐ Bonding req'd at hot gutters **F11,12** _____ [3609.2]   {250.92A}
☐ Service disconnects grouped in 1 location _____ [3601.7] {230.71A}
☐ Service conductors may not pass through interior
   of 1 building to another building _____ [3601.3]   {230.3}

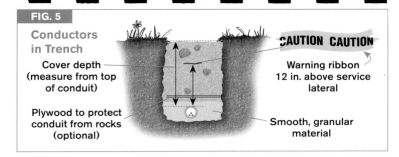

FIG. 5

Conductors in Trench

Cover depth (measure from top of conduit)

Plywood to protect conduit from rocks (optional)

CAUTION CAUTION

Warning ribbon 12 in. above service lateral

Smooth, granular material

TEMPORARY WIRING

Safety is the highest priority during construction, and GFCI protection is req'd for all 120V receptacles on construction sites. Some jurisdictions allow a limited number of temporary circuits from a service installation prior to the rough wiring stage (before weather protection).

General	11 NEC
☐ Allowed only during construction, repair, remodeling & similar _____	{590.3A}
☐ Service conductor clearances same as permanent **F2** _____	{590.4A}
☐ Support & brace pole to utility specifications _____	{utility}
☐ Provide overcurrent protection for branch circuits _____	{590.4C}
☐ No receptacles on branch circuits supplying temporary lighting _____	{590.4D}
☐ All multiwire circuits req handle ties _____	{590.4E}
☐ Lampholders req guards _____	{590.4F}
☐ Splices in NM cable or MC cable OK without splice box _____	{590.4G}
☐ Protect cords & cables from accidental damage _____	{590.4H}
☐ GFCI req'd on all 125V 15, 20 & 30A temporary receptacles EXC	{590.6A1}
• Listed GFCI cord-sets OK on existing permanent receptacles _ {590.6A2}[4]	

UNDERGROUND WIRING

Underground wiring methods include individual insulated conductors, cables rated for underground installation, and raceways. The most common method is PVC conduit. If there is a significant difference in elevation between the ends of an underground raceway, it may be necessary to install a pull-box for drainage near the downhill end.

General	09 IRC	11 NEC
☐ Burial depth must provide cover per **T1, F5** _____	[3803.1]	{300.5A}
☐ Warning ribbon in trench 12 in. above service laterals **F5** _____	[3803.2]	{300.5D3}
☐ Direct-buried cables or conductors must be protected by enclosures or raceways from req'd burial depth or 18 in. (whichever is less) to termination above grade or 8 ft. high (whichever is less) **F52** _____	[3803.3]	{300.5D1}
☐ Protect conductors & cables emerging from grade with RMC, IMC, PVC-80, or equivalent **F52** _____	[3803.3]	{300.5D4}
☐ OK to splice or tap direct-buried conductors without boxes with splicing means listed for the purpose _____	[3803.4]	{300.5E}
☐ Backfill smooth granular material—no rocks **F5** _____	[3803.5]	{300.5F}
☐ Boards or sleeves for protection where necessary **F5**	[3803.5]	{300.5F}
☐ Seal underground raceway entries (vapor protection)	[3803.6]	{300.5G}
☐ Bushing req'd between underground cables or individual conductors & protective conduit **F52** _____	[3803.7]	{300.5H}
☐ All conductors of circuit in same trench or raceway __	[3803.8]	{300.5I}
☐ Allow provision for earth movement (settlement or frost) using "S" loops, flexible connections &/or expansion fittings _	[3803.9]	{300.5J}

Load Calculation Steps (Long Form) (cont.) T2 — 11 NEC

7. Subtract 3,000 from amount in line 5 & enter difference in middle column. Multiply the middle column amount by 35% & enter in right column _____ {T220.42}
8. Range loads are calculated at nameplate rating. If a single range is > 8,000W & <12,000W, it still counts as 8,000W (8kW); >12,000W, add 5% of each additional 1,000W of nameplate load. Nameplates of a counter-mounted range & up to 2 wall ovens can be added together & computed as if they were 1 range. Enter in right column _____ {220.60}
9. Enter dryer circuit at 5,000W (or nameplate rating if greater) _____ {220.54}
10. Enter larger of fixed space heating or AC load _____ {220.60}
11–18. Enter nameplate ratings of appliances that are fixed in place. For appliances rated in amps, multiply amps times voltage to determine watts. If nameplate ratings unknown, use estimates in T4 _____ {220.53}
19. Enter total load of fixed appliances _____ {220.53}
20. If there are < 4 fixed appliances, enter number from line 19 in right column _____ {220.53}
21. If there are 4 fixed appliances, multiply line 19 by 75% & enter in right column _____ {220.53}
22. Add 25% of the largest motor load. Skip this step if a nameplate rated AC is largest load number since largest load has already been factored into nameplate min conductor ampacity _____ {220.18A}
23. Add numbers in third column _____ {220.40}
24. Divide line 23 by 240 to find req'd min amperage _____ {220.40}

SERVICE & FEEDER LOAD CALCULATIONS

The calculation methods in the codes take into account that not all of the possible electrical loads will be used at the same time. Each of the calculation methods allows the use of "demand factors." The "long form," shown below and in T2, is the most common calculation method. For information on multifamily load calculations, refer to the Code Check website at www.codecheck.com.

Load Calculation Steps (Long Form) T2 — 11 NEC

1. Determine the sq. ft. area of the residence & multiply by 3W (exclude garage & covered patios) _____ {220.12}
2. Min of 2 small-appliance circuits at 1,500W each _____ {220.52A}
3. Each additional small appliance circuit at 1,500W _____ {220.52A}
4. Minimum 1 laundry circuit at 1,500W _____ {220.52B}
5. Enter total of appliance circuits & general lighting _____ {220.42}
6. First 3,000W counted at 100% (carries to right column) _____ {220.42}

TABLE 1	MIN. COVER REQUIREMENTS IN TRENCH [T300.1] & {300.5}				
Cover	UF Cable	Rigid Metal	PVC	GFCI ≤ 20A Circuit	≤ 30V
General	24 in.	6 in.	18 in.	12 in.	6 in.
2 in. concrete	18 in.	6 in.	12 in.	6 in.	6 in.
Under building	0*	0	0	n/a	n/a
4 in. slab, no vehicles	18 in.	4 in.	4 in.	6 in.	6 in.
Street	24 in.	24 in.	24 in.	24 in.	24 in.
Driveway	18 in.	18 in.	12 in.	18 in.	18 in.

A. MC cable identified for direct burial also OK in 2011 NEC.

TABLE 2	LOAD CALCULATIONS [T3704.2(1)] & {220.40}		
General Lighting & Receptacle Loads			
1	Sq. ft. × 3W		
Small Appliance & Laundry Loads			
2	2 small appliance circuit	3,000	
3	Additional small appliance		
4	Laundry circuit	1,500	
5	Subtotal general light, small appliance & laundry		
6	First 3,000W @ 100%	3,000	3,000
7	Balance @ 35%	× .35	=
Special Appliance Loads			
8	Range	8,000 up to 12kW nameplate	
9	Dryer	5,000 (or nameplate if >)	
10	Heating or AC @ 100%		

TABLE 2	LOAD CALCULATIONS [T3704.2(1)] & {220.40} (CONT.)		
Appliances Fastened in Place			
11	Water heater		
12	Microwave		
13	Dishwasher		
14	Compactor		
15	Disposer		
16	Attic fan		
17	Spa—per manu.		
18	Other		
19	Subtotal		
20	If <4 appliances, enter subtotal @100% *or*		
21	If ≥4 appliances, enter subtotal × 75%		
22	Largest motor × 25%		
23	Total load		
24	Total load ÷ 240V = SERVICE AMPS		

Size Requirements–General | 09 IRC 11 NEC

- ☐ Min size for SFD 100A [3602.1] {230.79C}
- ☐ Service conductors adequate for load served [3602.1] {230.42}
- ☐ Feeders adequate for load served [3701.2] {215.2A1}
- ☐ Branch circuits adequate for load served [3701.2] {210.19A1}

The "optional" method is simpler & can be used to determine if an existing service is adequate for expansion. In the NEC, these methods apply to both services & feeders. In the IRC, the "long form" method, **T2**, is used for feeders per section E3604 & the "optional" method, **T3**, is used for services per section E3502. NEC 220.83 provides a specific method for evaluating the adequacy of an existing service for new air-conditioning loads.

TABLE 3	**MIN. SIZE OF ELECTRICAL SERVICE [T3602.2] & {220.82}**		
1.	Indoor sq. ft. × 3VA/ft.		
2.	Min. 2 small appliance circuits @ 1,500VA each	3,000	
3.	Laundry circuit @ 1,500VA	1,500	
4.	Nameplate VA of fixed appliances:		
	Dryer @ 5,000VA		
	Oven(s)		
	Cooktop		
	Water heater		
	Dishwasher		
	Disposer		
	Other		
5.	Subtotal		
6.	First 10,000VA @ 100%	10,000	10,000
7.	Balance @ 40% (subtract line 6 from line 5)	× .40	=
8.	Largest of heating or cooling load		
8a.	Nameplate rating(s) of air-conditioning & cooling equipment OR		
8b.	Heat pump nameplate if no supplemental electric heat OR		
8c.	Continuous electric thermal storage @ nameplate rating OR		
8d.	100% of heat pump nameplate rating plus 65% of supplemental electric heat or central electric heat OR		
8e.	Space heaters @ 65% of nameplate rating if < 4 units OR		
8f.	Space heaters @ 40% of nameplate rating if 4 units		
9.	Total load in VA		
10.	Divide by 240 = minimum service rating		

TABLE 4	TYPICAL APPLIANCE LOADS
Use actual nameplate ratings when known. This table is for estimating purposes when appliances are not yet specified.	

Appliance	Typical load (W)
Central AC or heat pump	1,800 per ton
Dishwasher	1,200
Food waste disposer	900
Trash compactor	1,200
Microwave	1,500
Central furnace	1,000
Central vacuum	1,500
Electric clothes dryer	5,000
Water heater	4,500
Electric cooktop	3,600
Single wall oven	4,800
Double wall oven	8,000
Pool pump	2,000
Well pump	2,000

Optional Method (Short Form)   11 NEC

1. 3W per ft. (exclude garage & covered patios) _____ {220.82B1}
2. Min 2 small-appliance circuits at 1,500W each, each
   additional small appliance circuit at 1,500W _____ {220.82B2}
3. Min 1 laundry circuit at 1,500W _____ {220.82B2}
4. Nameplate ratings of fixed appliances (see T4 if ratings not known);
   these include full nameplate rating of ranges & ovens without
   applying reductions allowed in the "long form" method _____ {220.82B3}
5. Enter sum of items 1–4 _____ {220.82B}
6. 100% of first 10,000VA_____ {220.82B}
7. Subtract line 6 from line 5, multiply by 40% & enter in
   right column _____ {220.82B}
8. Determine largest of the heating or cooling load. When using
   nameplate rating of heat pumps or AC, multiply
   "minimum circuit ampacity" times the voltage (240).
   If only size (tonnage) is known, refer to T4 _____ {220.82C}
9. Add numbers in right column & enter total_____ {220.82A}
10. Divide by 240 = amperage

GROUNDING ELECTRODES

Grounding electrodes are metal conducting objects through which a direct connection to earth is established. These electrodes provide a path for lightning and help reduce electrical noise on communications equipment. The most common grounding electrodes in residential construction are metal underground water piping, ground rods, and concrete-encased electrodes.

Grounding Electrode System (GES) F6   09 IRC   11 NEC
- ☐ Use all electrodes in F6 when present on premises__ [3608.1]   {250.50}
- ☐ Electrodes bonded together form a single system F6 __ [3608.1]   {250.50}
- ☐ Size electrode bonding conductors per GEC rules __ [3610.1]   {250.53C}
- ☐ Underground gas pipe not OK as electrode _____ [3608.6]   {250.52B1}

Water Pipe   09 IRC   11 NEC
- ☐ Metal water pipe if ≥10 ft. in direct contact with soil [3608.1.1]   {250.52A1}
- ☐ Bond around water meters, filters, etc. _____ [3608.1.1.1]   {250.53D1}
- ☐ Water pipe cannot be sole electrode _____ [3608.1.1.1]   {250.53D2}
- ☐ Metal well casing that is not bonded to metal pipe
  (e.g., plastic water service from well) OK as electrode [3608.1.1]   {250.52A8}

Pipes & Rods   09 IRC   11 NEC
- ☐ Rods min 8 ft. in contact with soil F6_____ [3608.1.4.1]   {250.53G}
- ☐ Pipe electrodes min 3/4 in. diameter _____ [3608.1.4]   {250.52A5}
- ☐ Unlisted ground rods min 5/8 in. diameter _____ [3608.1.4]   {250.52A5}
- ☐ Listed rods min 1/2 in. diameter_____ [3608.1.4]   {250.52A5}
- ☐ Drive rods vertical & fully below grade EXC _____ [3608.1.4.1]   {250.53G}
  - • If bedrock encountered, rod may be buried horizontally
    2 1/2 ft. deep or driven at 45° angle_____ [3608.1.4.1]   {250.53G}
  - • Clamp above grade OK if protected F6–10 ____ [3608.1.4.1]   {250.53G}
- ☐ If rod resistance > 25 ohms, install 2nd rod min 6 ft. from first
  & bond to 1st rod_____ [3608.4]   {250.56}

Recommended spacing 2× rod length, i.e., 16 ft.

Concrete-Encased Electrode F6   09 IRC   11 NEC
- ☐ Ufer = 20ft #4 or larger rebar near bottom of footing or
  20 ft. 4 AWG or larger Cu wire near bottom of footing[3608.1.2]   {250.52A3}
- ☐ Ufer must be used if present during construction____ [3608.1]   {250.50}
- ☐ Ufer not req'd in existing building if concrete would
  have to be disturbed to gain access _____ [3608.1X]   {250.50X}
- ☐ Ufer concrete encasement min 2 in._____ [3608.1.2]   {250.52A3}
- ☐ OK to bond sections of rebar with ordinary steel
  tie wires_____ [3608.1.2]   {250.52A3}
- ☐ Where multiple concrete-encased electrodes are present,
  only 1 req'd to be bonded to GES_____ [3608.1.2][5]   {250.52A3}
- ☐ Metal building frame OK as electrode if bonded to
  Ufer or if ≥ of steel 10 ft. in contact with earth with
  or without concrete encasement _____[n/a]   {250.52A2}

GROUNDING ELECTRODE CONDUCTORS (GECs)

A GEC connects the system of metal grounding electrodes in earth to the electrical system. It must have adequate size and protection to withstand the environmental and electrical forces imposed on it. Individual conductors can be run to each electrode of the GES, or a single conductor can be run to one of them or to the conductor that bonds the electrodes to each other.

Locations   09 IRC   11 NEC
- ☐ GEC must connect to EGCs, service entrance enclosures,
  service neutral & grounding electrodes _____ [3607.4]   {250.24D}
- ☐ Connect to service neutral anywhere from service point to
  bonded neutral in service disconnect _____ [3607.2]   {250.24A1}
- ☐ Bare Al not OK in masonry or earth _____ [3610.2]   {250.64A}
- ☐ Where outside, no Al ≤ 18 in. of earth _____ [3610.2]   {250.64A}
- ☐ Connection to metal water pipe that is part of GES not > 5 ft.
  after water entry to building F6_____ [3608.1.1]   {250.52A1}

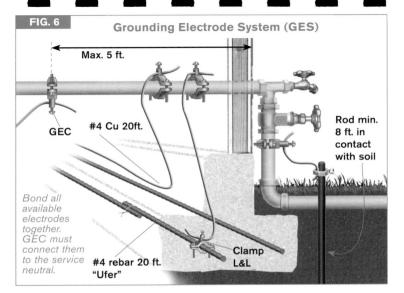

FIG. 6 — Grounding Electrode System (GES)

Max. 5 ft.

GEC

#4 Cu 20ft.

Rod min. 8 ft. in contact with soil

Bond all available electrodes together. GEC must connect them to the service neutral.

#4 rebar 20 ft. "Ufer"

Clamp L&L

Connections	09 IRC	11 NEC
☐ No splices between service & GES EXC _____	[3610.1]	{250.64C}
• Listed irreversible compression connectors or exothermic welding OK _____	[3610.1X]	{250.64C}
☐ GEC can connect to any electrode of GES _____	[3610.1]	{250.64F}
☐ Buried clamps L&L for direct burial (marked "DB") **F6**	[3611.1]	{250.70}
☐ Cu water tubing clamps L&L for Cu tubing _____	[3611.1]	{250.70}
☐ Ufer clamps L&L for rebar & encasement **F6** _____	[3611.1]	{250.70}
☐ Strap-type clamps suitable only for indoor telecommunications _____	[3611.1]	{250.70}
☐ Max 1 conductor per clamp unless listed for more ___	[3611.1]	{250.68A}
☐ Connections must be accessible EXC **F6** _____	[3611.2]	{250.68A}
Buried or encased connections **F6**	[3611.2]	{250.68AX}

Note: Rebar can be brought through the top of a foundation in a protected location, such as the garage, to provide an accessible point for the GEC to attach to the Ufer. The GEC can also be brought into the foundation and connect to the Ufer with L&L clamps or by exothermic welding.

Protection F7–10

	09 IRC	11 NEC
☐ 8 AWG must be protected by raceway or armor **F8,9**	[T3603.1]	{250.64B}
☐ 6 AWG OK unprotected if not subject to damage & following building contour **F7** _____	[T3603.1]	{250.64B}
☐ Bond each end of metal raceway enclosing GEC **F9**	[T3603.1]	{250.64E}

Size

	09 IRC	11 NEC
☐ Size per service conductor size **T5** EXC _____	[3603.4]	{250.66}
• 6 AWG Cu largest size GEC needed if ending at rod	[T3603.1]	{250.66A}
• 4 AWG Cu largest size GEC needed if ending at Ufer	[T3603.1]	{250.66B}

TABLE 5	GEC SIZING [T3603.1] & {250.66}	
Cu Service Wire (AWG)	Al Service Wire (AWG)	GEC Cu (AWG)
≤ 2	≤ 1/0	8
1 or 1/0	2/0 or 3/0	6
2/0 or 3/0	4/0 or 250kcmil	4
4/0–350kcmil	> 250–500kcmil	2
> 350–600kcmil	> 500–900kcmil	1/0

EQUIPMENT GROUNDING CONDUCTORS (EGCs)

EGCs limit the voltage on equipment enclosures and provide a path for fault current. Without EGCs, the conductive frame of an appliance could remain energized if there is a fault from an ungrounded "hot" conductor. Equipment grounding provides a low-impedance path so the overcurrent device will open the circuit. The equipment grounding system has a completely different purpose from the earth grounding system. In fact, earth plays no part in helping clear faults.

General   **IRC 60   NEC 11**

☐ EGC must provide effective ground-fault current path [3908.4]   {250.4A5}
☐ Earth is not an effective ground-fault current path [3908.5]   {250.4A5}
☐ Size EGCs per T6 [3908.12]   {250.122A}
☐ RMC, IMC, EMT, AC cable armor, electrically continuous raceways & surface metal raceways OK as EGC. [3908.8]   {250.118}
☐ Wire EGCs can be bare, covered, or insulated F16 [3908.8]   {250.118}
☐ Insulation on EGC green or green with yellow stripes [n/a]   {250.119}
☐ EGC > 6 AWG OK to strip bare for entire exposed length or use green tape or labels at termination of wire [n/a]   {250.119A}
☐ FMC & LFMC OK as EGC for non-motor circuits in combined lengths to 6 ft. with grounding fittings F60.61 [3908.8.1&2]   {250.118}
☐ Remove paint from threads & other contact surfaces for field-installed equipment such as ground terminal bars [3908.17]   {250.12}
☐ EGCs must run with other conductors of circuit EXC [3406.7]   {300.3B}
• Replacing nongrounding receptacles (see p.230) [n/a]   {250.130C}
☐ Neutral not to be used for grounding equipment EXC [3908.7]   {250.142B}
• Existing ranges & dryers [n/a]   {250.142BX1}

TABLE 6	EQUIPMENT GROUNDING CONDUCTORS (EGCs) [T3908.12] & {T250.122}	
Size of Breaker or Fuse Protecting Circuit (Amps)	Size of Cu EGC (AWG)	Size of Al EGC (AWG)
15	14	12
20	12	10
30-60	10	8
70-100	8	6
110-200	6	4
400	3	1

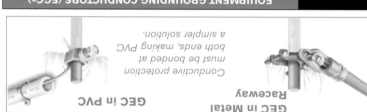

FIG. 7 Bare GEC

8 AWG must be protected. 6 AWG following the building contour does not need protection.

"Acorn" clamp

FIG. 8 Armor-clad GEC

Clamp must bond metal sheath to GEC.

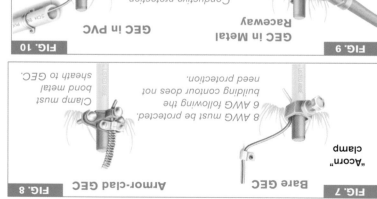

FIG. 9 GEC in Metal Raceway

Conductive protection must be bonded at both ends, making PVC a simpler solution.

FIG. 10 GEC in PVC

BONDING

Bonding ensures electrical continuity to limit voltage potential between conductive components. On the *line side* (ahead of the main disconnect **F15**), it provides a path back to the utility transformer for faults on service conductors and to limit voltage potential to other systems, such as telephones or CATV. On the *load side* (after the main overcurrent protection **F15**), bonding and equipment grounding provide a path to clear faults and protect against shocks.

Bonding & Equipment Grounding Methods   09 IRC   11 NEC

☐ Use listed connectors, terminal bars, exothermic welding, machine screws engaging 2 threads or secured with nut, or thread-forming machine screws engaging 2 threads. Not OK to use sheet metal or drywall screws _____ [3908.15][6]   {250.8A}

☐ Connections may not depend solely on solder _____ [3908.13]   {250.8B}

☐ Clean nonconductive coatings from contact surfaces [3908.17]   {250.12}

Line-Side Bonding F11, 12, 15   09 IRC   11 NEC

☐ Bond all service equipment, raceways & cable armor [3609.2]   {250.92A}

☐ Bond metal GEC enclosures at each end_____[T3603.1]   {250.64E}

☐ Threaded fittings OK for bonding service conduit _ [3609.4.2]   {250.92B2}

☐ Meyers hub OK for bonding service conduit **F11**__ [3609.4.2]   {250.92B2}

☐ Standard locknuts alone not sufficient
on line side of service **F11** _____ [3609.4.3]   {250.92B2}

☐ Bonding locknuts OK if no remaining concentrics **F11** [3609.4.4]   {250.92B4}

☐ Jumpers req'd around concentric knockouts or reducing washers
on line side of service **F12, 15** _____ [3609.4.4]   {250.92B4}[7]

☐ Service neutral can bond line-side equipment ____ [3609.4.1]   {250.142A}

☐ Size line-side bonding jumpers per **T5**_____ [3609.5]   {250.102C}

☐ Service enclosure main bonding jumper must connect enclosure, service neutral & equipment grounds **F15**_____ [3607.5]   {250.24B}

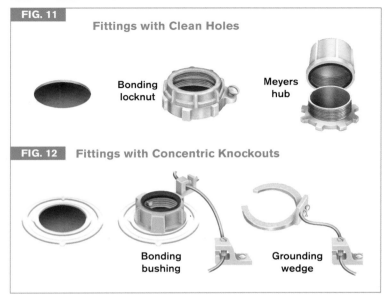

FIG. 11 Fittings with Clean Holes

Bonding locknut

Meyers hub

FIG. 12 Fittings with Concentric Knockouts

Bonding bushing

Grounding wedge

Load-Side Bonding   09 IRC   11 NEC

☐ Bond any metal piping system capable of becoming
energized, including hot & cold water & gas **F13** _ [3609.6&7]   {250.104}

☐ Size water pipe bonding per **T5** _____ [3609.6]   {250.104A1}

☐ Size gas pipe bonding per **T6**_____ [3609.7]   {250.104B}

☐ Bond metal well casings to EGC of pump motor_____[n/a]   {250.112M}

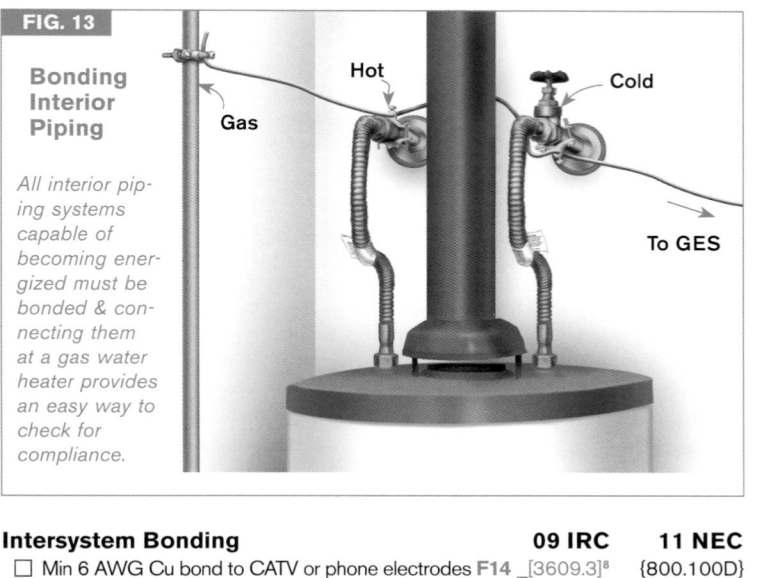

FIG. 13

Bonding Interior Piping

All interior piping systems capable of becoming energized must be bonded & connecting them at a gas water heater provides an easy way to check for compliance.

Gas

Hot

Cold

To GES

FIG. 14

Intersystem Bonding

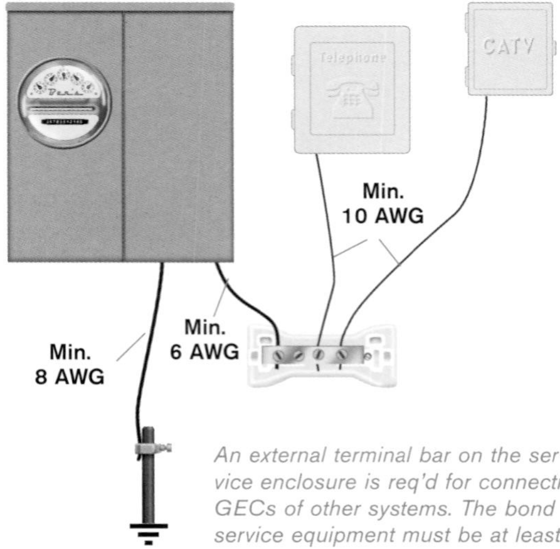

Min. 10 AWG

Min. 6 AWG

Min. 8 AWG

An external terminal bar on the service enclosure is req'd for connecting GECs of other systems. The bond to service equipment must be at least a 6 AWG conductor.

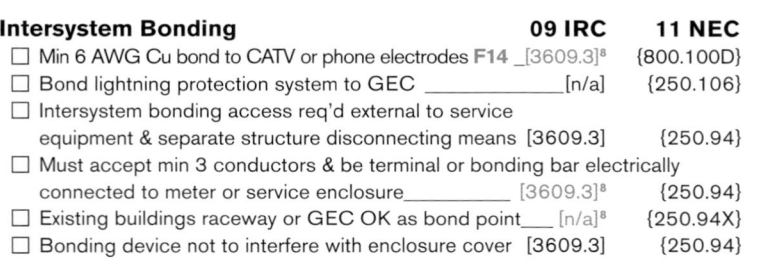

Intersystem Bonding	09 IRC	11 NEC
☐ Min 6 AWG Cu bond to CATV or phone electrodes **F14**	[3609.3][8]	{800.100D}
☐ Bond lightning protection system to GEC	[n/a]	{250.106}
☐ Intersystem bonding access req'd external to service equipment & separate structure disconnecting means	[3609.3]	{250.94}
☐ Must accept min 3 conductors & be terminal or bonding bar electrically connected to meter or service enclosure	[3609.3][8]	{250.94}
☐ Existing buildings raceway or GEC OK as bond point	[n/a][8]	{250.94X}
☐ Bonding device not to interfere with enclosure cover	[3609.3]	{250.94}

PANELBOARDS & CABINETS

What is commonly called an "electrical panel" is referred to as a panelboard (NEC 408) inside a cabinet (NEC 312). See **p.179** for working space requirements.

Clearances & Location
	09 IRC	11 NEC
☐ No panels or OCPDs in clothes closet or bathroom _	[3405.4]	{240.24D&E}
☐ No panels or OCPDs over steps of a stairway ____	[3405.4][9]	{240.24F}
☐ OCPDs readily accessible & max height 6 ft. 7 in. ___	[3705.7]	{240.24A}

Enclosures
	09 IRC	11 NEC
☐ Enclosures weatherproof in wet or damp locations __	[3907.2]	{312.2A}
☐ Surface-mounted wet or damp location metal enclosures min 1/4 in. air gap between enclosure & wall _____	[3907.2]	{312.2A}
☐ Equipment rated for dry or damp locations must be protected against damage from weather during construction___	[3404.4]	{110.11}
☐ Open knockouts & twistouts durably filled EXC _____	[3404.5]	{110.12A}
• Manu holes for mounting OK _____	[3404.6&3907.5]	{110.12A}
☐ Protect bus bars & other internal parts from contamination (paint or plaster) during construction _____	[3404.7]	{110.12B}
☐ Max setback in noncombustible wall 1/4 in. _____	[3907.3]	{312.3}
☐ Flush (no setback) in combustible (wood-frame) wall	[3907.3]	{312.3}
☐ Max plaster gap at side of flush mount panel 1/8 in. __	[3907.4]	{312.4
☐ Field labeling to distinguish each circuit from all others _	[3706.2]	{408.4}
☐ Labeling not based on transient conditions_____	[3706.2][10]	{408.4}
☐ Unused (spare) breakers labeled _____	[3706.2][11]	{408.4}

Grounding & Bonding
	09 IRC	11 NEC
☐ Bond neutral bar to enclosure & EGCs in service **F15**	[3607.5]	{250.24B}
☐ Isolate neutrals in subpanels **F16**_____	[3607.2 & 3908.6]	{250.24A5}
☐ Grounding terminal bar req'd if wire EGCs present **F16**__	[n/a]	{408.40}
☐ Continuity of neutral not to depend on enclosures	[3406.11][12]	{200.2B}
☐ Each neutral conductor req's individual terminal____	[3706.4]	{408.41}

OCPDs & Wiring
	09 IRC	11 NEC
☐ Panels req OCPD line side of bus **F15** _____	[3706.3]	{408.36}
☐ Breakers listed or classified AMI for panel _____	[3403.3]	{110.3B}
☐ Single-pole breakers with approved handle ties OK for 240V circuits **F16** _____	[n/a]	{240.15B2}
☐ All multiwire circuits req handle tie or single handle	[3701.5.1][13]	{210.4B}
☐ Handle tie req'd for 2 circuits to receptacles on same yoke	[n/a]	{210.7B}
☐ All conductors of multiwire circuit must be grouped (wire ties or other means) inside panel EXC **F16** _	[3701.5.1][14]	{210.4D}
• Cable systems where grouping is obvious **F16** _	[3701.5.2][14]	{210.4DX}
☐ Backfed breakers secured in place EXC_____	[3706.5]	{408.36D}
• Output circuits from utility interactive PV inverter_____	[n/a]	{705.12D6}
☐ Torque all breakers & terminals AMI _____	[3403.3]	{110.3B}
☐ Antioxidant on Al conductors AMI _____	[local]	{local}
☐ Secure each cable entering panel AMI **F15,16** _____	[3907.8]	{312.5C}
☐ Splices & taps in panels OK to 40% fill_____	[3907.1]	{312.8}
☐ Apply warning label to enclosure identifying power source of feed-through conductors_____	[n/a]	{312.8}[15]

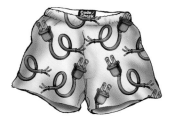

Beware of electrical shorts!

FIG. 15

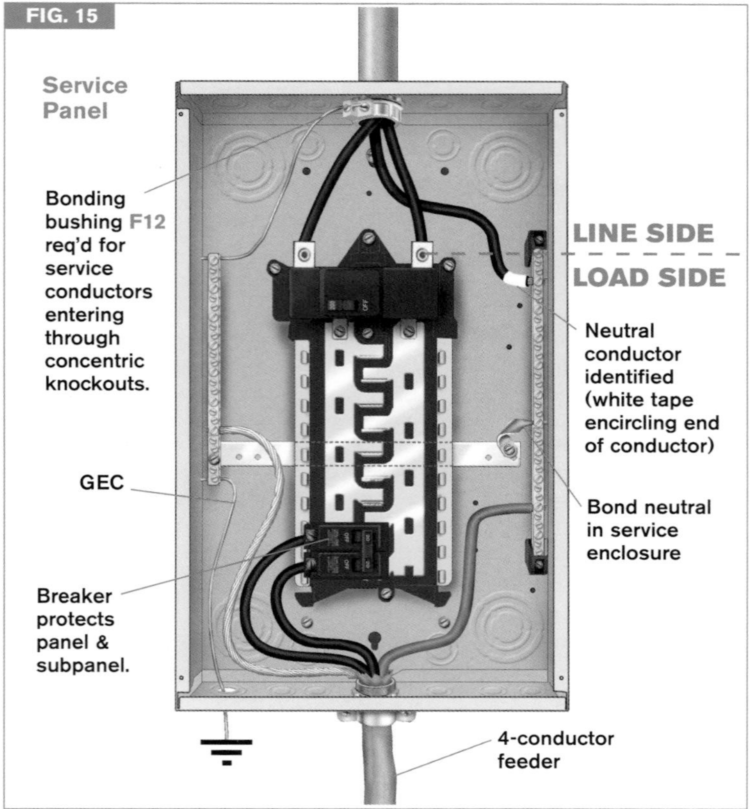

Service Panel

Bonding bushing **F12** req'd for service conductors entering through concentric knockouts.

GEC

Breaker protects panel & subpanel.

LINE SIDE
LOAD SIDE

Neutral conductor identified (white tape encircling end of conductor)

Bond neutral in service enclosure

4-conductor feeder

FIG. 16

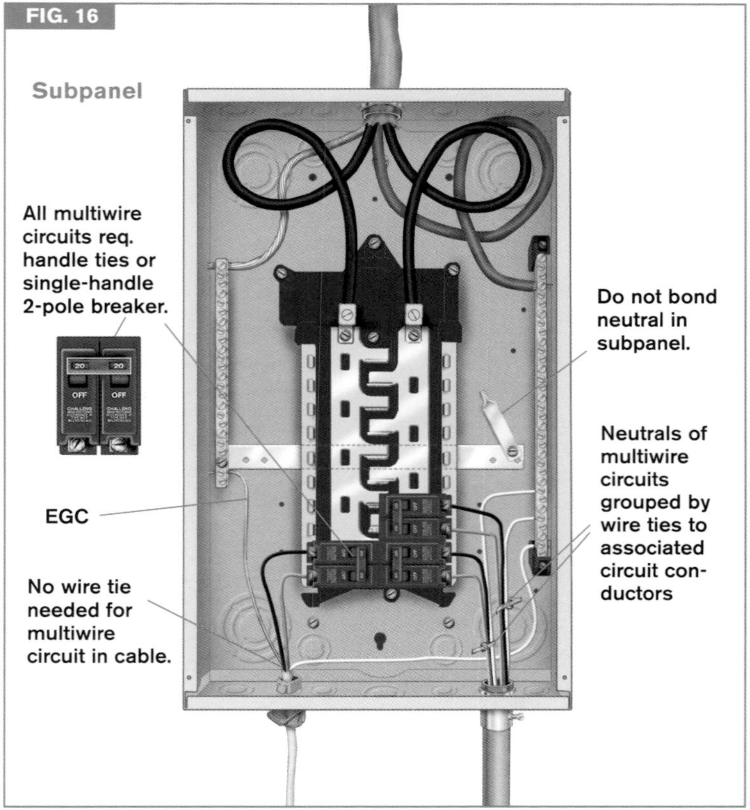

Subpanel

All multiwire circuits req. handle ties or single-handle 2-pole breaker.

EGC

No wire tie needed for multiwire circuit in cable.

Do not bond neutral in subpanel.

Neutrals of multiwire circuits grouped by wire ties to associated circuit conductors

OFF OFF
20 20

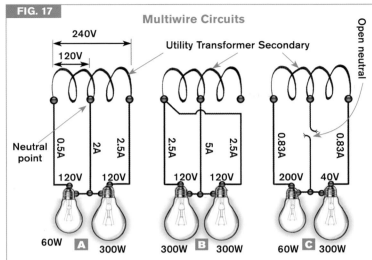

FIG. 17

Multiwire Circuits

240V

120V

Utility Transformer Secondary

Open neutral

Neutral point

| 0.5A | 2A | 2.5A | 2.5A | 5A | 2.5A | 0.83A | 0.83A |

| 120V | 120V | 120V | 120V | 200V | 40V |

60W **A** 300W   300W **B** 300W   60W **C** 300W

A **PROPER CIRCUIT** 2 unequal loads are fed by a 3-wire circuit. The neutral carries the imbalance between the 2 loads.

B **OVERLOADED NEUTRAL** Without voltage potential between the hot conductors, the neutral carries the sum of the loads. In a 3-conductor NM cable, the black & red wires must originate from different poles or the neutral can be overloaded because it carries the sum of the currents.

C **OPEN NEUTRAL** Two unequal loads in series across 240V from the transformer. The load with lowest resistance sees the greatest voltage drop. Voltage at each load depends on other loads and is unstable.

3-WIRE EDISON CIRCUITS (MULTIWIRE)

Standard electrical services to 1- and 2-family dwellings originate at a utility transformer with two ungrounded "hot" conductors and a neutral derived from the center of the transformer's secondary coil, as depicted in **F17**. The neutral is connected to earth and is referred to as the "grounded" conductor. The neutral limits the voltage on either of the hot conductors to 120V to ground. If the neutral is broken or loose, voltages become erratic, as in **F17** **C**. TV sets, motors, and computers don't do well with fluctuating voltages. The utility company should be notified if there are signs of unstable voltage, such as incandescent bulbs growing brighter or dimmer as other loads change. Not only is the service to the house a "3-wire" circuit, but 120V branch circuits are often installed with shared neutrals, which are then known as multiwire circuits.

Multiwire Circuits (also see p.191)	09 IRC	11 NEC
☐ Hot conductors must originate from opposite poles ___[3501]		{100}
☐ All conductors must originate from same panel _____ [3701.5]		{210.4A}
☐ Multiwire neutrals may not feed through devices such as receptacles (pigtail lead from neutral to device in box) _____ [3406.10.2]		{300.13B}

ARC-FAULT CIRCUIT INTERRUPTERS (AFCIs)

An AFCI provides fire protection by opening the circuit when an arcing fault is detected F18. They look similar to GFCI breakers F26, and AFCIs do provide some protection against shock hazards, though not at the level req'd for GFCIs. The 2008 NEC and 2009 IRC greatly expanded the areas that req AFCI protection. The time to plan for the AFCIs is during the rough wiring, so that separate cables are provided for the circuits requiring AFCI protection. Not all brands and models of AFCI are compatible with multiwire circuits.

Beginning January 1, 2008, all AFCIs were req'd to be "combination" type rather than the original "branch/feeder" type. Combination AFCIs provide a broader range of protection. Outlet types are mentioned in the codes, though at press time these were not yet available.

Acceptance of the AFCI code provisions varies widely by jurisdiction. Be sure to check with your local building department for their current AFCI requirements before beginning a wiring project.

AFCI Protection   09 IRC   11 NEC

☐ Combination-type AFCI req'd for 15A & 20A branch circuits supplying outlets in family rooms, dining rooms, living rooms, parlors, libraries, dens, bedrooms, sunrooms, recreation rooms, closets, hallways & similar rooms or areas EXC   [3902.11][16]   {210.12A}
• Not req'd on individual circuit for central station alarm in RMC, IMC, EMT, or steel-armored cable (type AC)   [3902.11X1]   {210.12AX2}

AFCI Protection (cont.)   09 IRC   11 NEC

☐ AFCI must protect entire branch circuit EXC   [3902.11]   {210.12A}
• OK to have protection at first outlet if wiring method between breaker & outlet is RMC, IMC, EMT, or steel-armored cable (type AC) & metal outlet or junction boxes are used   [3902.11X1]   {210.12AX1}
☐ Replacement or extension of branch circuit wiring req's AFCI breaker at origin of replacement circuit or AFCI outlet device at first receptacle of existing branch circuit   [n/a]   {210.12B}[17]

FIG. 18

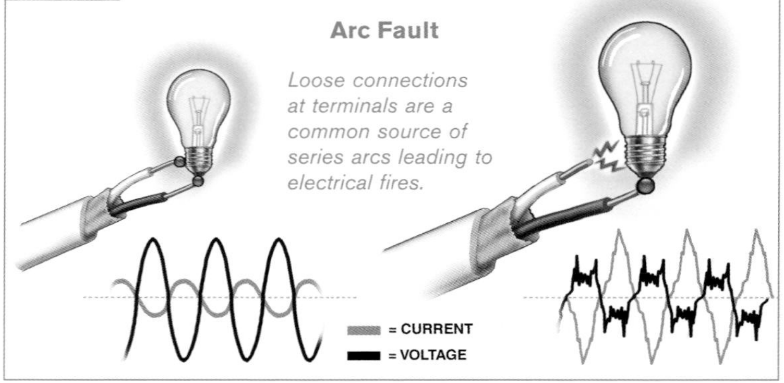

Arc Fault

Loose connections at terminals are a common source of series arcs leading to electrical fires.

= CURRENT   = VOLTAGE

BOXES

Boxes must be large enough to contain all the conductors and devices inside them, and sufficient wire must be brought into the box to safely make up connections. Luminaires that are supported from boxes are generally designed so their connections will be made inside the box, rather than inside the fixture canopy. Device boxes are threaded for 6–32 screws used to mount switches and receptacles. Lighting outlet boxes provide 8–32 (for luminaires) or 10–24 screws (for listed paddle fan boxes).

General

	09 IRC	11 NEC
☐ Metal boxes must be grounded _____	[3905.2]	{314.4}
☐ Box & conduit body covers must remain accessible	[3905.10]	{314.29}
☐ Max ¼ in. setback from noncombustible surface **F19**	[3906.5]	{314.20}
☐ Box extenders OK to correct excess setback _____	[3906.5]	{l314.20}
☐ Boxes flush with combustible surface **F19**	[3906.5]	{314.20}
☐ Plaster gap max ⅛ in. for flush cover boxes **F19** _____	[3906.6]	{314.21}
☐ Min 6 in. free conductor & 3 in. past box face _____	[3306.10.3]	{300.14}
☐ Luminaires only in boxes designed for luminaires EXC	[3905.6]	{312.27a}
• Wall sconces ≤ 6 lb. on device boxes		
with 2 #6 screws _____	[3905.6X]	{314.27A1X}
☐ Wall luminaire boxes max weight marked if ≤ 50 lb. __	[3905.6][18]	{314.27A1}
☐ Ceiling luminaire boxes req 50 lb. rating **F21** _____	[3905.7][18]	{314.27A2}
☐ Ceiling luminaires > 50 lb. req independent support	[3905.7][18]	{314.27A2}
☐ Smoke alarms OK on device boxes with 2 #6 screws____	[n/a]	{314.27DX}
☐ Paddle fans req L&L paddle fan box **F42** _____	[3905.9]	{314.27C}}
☐ Boxes must be rigidly supported _____	[3906.8]	{314.23}
☐ PVC & EMT not OK for box support _____	[3906.8.5]	{314.23E&F}
☐ PVC & EMT OK for conduit body support _____	[3906.8.5]	{314.23E&F}
☐ Wet location boxes & conduit bodies listed for wet _	[3905.12]	{314.15}
☐ Damp or wet location boxes must keep out water __	[3905.12]	{314.15}

Box Fill

	09 IRC	11 NEC
☐ Size sufficient to provide free space for conductors	[3905.13]	{314.16}
☐ Standard metal boxes per code tables _____	[3905.13.1.1]	{314.16A1}
☐ Include volume of marked mud rings & extensions	[3905.13.1]	{314.16A}
☐ Plastic boxes have volume marking _____	[3905.13.1.2]	{314.16A2}
☐ 4 in. (6 cu. in.) pancake OK only end of		
14/2 run **F21** _____	[3905.13.2]	{314.16B}
☐ 18 cu. in. box too small for 3 12/2 Romex **T8, F20**	[3905.13.2]	{314.16B}

TABLE 7	BOX FILL WORKSHEET [3905.13.2] & {314.16B}		
Item	**Size**	**#**	**Total**
#14 conductors exiting box	2.00		
#12 conductors exiting box	2.25		
#10 conductors exiting box	2.50		
#8 conductors exiting box	3.00		
#6 conductors exiting box	5.00		
Largest grounding conductor–count only 1		1	
Devices–2× times connected conductor size			
Internal clamps–1 based on largest wire present		1	
Fixture fittings–1 for each type based on largest wire			
		TOTAL	

Box Fill Factors T7,8

	09 IRC 60	11 NEC
☐ Count each conductor exiting box EXC	[3905.13.2.1]	{314.16B1}
EGCs from luminaires or up to 4 conductors < 14 AWG from luminaires with domed canopies	[3905.13.2.1X]	{314.16B1X}
☐ Unbroken conductors passing through box count as only 1 conductor EXC	[3905.13.2.1]	{314.16B1}
Looped unbroken conductors > 12 in. count as 2	[3905.13.2.1]	{314.16B1}
☐ Do not count pigtailed conductors to devices	[3905.13.2.1]	{314.16B1}
☐ All internal clamps count as 1, based on largest conductor in box	[3905.13.2.2]	{314.16B2}
☐ Support fittings count as 1 conductor for each fitting type based on largest conductor in box	[3905.13.2.3]	{314.16B3}
☐ Count devices as 2 conductors based on connected wire size	[3905.13.2.4]	{314.16B4}
☐ All EGCs count only as 1 based on largest	[3905.13.2.5]	{314.16B5}

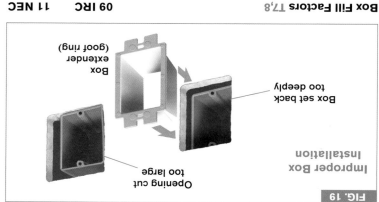

FIG. 19 — Improper Box Installation

- Opening cut too large
- Box set back too deeply
- Box extender (goof ring)

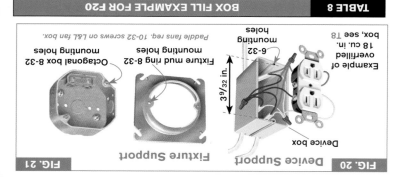

FIG. 21 — Fixture Support

- Octagonal box 8-32 mounting holes
- Fixture mud ring 8-32 mounting holes

FIG. 20 — Device Support

- Example of overfilled 18 cu. in. box
- 6-32 mounting holes
- $3^{9/32}$ in.
- Device box

Paddle fans req. 10-32 screws on L&L fan box. Paddle fans, see T8

TABLE 8 — BOX FILL EXAMPLE FOR F20

Item	Size	#	Total
#12 conductors exiting box	2.25	6	13.50
Largest grounding conductor–count only 1	2.25	1	2.25
Devices–2x times connected conductor size	4.50	1	4.50
Internal clamps–1 based on largest wire present	2.25	1	2.25
TOTAL			22.5

3 12/2+G Romex + device overfills 18 cu. in. box.

GROUND-FAULT CIRCUIT INTERRUPTERS (GFCIs)

A ground fault occurs when current leaks out of its normal path and finds a path back to the utility transformer through conductors that are not supposed to carry current. An example of such an abnormal path could include a human body. Ironically, even though the earth is not a sufficiently good conductor to provide a fault path that would trip a breaker, it is a good enough conductor to carry the low levels of current that can cause electrocution. GFCIs respond to very low levels of current imbalance in a circuit, such as those that occur when current returns through a person. GFCIs are designed to limit the duration of leaking current to safe levels.

*How does a GFCI work its magic? In **F22**, equal currents are flowing to & from the load. When any electrical current flows, it generates a magnetic field. The magnetic fields generated by the flow of electrons in these 2 conductors are of opposite polarity (north & south, leaving & returning). The forces are equal & opposite & their magnetic fields cancel each other. The circuit passes through a coil of wire inside the GFCI & the GFCI accounts for the electrons on each conductor. As long as the currents are balanced, GFCI allows current on the circuit.*

*During a ground fault—such as the flow of current through a person to something that is grounded—the circuit becomes unbalanced **F23**. Because the circuit is unbalanced, it produces a magnetic field that induces a small voltage on the sensing coil. The resulting current on the sensing coil signals the relay mechanism, which opens the circuit.*

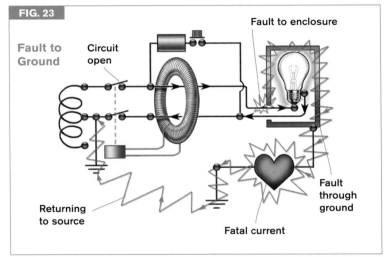

FIG. 23

Fault to Ground

Circuit open

Fault to enclosure

Returning to source

Fatal current

Fault through ground

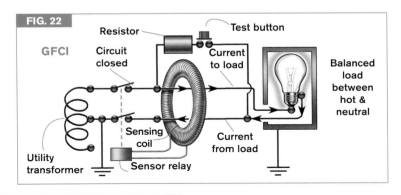

FIG. 22

GFCI

Resistor

Test button

Circuit closed

Current to load

Balanced load between hot & neutral

Sensing coil

Current from load

Utility transformer

Sensor relay

A GFCI also detects improper connections of the neutral (grounded conductor) to ground. A second "injector" coil **F24** surrounds the monitored circuit & induces a small current. Should the neutral have a downstream connection to the ground, the current will escape outside the circuit & the sensor coil circuit will be activated as described on p. 197.

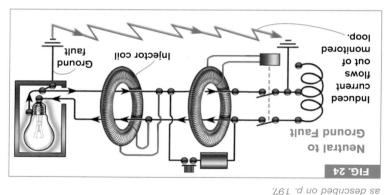

FIG. 24

Neutral to Ground Fault

Induced current flows out of monitored loop.

Injector coil

Ground fault

GFCIs take more space inside a box than do conventional receptacles. When adding GFCIs to old houses with shallow boxes, it might be necessary to first add an extension box, as in F25.

A GFCI will operate properly without an equipment ground. The receptacle should be labeled "no equipment ground" & any downstream protected receptacles should also have that label as well as a label stating that they are GFCI protected. Labels are not req'd for properly grounded GFCI-protected receptacles.

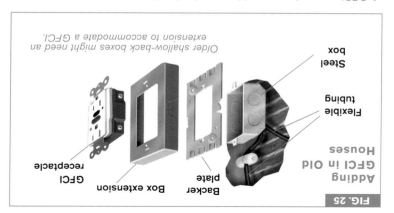

FIG. 25

Adding GFCI in Old Houses

GFCI receptacle

Box extension

Backer plate

Flexible tubing

Steel box

Older shallow-back boxes might need an extension to accommodate a GFCI.

A GFCI receptacle can provide protection for other receptacles downstream on the circuit. GFCI protection can be provided by GFCI breakers or GFCI receptacles **F26**.

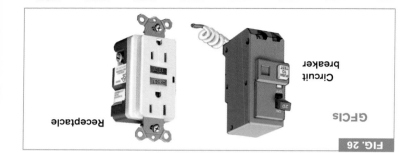

FIG. 26

Receptacle

Circuit breaker

GFCIs

Residential GFCI Protection   09 IRC   11 NEC

GFCI protection is req'd for 15A & 20A receptacles in the following locations. It is not req'd for 240V receptacles or 120V-30A receptacles.

- ☐ GFCIs req'd to be in readily accessible locations _____ [n/a]   {210.8A}[19]
- ☐ All bathroom receptacles_____ [3902.1]   {210.8A1}
- ☐ All garage & accessory building receptacles_____ [3902.2][20] {210.8A2}
- ☐ All receptacles in unfinished basements EXC_____ [3902.5] {210.8A5}
  - • Permanently installed fire or burglar alarm system _ [3902.5X3] {210.8A5X}

The 2005 NEC & 2006 IRC had exceptions for receptacles in garages & unfinished basements when those receptacles served appliances that are not easily moved, such as freezers. Those exceptions have been removed.

- ☐ All outdoor receptacles EXC _____ [3902.3] {210.8A3}
  - • GFPE circuit dedicated to nonreadily accessible receptacles for snow-melting or deicing equipment _____ [3902.3X] {210.8A3X}
- ☐ All receptacles in crawl spaces at or below grade level_[3902.4] {210.8A4}
- ☐ All receptacles serving kitchen counters **F30** _____ [3902.6] {210.8A6}
- ☐ Receptacles within 6 ft. of all non-kitchen sinks _____ [3902.7][21] {210.8A7}[22]

Pools, Spas, Whirlpool Tubs & Boathouses   09 IRC   11 NEC

- ☐ Receptacles ≤ 20 ft. of pools & outdoor hot tubs____ [4203.1.3] {680.22A4}
- ☐ Distance does not apply to cords that would have to pass through window or door _____ [4203.1] {680.22A5}
- ☐ Receptacles for 120V or 240V pool pump motors regardless of distance from pool_____ [4203.1.3] {680.22A4}
- ☐ Receptacles providing power to indoor spas or hot tubs __ [n/a] {680.43A3}
- ☐ Receptacles ≤ 10 ft. of indoor spas or hot tubs _____ [4203.1.5] {680.43A2}
- ☐ Pool cover motor & controller _____ [4206.11] {680.27B2}
- ☐ Hydromassage (whirlpool) tubs _____ [4209.1]   {680.71}
- ☐ Underwater pool lights > 15V **F68** _____ [4206.4] {680.23A3}
- ☐ Luminaires & lighting outlets < 10 ft. horizontally from outdoor pool or spa edge unless > 5 ft. vertically above water _____ [4203.4.5] {680.22B4}

Pools, Spas, Whirlpool Tubs & Boathouses (cont.) 09 IRC   11 NEC

- ☐ Existing luminaires allowed < 5 ft. horizontal if > 5 ft. vertical above water & GFCI protected_____ [4203.4.3] {680.22B3}
- ☐ Outlets supplying self-contained packaged spa/hot tub or field-assembled with heating < 50A EXC _____ [4208.1]   {680.44}
  - • Outlets supplying listed units with integral GFCIs____ [4208.1] {680.44A}
- ☐ Receptacles in boathouses_____ [3902.8] {210.8A8}
- ☐ 120V or 240V boat hoists_____ [3902.9][23] {210.8C}

UL 943–the standard of safety for GFCIs–was revised in 2003, requiring GFCIs to have greater resistance to corrosion & surges. GFCIs have become more reliable & do not have the problems of "nuisance tripping" that characterized these devices in the earlier stages of their development. Thanks to their increased reliability, it is no longer necessary to have the numerous exceptions that once existed for GFCIs associated with motor loads.

The new standard includes a line-load reversal test that req's the receptacle not be capable of resetting if it is miswired & a 2006 revision req's that there be no power to the face of a miswired receptacle. The contacts on newer GFCIs ensure proper resetting & prevent some miswiring that could appear from manipulation of the controls on the older GFCIs. In addition, manu installation instructions for GFCIs are now standardized for consistency. These instructions req specific methods for checking GFCI operation after installation to ensure that devices are properly wired & that they be tested on a regular basis for the life of the GFCI. As a result, these proven life savers have become more reliable than ever.

BRANCH CIRCUITS & OUTLETS

Branch circuits are the permanent wiring between the final overcurrent protective devices (fuses or breakers) and the lighting or receptacle outlets from which electrical equipment derives power. During rough-in of branch circuit wiring, care should be taken to ensure they are an adequate size for the load. Circuits for continuous loads and items such as water heaters or space heaters that are treated as continuous loads, must be sized to 125% of the load. There must be sufficient outlets for the needs of the occupants. An insufficient number of outlets could lead to the dangerous substitution of extension cords in place of permanent wiring. During rough-in, boxes are placed in the locations req'd for receptacle and lighting outlets, cables are run, and equipment grounds are connected.

Circuit Sizes, Number & Load Limitations	09 IRC	11 NEC
☐ Rule of thumb: min 1 general-purpose circuit per 500 sq. ft.	[3704.4]	{220.12}
☐ Load not to exceed rating of branch circuit	[3701.2]	{220.18}
☐ Min circuit size 125% of continuous load + 100% of noncontinuous load	[3701.2]	{210.19A}
☐ Continuous load = max current for 3 hours or more	[3501]	{100}
☐ Min size for branch circuit wiring 14 AWG	[3702.13]	{210.19A4}
☐ Branch circuit ratings for other than individual circuits must be 15A, 20A, 30A, 40A, or 50A.	[3702.2]	{210.3}
☐ Single piece of cord-&-plug-connected equipment not permanently fastened in place cannot exceed 80% of 15A or 20A branch circuit	[3702.3]	{210.23A1}
☐ Max single cord-&-plug-connected load on multi-receptacle circuit not to exceed 80% of circuit rating	[n/a]	{210.21B2}

Receptacle Locations—General	09 IRC	11 NEC
☐ Receptacles for specific appliances (laundry, garage door opener) within 6 ft. of appliance location	[3901.5]	{210.50C}
☐ Flexible cords not OK as fixed or concealed wiring	[3909.1]	{400.8}

TABLE 9	**RECEPTACLE RATINGS FOR MULTIPLE RECEPTACLES ON 1 CIRCUIT [4002.1.2] & {210.21.B3}**

Circuit Rating	Receptacle Rating
15A	not over 15A
20A	15 or 20A
30A	30A
40A	40 or 50A
50A	50A

For the purposes of these rules, a duplex receptacle is 2 receptacles, not a "single" receptacle.

Receptacles	09 IRC	11 NEC
☐ All receptacles on 15A & 20A circuits grounding type	[4002.2]	{406.4A}
☐ Receptacles for direct Al connection marked "CO/ALR"	[4002.3]	{406.3C}
☐ All req'd receptacles listed TR type EXC.	[4002.14]	{406.12}
• Receptacles located > 5½ ft. above floor	[4002.14][24]	{406.12X}[25]
• Receptacles that are part of a luminaire	[4002.14][24]	{406.12X}[25]
• Single receptacles within dedicated space for an appliance not easily moved or duplex receptacle for 2 such appliances	[n/a]	{406.12X}[25]
• Replacement nongrounding receptacles (see P. 230)	[n/a]	{406.12X}[25]
☐ Single receptacles rated not less than branch circuit	[4002.1.1]	{210.21B1}
☐ Multiple receptacles on branch circuit per T9	[4002.1.2]	{210.21B3}

FIG. 27

6 ft. & 12 ft. Rule

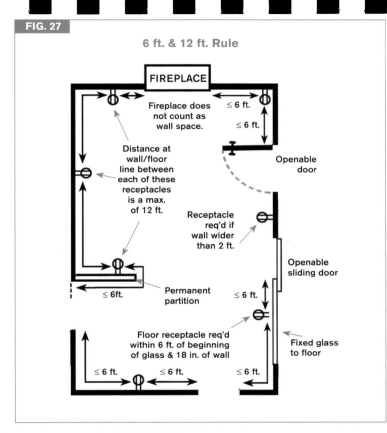

FIREPLACE

Fireplace does not count as wall space.

≤ 6 ft.

≤ 6 ft.

Distance at wall/floor line between each of these receptacles is a max. of 12 ft.

Openable door

Receptacle req'd if wall wider than 2 ft.

Openable sliding door

≤ 6ft.

≤ 6 ft.

Permanent partition

Floor receptacle req'd within 6 ft. of beginning of glass & 18 in. of wall

Fixed glass to floor

≤ 6 ft.   ≤ 6 ft.   ≤ 6 ft.

Receptacle Outlets–General Purpose F27,28   09 IRC   11 NEC

☐ Walls ≥ 2 ft. wide req receptacle _____ [3901.2.2]   {210.52A2}
☐ Partitions & bar-type counters count as walls F30 _ [3901.2.2]   {210.52A2}
☐ Doorways & fireplaces not counted as walls_____ [3901.2.2]   {210.52A2}
☐ Receptacle req'd within 6 ft. measured horizontally of
   any point along floor line _____ [3901.2.1]   {210.52A1}
☐ Receptacle req'd for hallways ≥ 10 ft. in length F28 [3901.10]   {210.52H}
☐ Receptacles that are part of electric baseboard
   heaters OK as req'd outlets _____ [3901.1]   {210.52}
☐ Receptacles > 5½ ft. high not OK as req'd outlets__ [3901.1]   {210.52}
☐ Floor receptacles > 18 in. from wall not OK as
   req'd outlets _____ [3901.2.3]   {210.52A3}
☐ Switched receptacles installed as req'd lighting do not count
   as part of req'd receptacle outlets unless "half hot" [3901.1][26]   {210.52}
☐ Receptacles req'd each wall ≥ 3 ft. in foyers > 60 sq. ft.__[n/a]   {210.52I}[27]
☐ Garages & unfinished basements req min 1 receptacle
   in addition to any for specific equipment _____ [3901.9]   {210.52G}

FIG. 28

6 ft. & 12 ft. Rule Explained

Wall receptacles serve spaces for 6 ft. on each side of receptacle. Therefore, max. spacing between wall receptacles is 12 ft.

6 ft.   6 ft.

12 ft. max.

Bathrooms | 09 IRC | 11 NEC

- ☐ Receptacle req'd on wall or partition within 3 ft. of each basin or in side or face of cabinet ≤ 12 in. below countertop __[3901.6] {210.52D}
- ☐ No face-up outlets on vanity countertop _____ [3901.6] {406.4E}
- ☐ Listed countertop-mounted receptacles OK _____ [n/a] {210.52D}[28]
- ☐ No receptacles within or directly over tub or shower [4002.11] {406.8C}
- ☐ Separate 20A circuit for bath receptacles only OR __ [3703.4] {210.11C3}
  - Dedicated 20A circuit to each bathroom _____ [3703.4X] {210.11C3X}
- ☐ Max rating of fixed space heater on general lighting circuit 15A circuit: 900W; 20A circuit: 1,200W _____ [3702.5] {210.23A2}

Laundry | 09 IRC | 11 NEC

- ☐ Min 1 20A circuit for laundry receptacles _____ [3703.3] {210.11C2}
- ☐ No other outlets on laundry receptacle circuit_____ [3703.3] {210.11C2}
- ☐ Receptacle within 6 ft. of intended appliance location [3901.5] {210.50C}
- ☐ Electric dryer min 30A circuit (10 AWG Cu, 8 AWG Al)[T3704.2(1)] {220.54}
- ☐ Electric dryer req's 4-conductor branch circuit EXC _ [3908.7] {250.140}
  - Existing 3-wire circuits allowed to remain in use _____ [n/a] {250.140X}

Outdoors | 09 IRC | 11 NEC

- ☐ Receptacle accessible from grade req'd at front & rear of dwelling max 6½ ft. above grade _____ [3901.7] {210.52E1}
- ☐ Receptacle req'd at balconies with interior access EXC [3901.7][29] {210.52E3}
  - Not req'd if balcony < 20 sq. ft. _____ [3901.7] {n/a}[30]
- ☐ Receptacles in damp or wet locations req'd to be listed weather-resistant type _____ [4002.8][31] {406.8A&B}
- ☐ Outdoor damp location receptacle (e.g., protected porch) req's weatherproof cover F29 _____ [4002.8] {406.8A}
- ☐ Wet location 15A & 20A receptacles req in-use covers F29 [4002.9] {406.8B1}

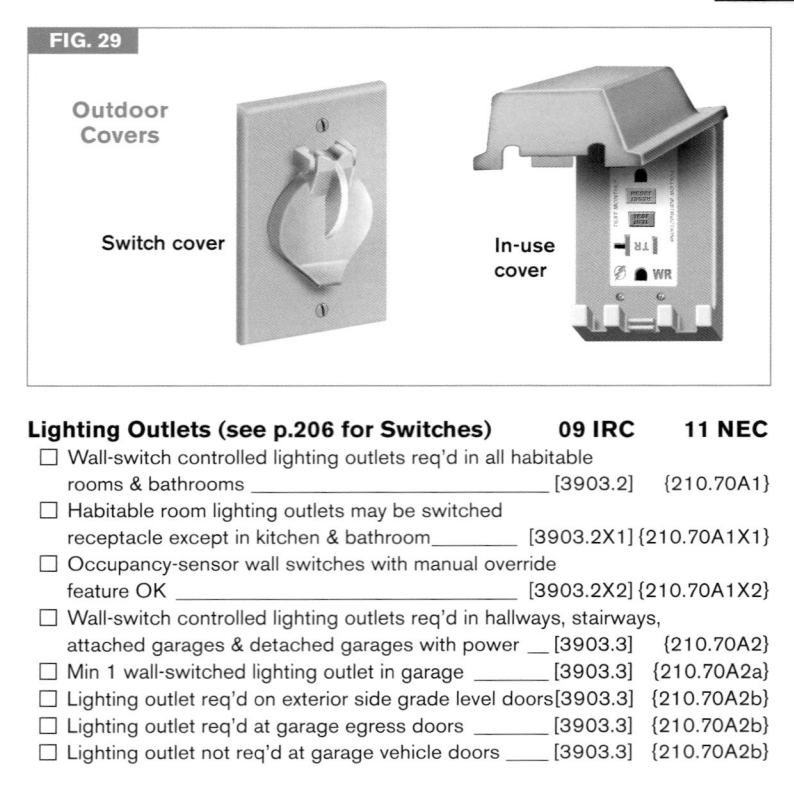

FIG. 29

Outdoor Covers

Switch cover

In-use cover

Lighting Outlets (see p.206 for Switches) | 09 IRC | 11 NEC

- ☐ Wall-switch controlled lighting outlets req'd in all habitable rooms & bathrooms _____ [3903.2] {210.70A1}
- ☐ Habitable room lighting outlets may be switched receptacle except in kitchen & bathroom_____ [3903.2X1] {210.70A1X1}
- ☐ Occupancy-sensor wall switches with manual override feature OK _____ [3903.2X2] {210.70A1X2}
- ☐ Wall-switch controlled lighting outlets req'd in hallways, stairways, attached garages & detached garages with power __ [3903.3] {210.70A2}
- ☐ Min 1 wall-switched lighting outlet in garage _____ [3903.3] {210.70A2a}
- ☐ Lighting outlet req'd on exterior side grade level doors[3903.3] {210.70A2b}
- ☐ Lighting outlet req'd at garage egress doors _____ [3903.3] {210.70A2b}
- ☐ Lighting outlet not req'd at garage vehicle doors ____ [3903.3] {210.70A2b}

KITCHENS

A minimum of 2 20A small-appliance branch circuits are req'd for portable appliances that are used in kitchens and dining areas. These circuits are in addition to those that supply lighting or permanently installed appliances. Portable kitchen appliances have short cords so they are not as likely to be run across cooktops or sinks or to hang down in the reach of children. A receptacle is needed to serve every countertop 1 ft. or more in width.

Branch Circuits	09 IRC	11 NEC
☐ Min 2 20A small-appliance circuits req'd_____	[3703.2]	{210.11C}
☐ Small-appliance circuits must serve refrigerator & all countertop & exposed wall receptacles in kitchen, dining room & pantry EXC _____	[3703.2]	{210.52B1}
• Refrigerator OK on individual branch circuit ≥ 15A	[3703.2X]	{210.52B1X2}
☐ Switched receptacle for dining room light OK on non-small-appliance circuit _____	[3901.3X1]	{210.52B1X1}
☐ No other outlets (including lights) on small appliance branch circuits EXC_____	[3901.3.1]	{210.52B2}
• Receptacles for clock or gas range ignition OK_	[3901.3.1X]	{210.52B2X}
☐ Dishwasher & disposer req separate circuits if combined rating exceeds branch circuit rating _____	[3701.2]	{210.19A1}
☐ Circuits for ranges ≥ 8.75kW min 40A 240V _____	[3702.9.1]	{210.19A3}

Receptacles for Countertop Spaces	09 IRC	11 NEC
☐ Receptacles req'd for wall counter spaces ≥ 12 in. wide	[3901.4.1]	{210.52C1}
☐ Countertop spaces separated by sinks or ranges considered separate countertop spaces F30 _____	[3901.4.4]	{210.52C4}
☐ Spacing so no point > 24 in. from receptacle F31	[3901.4.1]	{210.52C1}
☐ Area behind sink or range considered countertop space if ≥ 12 in. to wall F32 or ≥ 18 in. to corner F33 __	[3901.4.1X]	{210.52C1X}
☐ Max 20 in. above countertop_____	[3901.4.5]	{210.52C5}
☐ Peninsulas req receptacle if long dimension ≥ 24 in. & short dimension > 12 in., measured from connecting edge F30 _____	[3901.4.3]	{210.52C3}
☐ Island & peninsula countertop spaces min 1 receptacle per space—no 24 in. rule F30 _____	[3901.4.2&3]	{210.52C2&3}
☐ Sink or range with < 12 in. behind divides counters into separate spaces for above rule_____	[3901.4]32	{210.52C4}
☐ Island & peninsula receptacles OK ≤ 12 in. below counter overhanging ≤ & no means of installing receptacle in overhead cabinet F30 _____	[3901.4.5X]	{210.52C5X}
☐ No face-up countertop receptacles _____	[3901.4.5]	{406.4E}
☐ GFCI protection for all receptacles serving countertops	[3902.6]	{210.8A6}

FIG. 30

Kitchen Receptacles

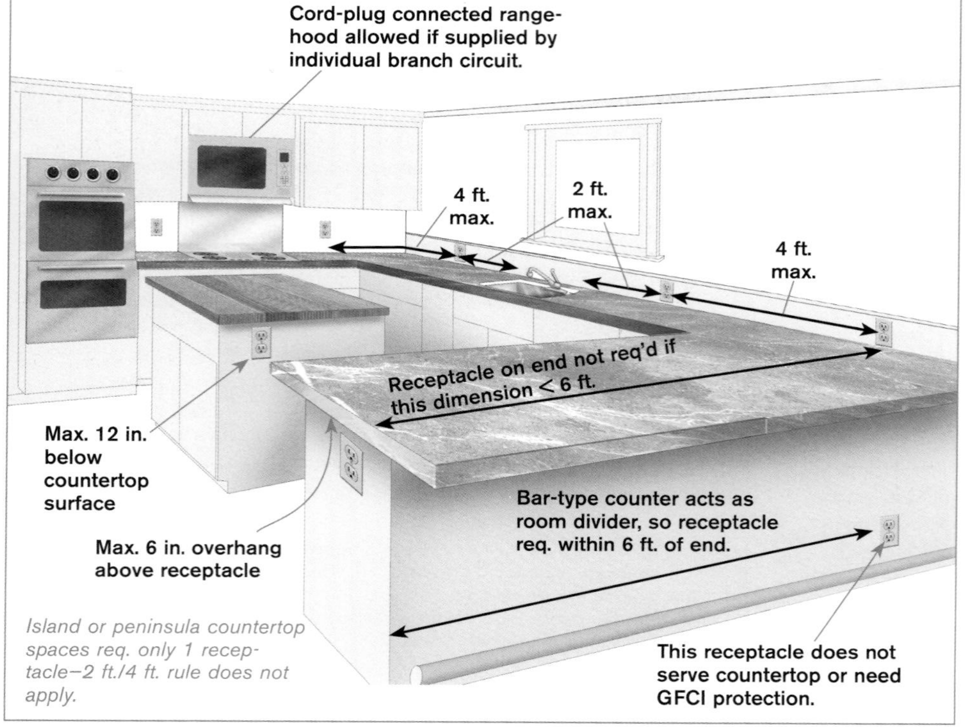

Cord-plug connected range-hood allowed if supplied by individual branch circuit.

4 ft. max.

2 ft. max.

4 ft. max.

Receptacle on end not req'd if this dimension < 6 ft.

Max. 12 in. below countertop surface

Max. 6 in. overhang above receptacle

Bar-type counter acts as room divider, so receptacle req. within 6 ft. of end.

This receptacle does not serve countertop or need GFCI protection.

Island or peninsula countertop spaces req. only 1 receptacle—2 ft./4 ft. rule does not apply.

FIG. 31

2 ft./4 ft. Rule Explained

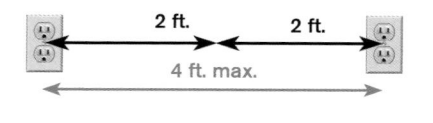

2 ft.

2 ft.

4 ft. max.

Wall countertop receptacles serve spaces for 2 ft. on each side of the receptacle. Therefore, the max. spacing between receptacles on the same countertop space is 4 ft.

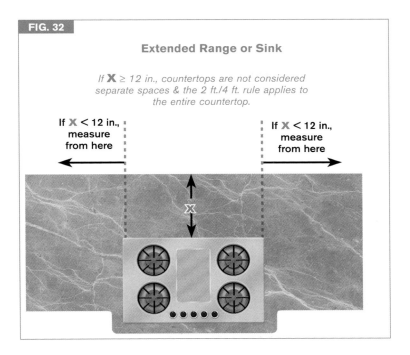

FIG. 32

Extended Range or Sink

*If **X** ≥ 12 in., countertops are not considered separate spaces & the 2 ft./4 ft. rule applies to the entire countertop.*

If **X** < 12 in., measure from here

If **X** < 12 in., measure from here

X

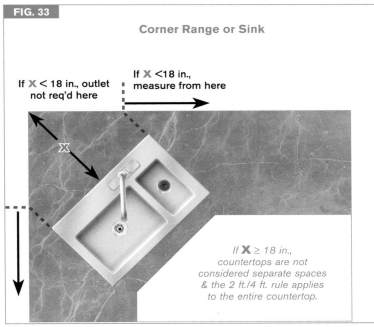

FIG. 33

Corner Range or Sink

If **X** < 18 in., outlet not req'd here

If **X** <18 in., measure from here

X

*If **X** ≥ 18 in., countertops are not considered separate spaces & the 2 ft./4 ft. rule applies to the entire countertop.*

FIG. 34

3-Way Switch

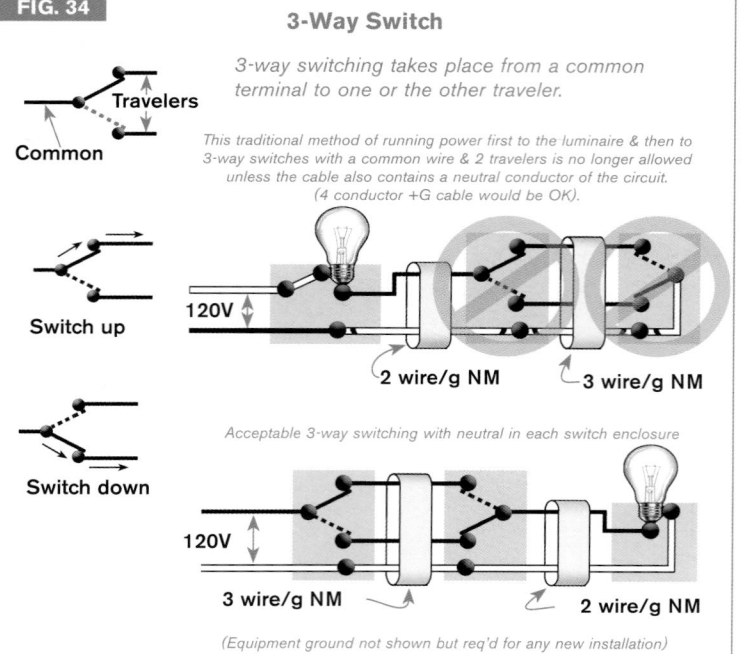

3-way switching takes place from a common terminal to one or the other traveler.

Travelers

Common

This traditional method of running power first to the luminaire & then to 3-way switches with a common wire & 2 travelers is no longer allowed unless the cable also contains a neutral conductor of the circuit. (4 conductor +G cable would be OK).

Switch up

120V

2 wire/g NM   3 wire/g NM

Switch down

Acceptable 3-way switching with neutral in each switch enclosure

120V

3 wire/g NM   2 wire/g NM

(Equipment ground not shown but req'd for any new installation)

SWITCHES

General	09 IRC	11 NEC
☐ All switching in ungrounded conductors **F34,35** __[4001.8&9]		{404.2A&B}
☐ Provide neutral in switchbox EXC _____[n/a]		{404.2C}[33]
• In raceway with sufficient room to add neutral _____[n/a]		{404.2CX}[34]
• Where switch not enclosed by building finishes_____[n/a]		{404.2CX}[34]
☐ Snap switches & dimmers req grounding EXC___ [4001.11.1]		{404.9B}
• Replacements where no grounding means present OK with plastic faceplate or GFCI protection _____[4001.11.1X]		{404.9BX}
☐ Grounding OK by screws to grounded metal box [4001.11.1]		{404.9B1}
☐ Metal faceplates must be grounded to switch ___ [4001.11.1]		{404.9B}
☐ Faceplate must completely cover wall opening_____[4001.11]		{404.9A}
☐ Switch at each entrance of stairs with ≥ 6 risers ____ [3903.3]		{210.70A2c}
☐ Dimmers only for incandescent lights not receptacles [4001.12]		{404.14E}
☐ Current-carrying conductors of circuit grouped **F34** _ [3406.7]		{300.3B}
☐ Re-identify ungrounded white or gray wires **F34** ___ [3407.3X]		{200.7C}
☐ "CO/ALR" switch req'd if direct Al wire connection__ [4001.2]		{404.14C}

FIG. 35

4-Way Switch

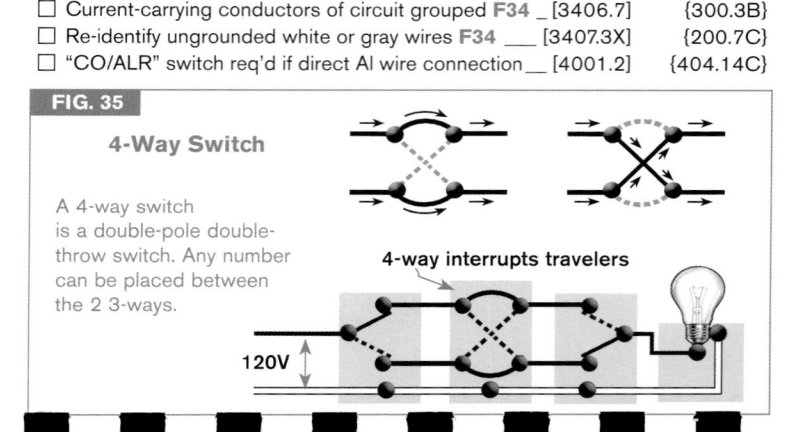

A 4-way switch is a double-pole double-throw switch. Any number can be placed between the 2 3-ways.

4-way interrupts travelers

120V

LIGHTING

Lighting outlets and luminaires must be installed with no exposed live parts that could pose a shock hazard. The heating effect of luminaires must be considered, especially around thermal insulation. Lights rated "type IC" are suitable for insulated ceilings. See **p.202** for req'd locations.

General	09 IRC	11 NEC
☐ All luminaires & lampholders listed _____	[3403.3]	{410.6}
☐ Exposed metal parts grounded EXC _____	[4003.3]	{410.42A}
• Incidental metal parts such as mounting screws ___	[4003.3]	{410.42A}
☐ Wet location luminaires L&L for wet location_____	[4003.8]	{410.10A}
☐ Damp location luminaires L&L for damp or wet location	[4003.8]	{410.10A}
☐ Screw shells for lampholders only–no adapters _____	[4003.4]	{410.90}

Recessed Lights	09 IRC	11 NEC
☐ Non-Type IC min 1/2 in. from combustibles _____	[4004.8]	{410.116A1}
☐ Non-Type IC min 3 in. from insulation_____	[4004.9]	{410.116B}
☐ Type IC OK in contact with combustible material ____	[4004.8]	{410.116A2}
☐ Type IC OK in contact with insulation _____	[4004.9]	{410.116B}
☐ Luminaires that req > 60°C wire must be marked _____	[n/a]	{410.74}
☐ Connect proper temp-rated wire to luminaire _____	[n/a]	{410.117A}
☐ Tap conductors to 60°C wire min 18 in. max 6 ft. **F36**____	[n/a]	{410.117C}

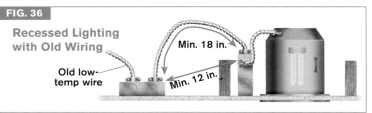

FIG. 36

Recessed Lighting with Old Wiring

Old low-temp wire

Min. 18 in.

Min. 12 in.

Closet Lights F37

	09 IRC	11 NEC
☐ Incandescent bulbs req'd to be fully enclosed _____	[4003.12]	{410.16A1}
☐ Partially enclosed incandescent bulbs prohibited___	[4003.12]	{410.16B}
☐ Surface-mounted only on ceiling or wall above door	[4003.12]	{410.16C1&2}
☐ Surface incandescents min 12 in. from storage ____	[4003.12]	{410.16C1}
☐ Surface fluorescents min 6 in. from storage _____	[4003.12]	{410.16C2}
☐ Recessed (wall or ceiling) min 6 in. from storage ___	[4003.12]	{410.16C3&4}
☐ Surface fluorescent or LED (light-emitting diode) OK in storage area if listed for same_____	[4003.12][35]	{410.16C5}

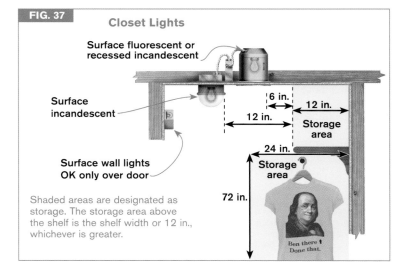

FIG. 37

Closet Lights

Surface fluorescent or recessed incandescent

Surface incandescent

Surface wall lights OK only over door

6 in.

12 in.

12 in.

24 in.

72 in.

Storage area

Storage area

Ben there. Done that.

Shaded areas are designated as storage. The storage area above the shelf is the shelf width or 12 in., whichever is greater.

Track Lighting

	09 IRC	11 NEC
☐ Branch circuit rating ≤ track rating	[4005.1]	(410.151B)
☐ Connected load ≤ track rating	[4005.3]	(410.151B)
☐ No track concealed, extended through walls or partitions, or	[4005.4]	
☐ in damp or wet locations	[4005.4]	(410.151C)
☐ Track must be securely fastened	[4005.5]	(410.154)
☐ Track must be grounded	[4005.6]	(410.155B)

Tub & Shower Areas F38

	09 IRC	11 NEC
☐ No cord-connected or pendant luminaires, lighting track, or ceiling-suspended paddle fans 1st 8 ft. above tub rim or shower threshold & for zone extending 3 ft. outside.	[4003.10]	(410.10D)
☐ Luminaires directly above tub & shower listed for damp locations (or wet locations if subject to shower spray)	[4003.10]	(410.10D)

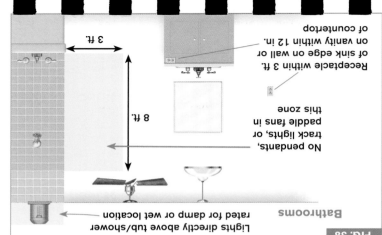

FIG. 38

Lights directly above tub/shower rated for damp or wet location

Bathrooms

No pendants, track lights, or paddle fans in this zone

8 ft.

3 ft.

Receptacle within 3 ft. of sink edge on wall or on vanity within 12 in. of countertop

---

APPLIANCES

The term *appliances* is a generic term for standardized manufactured equipment that uses electricity (other than lighting). Whether portable or permanent, all appliances req a means of disconnecting the power source so the appliance can be safely serviced or replaced. The codes provide general rules for disconnecting appliances as well as specific rules for common built-in (fixed in place) appliances.

Disconnecting Devices

	09 IRC	11 NEC
☐ All appliances req disconnecting means	[410.1.5]	(422.30)
☐ Cord-connected appliances req attachment plug	[3909.4]	(410.7B)
☐ Accessible attachment plug OK as disconnect	[T410.1.5]	(422.33A)
☐ Additional disconnect req'd if plug not accessible	[T410.1.5]	(422.33A)
☐ Breaker alone OK for appliances <300VA or 1/8hp	[T410.1.5]	(422.31A)
☐ In-sight switch or breaker req'd if ≥300VA or 1/8hp,	[T410.1.5]	
☐ or lockable breaker OK when not in sight F39	[T410.1.5]	(422.31B)
☐ Breaker lockouts req permanent hasp F39	[T410.1.5]	(422.31B)
☐ Unit switch opening all ungrounded conductors OK	[T410.1.5]	(422.34)

FIG. 39

Breaker Lockout

Hydromassage Tub (Whirlpool Bathtub)

	09 IRC	11 NEC
☐ Readily-accessible GFCI protection req'd_____	[4209.1]	{680.71}
☐ Individual branch circuit req'd _____	[4209.1][36]	{680.71}
☐ Electrical equipment (pump motor) must be accessible	[4209.3]	{680.73}
☐ Disconnecting means req'd in sight of motor _____	[T4101.5]	{430.102B}
☐ Bond metal parts in contact with circulating water___	[4209.4]	{680.74}
☐ Bonding conductor min solid 8 AWG Cu **F40** _____	[4209.4]	{680.74}
☐ Bond metal piping system to motor lug EXC **F40** __	[4209.4][37]	{680.74}
• Double-insulated motor_____	[4209.4]	{680.74}
☐ Bonding conductor need not connect to panelboards	[4209.4]	{680.74}

FIG. 40

Hydromassage Tub (Whirlpool)

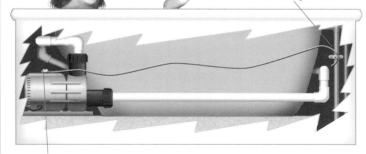

Bond to metal piping systems & any grounded metal parts in contact with circulating water

8 AWG conductor bonded to motor lug intended for bonding

Kitchens

	09 IRC	11 NEC
☐ Cords must be L&L (no NM cable)_____	[4101.3]	{422.16A}
☐ Garbage disposer cord min 18 in. max 36 in. _____	[T4101.3]	{422.16B1}
☐ Dishwasher or trash compactor cord min 3 ft. max 4 ft. measured from back _____	[T4001.3]	{422.16B2}
☐ Dishwasher & compactor receptacles in same space as appliance or in adjacent space_____	[n/a]	{422.16B2}
☐ Range hoods can be cord & plug connected if L&L for cord & on individual branch circuit _____	[4101.3]	{422.16B4}
☐ Range hood cords min 18 in. max 36 in. _____	[T4101.3]	{422.16B4}
☐ Cord & plug ovens & cooking units OK if L&L_____	[4101.3]	{422.16B3}

Central Furnace

	09 IRC	11 NEC
☐ In-sight disconnect req'd_____	[T4101.5]	{422.31B&C}

Refer to manu instructions for possible supplemental OCPD requirements **F41**.

☐ Lighting outlet switched at entry to equipment space	[3903.4]	{210.70A3}
☐ Central furnace must be on individual circuit EXC ___	[3703.1]	{422.12}
• Associated equipment (electrostatic filters, pumps, etc.)	[3703.1]	{422.12X1}
☐ 120V receptacle req'd within 25 ft. on same elevation	[3901.11]	{210.63}

FIG. 41

"SSU" Switch

A fused disconnect provides supplementary overcurrent protection & is sometimes a manufacturer's instruction.
An example might be a furnace requiring 15A overcurrent protection installed on a 20A circuit.

Electric Furnaces & Space Heaters — 09 IRC — 11 NEC

- ☐ Branch circuit 125% load (heat watts + motor) ____ [3702.10] — {424.3B}
- ☐ Disconnect in sight or lockable breaker **F39** _____ [T4101.5] — {424.19}
- ☐ Unit switch that opens all ungrounded conductors OK as disconnect for space heater with no motor > 1/8hp _____ [T4101.5] — {424.19C}

Central Vacuum — 09 IRC — 11 NEC

- ☐ Max 80% individual branch circuit rating, 50% of multi-outlet branch circuit rating _____ [3702.3] — {210.23A}
- ☐ Cord must have same ampacity as branch circuit _____[n/a] — {422.15B}
- ☐ Bond all non-current-carrying metal parts _____ [3908.2] — {422.15C}

Water Heater — 09 IRC — 11 NEC

- ☐ In-sight or lockable breaker or switch OK **F39** _____ [T4101.5] — {422.31B}
- ☐ Breaker lockout hasp req'd to remain in place with lock removed **F39** _____ [T4101.5] — {422.31B}
- ☐ Bond hot, cold & gas pipes **F13**_____ [3609.7] — {250.104}

Outdoor De-icing & Snow Melting Equipment — 09 IRC — 11 NEC

- ☐ GFPE protection req'd for de-icing equipment_____ [4101.7] — {426.28}

Some jurisdictions allow the GFPE function of an AFCI to meet this rule.

Air-Conditioning — 09 IRC — 11 NEC

- ☐ Wiring & OCPD per nameplate of L&L equipment __ [3702.11] — {440.4B}
- ☐ Disconnect on or within sight of condenser **F43** ___ [T4101.5] — {440.14}
- ☐ Disconnect not OK on compressor access panel _____[n/a] — {440.14}
- ☐ Working space req'd in front of disconnect **F43** ____ [3405.1] — {110.26A}
- ☐ Room AC plug disconnect OK if controls ≤6 ft. of floor___[n/a] — {440.63}
- ☐ Max cord length 120V = 10 ft., 240V = 6 ft._____[n/a] — {440.64}
- ☐ AFCI or leakage current detection interrupter (LCDI) in cord or plug for room AC units_____[n/a] — {440.65}

FIG. 42

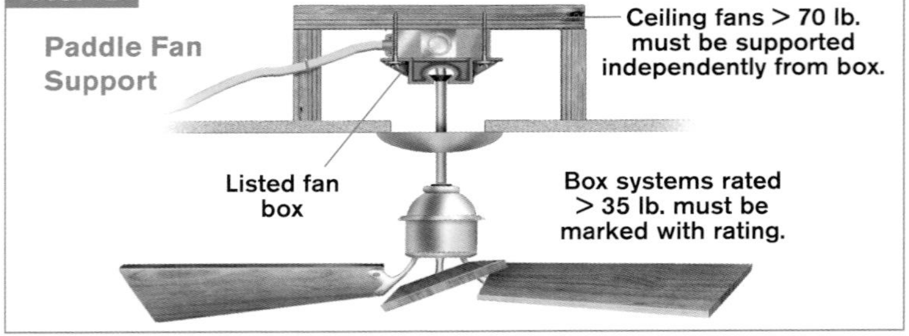

Paddle Fan Support

Ceiling fans > 70 lb. must be supported independently from box.

Listed fan box

Box systems rated > 35 lb. must be marked with rating.

Paddle Fans F42 — 09 IRC — 11 NEC

- ☐ Listed box for fan support (no standard boxes)_____ [3905.9] — {314.27C}
- ☐ Listed fan boxes without weight marking OK to 35 lb. [3905.9] — {314.27C}
- ☐ > 35 lb. & < 70 lb., fan box L&L for suitable weight __ [3905.9] — {314.27C}
- ☐ Independent support for fans > 70 lb. _____ [3905.9] — {314.27C}

FIG. 43

Air-Conditioning Condenser

All ACs req. an in-sight disconnect.

Switch not to be installed directly behind condenser.

Smoke Alarms   09 IRC

- ☐ NFPA 72 systems OK if permanent part of property _____ [314.2][38]
- ☐ Alarms req'd in each sleeping room & adjoining areas **F44**_____ [314.3]
- ☐ Req'd each story including basements & habitable attics **F44** _____ [314.3]
- ☐ Interconnect so activation of 1 alarm sets off all alarms _____ [314.3]
- ☐ Power from building wiring & battery backup EXC_____ [314.4]
  - • Battery-only OK alterations or repairs with no access to wire path [314.4X2]

Carbon Monoxide Alarms   09 IRC

- ☐ Req'd outside sleeping areas in dwellings with fuel-fired appliances or with attached garages **F44**_____ [315.1][39]
- ☐ Req'd when remodeling requiring permit is performed _____ [315.2][39]
- ☐ Install AMI & in compliance with UL 2034 _____ [315.3][39]

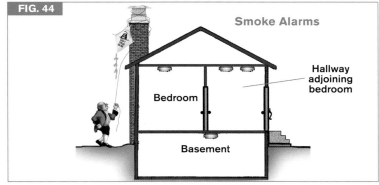

FIG. 44

Smoke Alarms

Bedroom

Basement

Hallway adjoining bedroom

T10 is a "quick reference" guide to the maximum size breaker for a given size of wire. It is an abbreviated version of T11–14. Always consider if the conductors must be "derated" for ambient temperature, grouping, or the other factors on the next page. The sizes given for service entrance conductors apply for wires only with insulation types RHH, RHW, RHW-2, THHN, THHW, THW, THW-2, THWN, THWN-2, XHHW, XHHW-2, SE, USE, and USE-2.

TABLE 10	SIZING CONDUCTORS			
Fuse or Breaker	Branch Circuits or Feeders Wire Size		Service Conductors (AWG)	
	Cu	Al	Cu	Al
15	14	12	n/a	n/a
20	12	10	n/a	n/a
30	10	8	n/a	n/a
40	8	6	n/a	n/a
50	6	4	n/a	n/a
60	6	3	n/a	n/a
70	4	2	n/a	n/a
80	3	1	n/a	n/a
90	2	1/0	n/a	n/a
100	2	1/0	4	2
110	1	1/0	3	1
125	1/0	1/0	2	1/0
150	1/0	2/0	1	2/0
200	3/0	4/0	2/0	4/0
225	4/0	250kcmil	3/0	250kcmil
400	500kcmil	700kcmil	400kcmil	600kcmil

AMPACITY OF WIRE

When wire overheats, its insulation begins to break down, and we say the wire has exceeded its ampacity. Protecting conductors and equipment from overheating and insulation failure is one of the main principles of electrical safety.

General

	IRC 60	11 NEC
☐ Protect conductors at their ampacity EXC.	[3705.5]	{240.4}
• Small conductors protected per note A in **T11**	[3705.5.3]	{240.4D}
• AC protected AMI	[3705.5.4]	{240.4G}
☐ OCPD for NM cable not to exceed 60°C ampacity	[3705.4.4]	{334.80}

Derating

	IRC 60	11 NEC
☐ Apply temp-correction factor **T12**	[3705.15B2]	{310.15B2}
☐ Add correction for rooftop conduits per **T13**	[n/a]	{310.15B3c}
☐ Derate for > 3 current-carrying conductors in raceway or cables grouped without spacing > 24 in. in length	[3705.3]	{310.15B3a}
☐ Derate > 2 NM cables in caulked (fireblocked) hole	[3705.4.4]	{334.80}
☐ Derate > 2 NM cables installed without spacing	[3705.4.4][40]	{334.80}

The first step in determining the allowable ampacity of a conductor is to look it up in **T11** based on the wire size & insulation type. The most common ratings of conductor insulation are 60°C, 75°C & 90°C. We use the 90°C column only for derating (temp. corrections), not for selection of the breaker or fuse. Conductors can be dual rated, with 75°C ratings in wet locations & 90°C ratings in dry locations, such as THWN/THHN.

Breaker & equipment terminations have a temp. rating, typically 60°C and/or 75°C. The overall ampacity of a circuit is limited by the lowest-rated device or conductor in the circuit. The final choice of breaker is, therefore, usually limited by the temp. rating of the breaker terminals; & the insulation rating is used in the derating calculations. Nonmetallic sheathed cable & SE cable as interior wiring are restricted to a 60°C rating despite containing 90°C rated conductors.

TABLE 11 — WIRE AMPACITIES [T3705.1] {310.15B16}

	Cu			Al			
	60°C	75°C	90°C	60°C	75°C	90°C	
	140°F	167°F	194°F	140°F	167°F	194°F	
Cu (AWG)	TW UF	THHW THW THWN USE	THHN THHW THW-2 THWN2 USE-2	UF	USE XHHW	USE-2 XHHW-2	**Al (AWG)**
14*	20	20	25	—	—	—	—
12*	25	25	30	20	20	25	12
10*	30	35	40	25	30	35	10
8	40	50	55	30	40	45	8
6	55	65	75	40	50	60	6
4	70	85	95	55	65	75	4
3	85	100	110	65	75	85	3
2	95	115	130	75	90	100	2
1	110	130	150	85	100	115	1
1/0	125	150	170	100	120	135	1/0
2/0	145	175	195	115	135	150	2/0
3/0	165	200	225	130	155	175	3/0
4/0	195	230	260	150	180	205	4/0
250	215	255	290	170	205	230	250

A. For Cu wire: max OCPD 30A for 10 AWG, 20A for 12 AWG & 15A for 14 AWG.
For Al wire, max OCPD 25A for 10 AWG & 20A for 12 AWG.

In addition to size, material & insulation type, other factors must be considered. These are ambient temp. **T12**, the rate of heat dissipation into the ambient medium & the adjacent load-carrying conductors **T14**. Heat dissipates more readily to free air than to water, such as found in underground conduits. Thermal insulation traps heat, as do adjacent conductors when they are grouped together.

To determine the ambient temp. correction; apply the factors of **T12** to the ampacity listed in the appropriate column of **T11**. The heating effect of reflected sunlight must also be added to the temp. correction, per **T13**.

TABLE 12	AMBIENT TEMPERATURE CORRECTION [T3705.2] & {310.15B2a}			
Ambient Temp. °C	For Ambient Temp. > 30°C (86°F), Multiply the Allowable Ampacities in T11 by the Following Percentages:			Ambient Temp. °F
	60°C	75°C	90°C	
31–35	0.91	0.94	0.96	87–95
36–40	0.82	0.88	0.91	96–104
41–45	0.71	0.82	0.87	105–113
46–50	0.58	0.75	0.82	114–122
51–55	0.41	0.67	0.76	123–131
56–60	–	0.58	0.71	132–140
61–70	–	0.33	0.58	141–158

This table may have little effect on post-1984 90°C-based NM-B wiring. It can be important in remodels with older 60°C wire.

TABLE 13	TEMPERATURE ADJUSTMENT FOR CONDUITS EXPOSED TO SUNLIGHT ABOVE ROOFTOPS {T310.15B2c}	
Distance between Roof & Conduit	Temp. Added to T12	
0 – 1/2 in.	33°C	60°F
> 1/2 in. – 3 1/2 in.	22°C	40°F
> 3 1/2 in. – 12 in.	17°C	30°F
> 12 in.	14°C	25°F

Another consideration is conductor proximity, which traps heat & prevents heat dissipation when conductors are grouped. When there are more than 3 current-carrying conductors in a raceway, the derating factors of **T14** must be applied, in addition to any ambient temp. correction. These same derating factors also apply to a grouping of cables installed without spacing for a length of 24 in. or more & for groups >2 NM cables passing through an opening in wood framing that is fireblocked with thermal insulation, caulk, or foam & to NM cables installed without spacing & in contact with thermal insulation.

TABLE 14	DERATING FOR CONDUCTOR PROXIMITY [T3705.3] {310.15B3a}
Number of Current-Carrying Wires	Ampacity Correction
4–6	80
7–9	70
10–20	50

With modern 90°C small conductors this table becomes significant when there are > 9 current-carrying conductors in a conduit or cable group, or when compounded by temp. corrections. Cables installed without spacing > 2 ft. are subject to the above derating. When newer 90°C wire is connected to older 60°C wire, such as pre-1984 NM, the ampacity of the lower-rated conductors applies to the entire circuit.

CABLE SYSTEMS

Cable systems are the most common residential wiring methods. Cables contain all conductors of the circuit inside a protective outer sheath of metal or plastic Starting with the 2005 edition, the NEC uses a parallel numbering system for rules pertaining to cables and raceways. See the common numbering system table (**T24**) on **p.233**.

Cable Protection Indoors (NM, AC, MC, UF, SE) 09 IRC   **11 NEC**

☐ Bored holes & standoff clamps 1¼ in. setback **F56** _ [3802.1]   {300.4A&D}

☐ Protect cables with ¹⁄₁₆ in. steel plate {or L&L plate}
   if closer than 1¼ in. to framing surfaces **F45** _____ [3802.1]   {300.4A&D}

☐ Cables min 1½ in. below sheet steel roof decks _____[n/a]   {300.4E}[41]

☐ Provide guard strips within 6 ft. of attic scuttle
   (& up to 7 ft. high if attic has permanent access) __ [3802.2.1]   {334.23}

FIG. 45

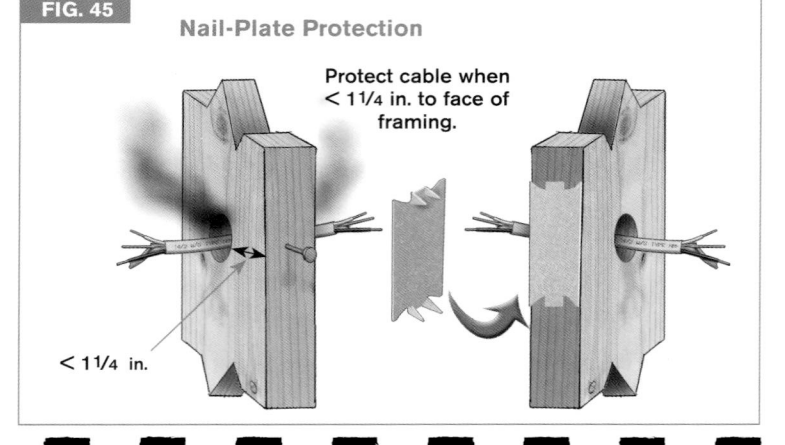

Nail-Plate Protection

Protect cable when < 1¼ in. to face of framing.

< 1¼ in.

FIG. 46   Cable in an Attic with No Permanent Stair

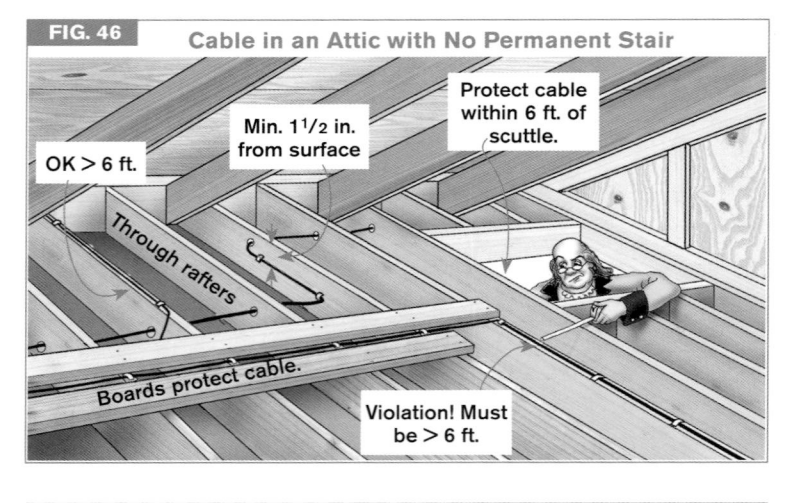

OK > 6 ft.

Min. 1½ in. from surface

Protect cable within 6 ft. of scuttle.

Through rafters

Boards protect cable.

Violation! Must be > 6 ft.

FIG. 47   Underfloor Cable in Basement or Crawl Space

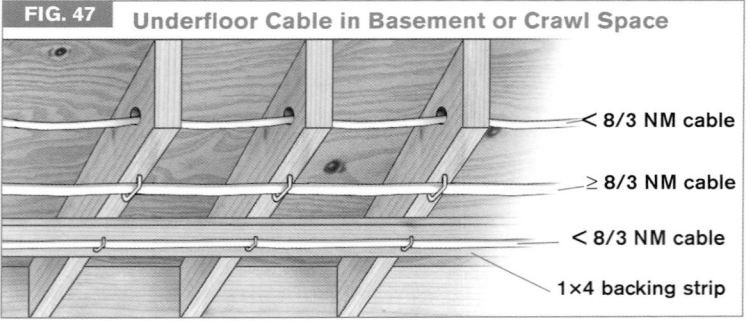

< 8/3 NM cable

≥ 8/3 NM cable

< 8/3 NM cable

1×4 backing strip

NM–Nonmetallic Sheathed Cable F48

	09 IRC	11 NEC
☐ OK in dry locations only	[3801.4]	{334.12B4}
☐ Protect exposed cable from damage where necessary	[3802.3.2]	{334.15B}
☐ Listed grommets for holes through metal framing	[3802.1]	{300.4B1}
☐ OCPD selection based on 60° column **T11**	[3705.4.4]	{334.80}
☐ Derating & temp correction based on 90° rating	[3705.4.4]	{334.80}
☐ Derate > 2 NM cables in same caulked (fireblocked) hole	[3705.4.4]	{334.80}
☐ Derate > 2 NM cables installed without spacing in contact with thermal insulation	[3705.4.4][40]	{334.80}
☐ Secure to box with approved NM clamp EXC **F49**	[3905.3.2]	{314.17B&C}
• Single gang (2¼ × 4 in.) plastic box stapled within 8 in.	[3905.3.2]	{314.17CX}
☐ Min ¼ in. sheathing into plastic boxes	[3905.3.1]	{314.17C}
☐ Secure within 12 in. of box & max 4½ ft. intervals	[3802.1]	{334.30}
☐ Do not overdrive staples or staple flat cable on edge	[3802.1]	{334.30}
☐ Bends gradual (min 5× cable diameter)	[3802.5]	{334.24}
☐ Running board for small cable under joists **F47**	[3802.4]	{334.15C}

FIG. 48

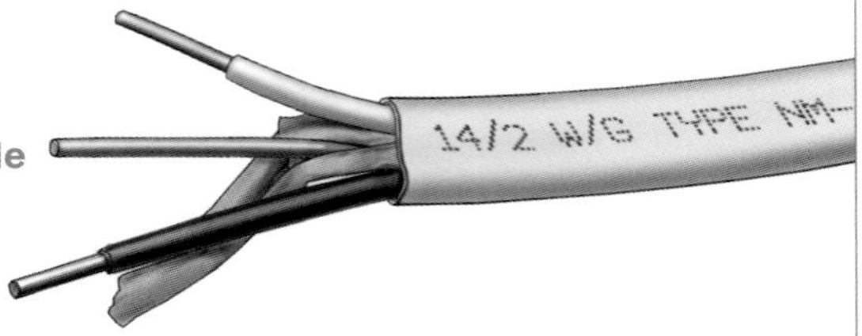
NM –
Nonmetallic-
Sheathed Cable
14/2 W/G TYPE NM-

FIG. 49

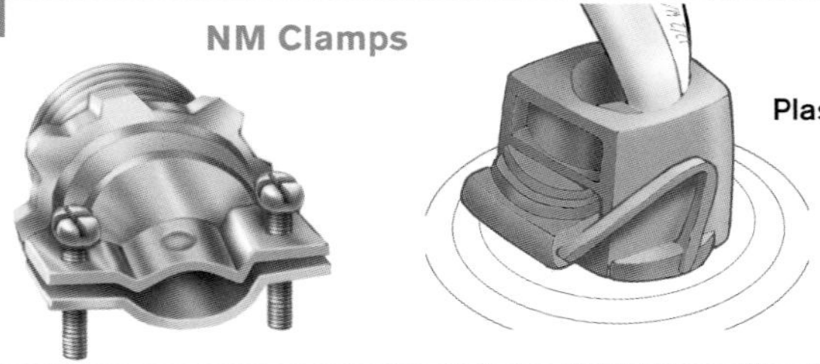
NM Clamps
Metal
Plastic

AC–Armored Cable (BX™) F50

	09 IRC	11 NEC
☐ Dry locations only	[3801.4]	{320.10}
☐ Secure within 12 in. of box & max 4½ ft. intervals EXC	[3802.1]	{320.30B}
• 2 ft. where flexibility needed (motors)	[3802.1]	{320.30D}
☐ Insulated (anti-short) bushing at terminations **F50**	[3802.1]	{320.40}
☐ Armor is EGC–don't bring bond wire into box **F50**	[3908.8]	{250.118}
☐ Underside of joists–secure at each joist	[n/a]	{320.15}

FIG. 50

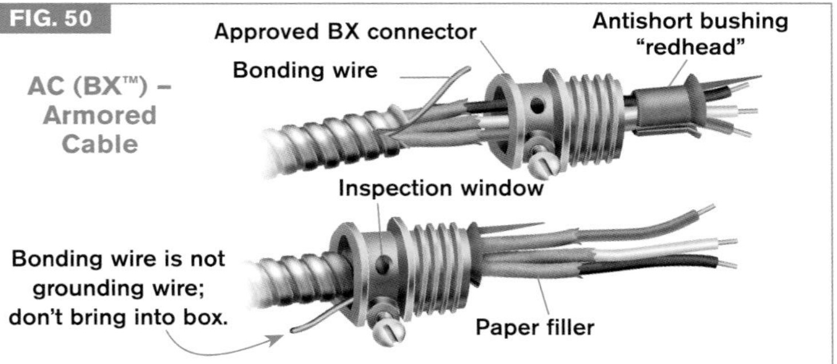
AC (BX™) –
Armored
Cable
Approved BX connector
Antishort bushing
"redhead"
Bonding wire
Inspection window
Bonding wire is not
grounding wire;
don't bring into box.
Paper filler

UF–Underground Feeder Cable F51

	09 IRC	11 NEC
☐ Interior installation same rules as NM	[3801.4]	{340.10}
☐ May be buried in earth with cover per **T1, F52**	[3801.4]	{340.10}
☐ Protect where emerging from earth from 18 in. below grade to 8 ft. above **F52**	[3803.3]	{300.5D1}
☐ Single conductors in trench must be grouped	[3803.8]	{340.10}
☐ UV-resistant type OK exposed to sunlight	[3802.3.3]	{340.12}
☐ Not OK strung through air without support messenger	[3802.1]	{340.12}

FIG. 51

UF – Underground Feeder Cable

UF 14/2

FIG. 52

Protecting Underground Cable

UF cable req's protection where it emerges from the ground & to a height of at least 8 ft.

The protection should extend underground to the burial depth or 18 in., whichever is less.

PVC

8 ft. min.

Per T1

UF

Must have bushing

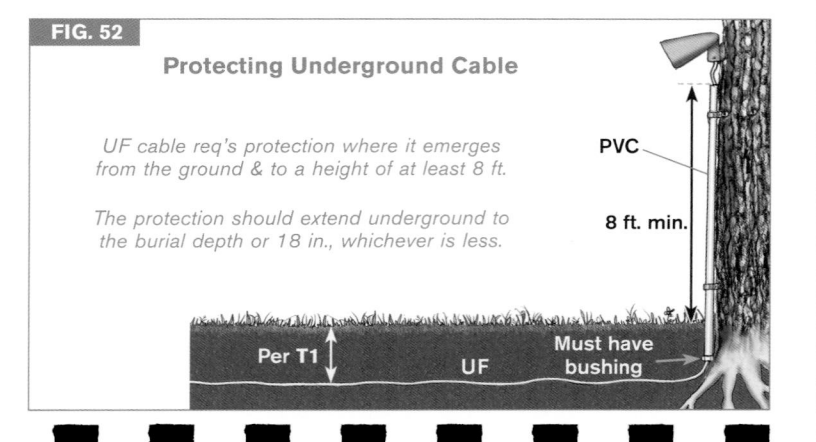

SE–Service Entrance Cable F53 & USE–Underground Service Entrance Cable

	09 IRC	11 NEC
☐ OK as service entrance conductor (see **p.178**)	[3801.4]	{338.10A}
☐ Type SE interior installation same rules as NM	[3802.1][42]	{338.10B4a}
☐ Type USE not OK for interior wiring	[3801.4]	{338.12B}
☐ SE not OK for direct burial, USE OK for direct burial	[3801.4]	{338.12B}
☐ Bare neutral OK for EGC of 240V branch circuit	[3801.4]	{338.10B2}
☐ Insulated neutral (type SE-R) req'd for feeders except to separate existing building with no other continuous metal path (see **p.180,192**)	[3801.4]	{338.10B2X}
☐ Bends gradual (min 5× cable diameter)	[3802.5]	{338.24}

FIG. 53

SE Cable

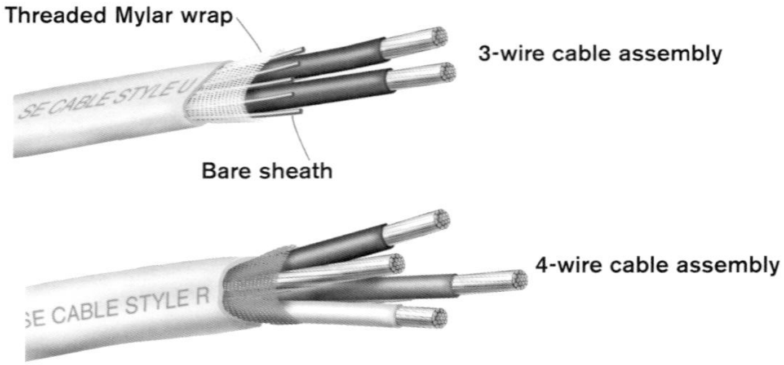

Threaded Mylar wrap

3-wire cable assembly

SE CABLE STYLE U

Bare sheath

4-wire cable assembly

SE CABLE STYLE R

MC–Metal-Clad Cable F54,55   09 IRC   11 NEC

☐ Support or secure at max 6 ft. intervals _____ [3802.1] {330.30B&C}
☐ Secure within 12 in. of box or other termination EXC_ [3802.1]   {330.30B}
  • Unsupported whip ≤ 6 ft. to luminaire in accessible
    ceiling _____ [3802.1]   {330.30D}
  • Where fished _____ [n/a]   {330.30D}
☐ Bends gradual (min 7× interlocked armor diameter) _ [3802.5]   {330.24}

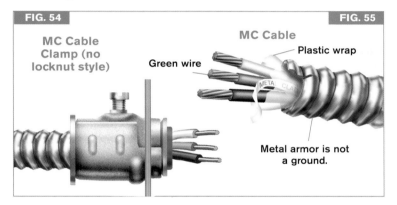

FIG. 54   **FIG. 55**

MC Cable Clamp (no locknut style)

MC Cable

Green wire

Plastic wrap

META CLAD

Metal armor is not a ground.

FIG. 56

Stand-Off Clamp

Used to maintain clearances to stud or joist edge

VOLTAGE DROP

When laying out wiring, consider the voltage drop caused by long runs of wire. Fine-print note #4 of 210.19 of the NEC recommends (though it does not req) a maximum voltage drop of 3% on branch circuits and a 5% overall voltage drop, including the feeders. Excessive voltage drop can cause problems in connected equipment and adds to the monthly utility costs. One way to overcome a voltage drop problem is to use larger wire than the minimum size and to make sure that all connections are tight. Voltage drop increases proportionately to the load on the circuit. Adding more than the minimum number of circuits helps prevent individual circuits from overloading. The added cost of more wiring will pay for itself over time in reduced utility costs and greater equipment efficiency.

In the table below, the distances shown are the maximum length of cable to stay within a 3% branch circuit voltage drop at 80% of the allowable load on the circuit. Multiwire circuits (**p.193**) act as 240V circuits for voltage drop to the extent that the load on them is balanced. When only one side of the multiwire circuit has a load, the voltage drop is the same as for any other 120V circuit.

TABLE 15	CABLE LENGTH TO LIMIT VOLTAGE DROP TO 3%			
Wire Size (AWG)	Cu Distance (ft.)		Al Distance (ft.)	
	120V	**240V**	**120V**	**240V**
14	50	100	N/A	N/A
12	60	120	36	72
10	64	128	38	76
8	76	152	46	92
6	94	188	57	114

Based on 80% circuit loading for normal OCPD.

RACEWAYS

Raceways are complete systems of conduit or tubing through which conductors are installed. In the NEC numbering system, all articles pertaining to raceways have a parallel numbering system so the portion after the article number is the same for all types. Article numbers are the first 3 digits inside the period before the period. See the common numbering system table T24 on p. 233.

General	09 IRC	11 NEC
☐ Conductors in raceways stranded if ≥ 8 AWG	[3406.4]	{310.3}
☐ Wet-rated conductors req'd in raceways above grade in wet locations	[3802.7][43]	{300.9}
☐ Raceway req to be complete before wiring EXC	[3904.5]	{300.18A}
• Short sections of raceway for cable protection	[3904.5X]	{300.18AX}
☐ Bends req'd to have even radius—no kinks	[3802.5]	{***.24}
☐ 360° max bends between pull points F57	[3802.1]	{***.26}
☐ Raceway must be reamed after cutting	[3802.1]	{***.28}
☐ Plastic bushing/liner req'd if conductors 24 AWG	[3906.1.1]	{300.4G}
☐ Box & conduit body covers must remain accessible	[3905.10]	{314.29}
☐ No plastic boxes with metal cables or raceways unless bonded through box	[3905.3X]	{314.3X}
☐ No splicing in conduit bodies except conduit bodies with sufficient volume per marking	[3905.12.3.1]	{314.16C2}
☐ Max 40% fill if > 2 conductors T21, 22	[3904.6]	{***.22}
☐ Derate conductors as needed T11-14	[3705.283]	{310.15B2a}

EMT-Electrical Metallic Tubing F58	09 IRC	11 NEC
☐ Direct burial or embedment not OK	[3801.4]	{358.10B}
☐ In dry/wet locations L&L wet fittings	[3905.11]	{358.42}
☐ Secure in place max 10 ft. intervals & 3 ft. from each box, conduit body or cabinet	[3802.1]	{358.30A}

EMT-Electrical Metallic Tubing (cont.) F58	09 IRC	11 NEC
☐ Horizontal runs supported by holes in framing OK if securely fastened within 3 ft. of box, conduit body, or cabinet	[3802.1]	{358.30B}
☐ Not OK as support for boxes, but OK for conduit bodies	[n/a]	{358.12}

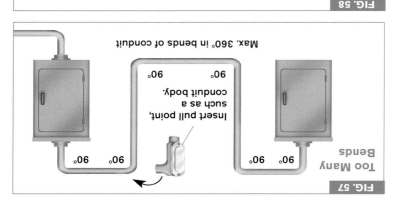

FIG. 57

Too Many Bends

90° 90° 90° 90° 90° 90° 90° 90°

Insert pull point, such as a conduit body.

Max. 360° in bends of conduit

FIG. 58

EMT-Electrical Metallic Tubing

Dry location / Raintight wet location

Compression ring / Sealing ring

Older style EMT connectors with only compression ring were not listed.

RMC–Rigid Metal Conduit F59   09 IRC   11 NEC

- ☐ Galvanized RMC typically sufficient corrosion protection
for direct burial or embedment _____ [3801.4] {344.10B}
- ☐ Coat buried field cut threads with L&L compound ___ [3801.4] {300.6A}
- ☐ Provide bushing or fitting at box connection F59 ____ [3802.1] {344.46}
- ☐ No threadless connectors on threaded conduit ends ____[n/a] {344.42}
- ☐ Secure in place within 3 ft. of termination _____ [3802.1] {344.30A}
- ☐ Horizontal support spacing max 10 ft. _____ [3802.1] {344.30B}

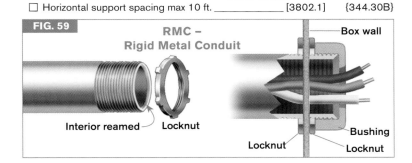

FIG. 59

RMC –
Rigid Metal Conduit

Interior reamed — Locknut

Box wall

Bushing

Locknut

Locknut

Locknut

FMC–Flexible Metal Conduit ("Greenfield") F60   09 IRC   11 NEC

- ☐ Dry locations only _____ [3801.4][44] {348.12}
- ☐ Support max spacing 4 1/2 ft. & 12 in. from boxes EXC [3802.1] {348.30A}
  - • Lighting whip in accessible ceiling OK to 6 ft. OR__ [3802.1] {348.30AX4}
  - • 36 in. where flexibility is needed_____ [3802.1] {348.30AX1}
- ☐ Armor is OK as EGC if fittings listed, circuit ≤ 20A,
no flexibility needed & ≤ 6 ft. long_____ [3908.8.1] {250.118}
- ☐ Angle connections may not be concealed F60_____[n/a] {348.42}

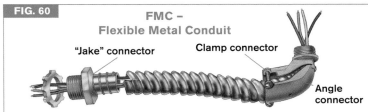

FIG. 60

FMC –
Flexible Metal Conduit

"Jake" connector

Clamp connector

Angle connector

LFMC–Liquidtight Flexible Metal Conduit F61   09 IRC   11 NEC

- ☐ OK for wet locations _____ [3801.4] {350.10}
- ☐ OK for direct burial if L&L _____ [3801.4] {350.10}
- ☐ OK as EGC up to 6 ft. if fittings listed, circuit ≤20A or ≤60A
for sizes 3/4–1 1/4 in. & no flexibility needed _____ [3908.8.1] {250.118}
- ☐ Support max spacing 4 1/2 ft. & 12 in. from boxes EXC[3802.1] {350.30A}
  - • 36 in. where flexibility is needed_____ [3802.1] {350.30AX2}

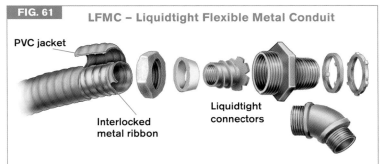

FIG. 61

LFMC – Liquidtight Flexible Metal Conduit

PVC jacket

Interlocked metal ribbon

Liquidtight connectors

LFNC–Liquidtight Flexible Nonmetallic Conduit F62

	09 IRC	11 NEC
☐ OK in lengths > 6 ft. if secured every 3 ft.	[n/a]	{356.10}
☐ Securing or supporting not req'd up to 3 ft. for motors	[3802.1]	{356.30}
☐ OK for direct burial or encasement when L&L	[3801.4]	{356.10}
☐ EGC req'd	[3908.4]	{250.4A5}

PVC–Rigid Polyvinyl Chloride Conduit F63

	09 IRC	11 NEC
☐ Burial depth per **T1**	[3803.1]	{300.5A}
☐ Support to prevent sags per **T16** & within 3 ft. of box	[3802.1]	{352.30}
☐ Expansion joints req'd if subject to ≥ 1/4 in. shrinkage	[n/a]	{352.44}
☐ Not OK for support of luminaires or boxes	[n/a]	{352.12B}
☐ Not permitted in environments > 50°C (122°F)	[n/a]	{352.12D}

ENT–Electrical Nonmetallic Tubing F64

	09 IRC	11 NEC
☐ OK embedded in concrete with approved fittings	[3801.4]	{362.10}
☐ Not OK in environments > 50°C (122°F)	[n/a]	{362.12}
☐ Not OK for direct earth burial	[3801.4]	{362.12}
☐ Must be identified as sunlight resistant if outdoors	[3801.4]	{362.12}
☐ Secure or support every 3 ft. EXC	[3802.1]	{362.30A&B}
• 6 ft. unsupported OK to luminaires in accessible ceiling	[3802.1]	{362.30AX2}

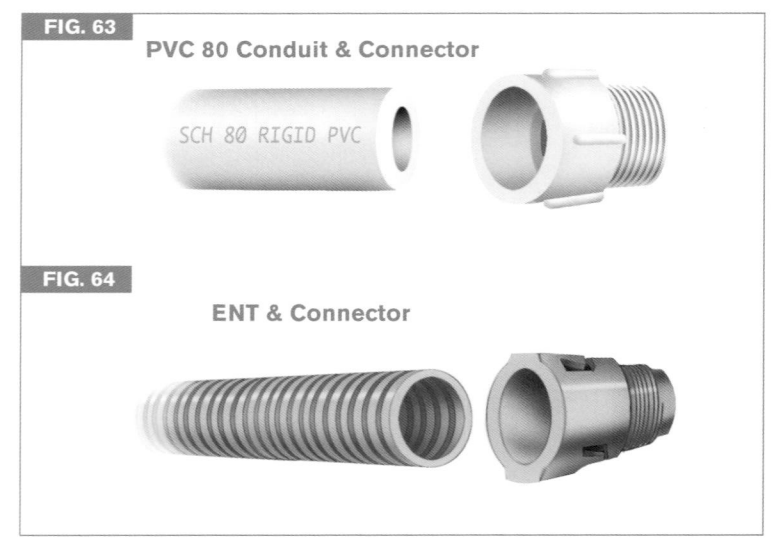

FIG. 63

PVC 80 Conduit & Connector

SCH 80 RIGID PVC

FIG. 64

ENT & Connector

FIG. 62

LFNC–Liquidtight Flexible Nonmetallic Conduit

NON-METALLIC LIQUID

TABLE 16	PVC CONDUIT SUPPORT MAX. SPACING [T3802.1] {T352.30}	
Conduit Trade Size	**2009 IRC**	**2011 NEC**
1/2 in.–1 in.	3 ft.	3 ft.
1 1/4 in.–2 in.	5 ft.	5 ft.
2 1/2 in.–3 in.	5 ft.	6 ft.
3 1/2 in.–5 in.	5 ft.	7 ft.

FILL TABLES FOR ALL CONDUCTORS OF THE SAME SIZE (BASED ON NEC CHAPTER 9 & ANNEX C & [E3804.6(1) & (9)])

Size (AWG)	**TABLE 17** EMT FILL — Number of Conductors in THHN, THWN						Size (AWG)	**TABLE 18** EMT FILL — Number of Conductors in XHHW (Compact Stranding)						Size (AWG)	**TABLE 19** SCHEDULE 80 PVC FILL — Number of Conductors in THHN, THWN						Size (AWG)	**TABLE 20** SCHEDULE 80 PVC FILL — Number of Conductors in XHHW (Compact Stranding)					
•	½	¾	1	1¼	1½	2	•	½	¾	1	1¼	1½	2	•	½	¾	1	1¼	1½	2	•	½	¾	1	1¼	1½	2
14	12	22	35	61	84	138	14	8	15	25	43	58	96	14	9	17	28	51	70	118	14	6	11	20	35	49	82
12	9	16	26	45	61	101	12	6	11	19	33	45	74	12	6	12	20	37	51	86	12	5	9	15	27	38	63
10	5	10	16	28	38	63	10	5	8	14	24	33	55	10	4	7	13	23	32	54	10	3	6	11	20	28	47
8	3	6	9	16	22	36	8	3	5	8	15	20	34	8	2	4	7	13	18	31	8	3	5	8	14	20	33
6	2	4	7	12	16	26	6	1	4	6	11	15	25	6	1	3	5	9	13	22	6	1	4	6	11	15	25
4	1	2	4	7	10	16	4	1	3	4	8	11	18	4	1	1	3	6	8	14	4	1	2	4	8	11	18
3	1	1	3	6	8	13	3	1	1	4	6	8	14	3	1	1	3	5	7	12	3	1	1	3	5	7	12
2	1	1	3	5	7	11	2	1	1	3	6	8	13	2	1	1	2	4	6	10	2	1	1	3	5	8	13
1	1	1	1	4	5	8	1	1	1	2	4	6	10	1	0	1	1	3	4	7	1	1	1	2	4	6	9
1/0	1	1	1	3	4	7	1/0	1	1	1	3	5	8	1/0	0	1	1	2	3	6	1/0	1	1	1	3	5	8
2/0	0	1	1	2	3	6	2/0	1	1	1	3	4	7	2/0	0	1	1	1	3	5	2/0	1	1	1	3	4	7
3/0	0	1	1	1	3	5	3/0	0	1	1	2	3	6	3/0	0	1	1	1	2	4	3/0	0	1	1	2	3	5
4/0	0	1	1	1	2	4	4/0	0	1	1	1	3	5	4/0	0	0	1	1	1	3	4/0	0	1	1	1	3	5
250	0	0	1	1	1	3	250	0	1	1	1	2	4	250	0	0	1	1	1	3	250	0	0	1	1	1	4

Conduit Fill Calculations

When all conductors are the same size, use T17–20. When different sized conductors are used, use T21 to find the wire areas, add them up, and use T22 to find the minimum size conduit. Example: 3 2 AWG THHN + 3 8 AWG XHHW in FMC: (3 × 0.1158) + (3 × 0.0437) = 0.4785, and the next greater size in the 40% column is 0.511. Therefore, a 1¼ in. FMC conduit meets code. When conductor calculation is close to conduit table values, one size larger is recommended.

TABLE 21	SQ. IN. AREA OF CONDUCTORS (BASED ON NEC T5 CHAPTER 9)												
	14	12	10	8	6	4	2	1	1/0	2/0	3/0	4/0	250
TW	.0139	.0181	.0243	.0437	.0726	.0973	.1333	.1901	.2223	.2624	.3117	.3718	.4596
THHN	.0097	.0133	.0211	.0366	.0507	.0824	.1158	.1562	.1855	.2223	.2679	.3237	.3970
XHHW	.0139	.0181	.0243	.0437	.0590	.0814	.0962	.1146	.1534	.1825	.2190	.2642	.3197

TABLE 22								CONDUIT & TUBING FILL (BASED ON NEC T4 CHAPTER 9)																
Trade Size	Internal Diameter								2 wire sq. in. Fill 31%								> 2 Wire sq. in. Fill 40%							
	EMT	ENT	FMC	LFNMC	IMC	RMC	PVC80	PVC40	EMT	ENT	FMC	LFNMC	IMC	RMC	PVC80	PVC40	EMT	ENT	FMC	LFNMC	IMC	RMC	PVC80	PVC40
⅜	–	–	.384	.494	–	–	–	–	–	–	.036	.059	–	–	–	–	–	–	.046	.077	–	–	–	–
½	622	.560	.635	.632	.660	.632	.526	.602	.094	.076	.098	.097	.106	.097	.067	.088	.122	.099	.127	.125	.137	.125	.087	.114
¾	.824	.760	.824	.830	.864	.836	.722	.804	.165	.141	.165	.168	.182	.170	.127	.157	.213	.181	.213	.216	.235	.220	.164	.203
1	1.049	1.000	1.020	1.054	1.105	1.063	.936	1.029	.268	.243	.253	.270	.297	.275	.213	.258	.346	.314	.327	.349	.384	.355	.275	.333
1¼	1.380	1.340	1.275	1.395	1.448	1.394	1.255	1.360	.464	.437	.396	.474	.510	.473	.383	.450	598	.564	.511	.611	.658	.610	.495	.581
1½	1.610	1.570	1.538	1.588	1.683	1.624	1.476	1.590	.631	.600	..576	.614	.689	1.056	.530	.616	.814	.774	.743	.792	.889	.829	.684	.794
2	2.067	2.020	2.040	2.033	2.150	2.083	1.913	2.047	1.040	.994	1.013	1.006	1.125	1.508	.891	1.020	1.342	1.282	1.307	1.298	1.452	1.363	1.150	1.316

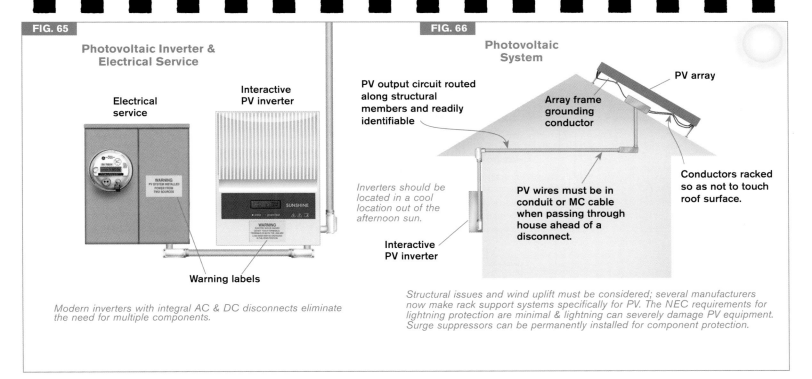

FIG. 65

Photovoltaic Inverter & Electrical Service

Electrical service

Interactive PV inverter

WARNING
PV SYSTEM INSTALLED
POWER FROM
TWO SOURCES

SUNSHINE

WARNING
ELECTRIC SHOCK HAZARD
DO NOT TOUCH TERMINALS
TERMINALS ON BOTH THE LINE AND
LOAD SIDES MAY BE ENERGIZED
IN THE OPEN POSITION

Warning labels

Inverters should be located in a cool location out of the afternoon sun.

Modern inverters with integral AC & DC disconnects eliminate the need for multiple components.

FIG. 66

Photovoltaic System

PV output circuit routed along structural members and readily identifiable

Array frame grounding conductor

PV array

Conductors racked so as not to touch roof surface.

PV wires must be in conduit or MC cable when passing through house ahead of a disconnect.

Interactive PV inverter

Structural issues and wind uplift must be considered; several manufacturers now make rack support systems specifically for PV. The NEC requirements for lightning protection are minimal & lightning can severely damage PV equipment. Surge suppressors can be permanently installed for component protection.

PHOTOVOLTAICS

In most states, the utility will rebate a portion of the cost of a PV system. Time-of-use and net metering can reduce or eliminate monthly utility costs. The quality and efficiency of PV equipment have improved greatly in the last few years. What once req'd numerous separate components is often integrated into a single piece of equipment. Contact the utility and building department before beginning any project involving renewable energy sources.

Definitions

Array: An assembly of panels that forms the power-producing unit F66.

Combiner: The location where parallel PV source circuits are connected to create a PV output circuit.

Hybrid system: A system with multiple power sources (not including the utility or batteries). An example would be a system with a generator & a PV source.

Interactive system: A solar PV system that operates in parallel to the utility.

Inverter: Equipment that converts the DC current & voltage of a PV output circuit to an AC waveform F65.

Inverter output circuit: The AC conductors from an inverter to an AC panelboard or service F65.

Module: A group of PV cells connected together & encapsulated in an environmentally protective laminate—usually tempered glass—to generate DC power when exposed to the sun.

Panel: A group of modules preassembled onto a common frame & designed to be field installed.

PV output circuit: Conductors between the PV source circuits & the inverter F66.

PV source circuits: Circuits between modules & circuits from modules to the common connection points (combiners) of the DC system.

Stand-alone system: Solar PV system supplying power independent of the utility.

General — 11 NEC

- ☐ Inverters, modules, panels, source circuit combiners L&L for PV {690.4D}
- ☐ PV req'd to be installed by only qualified persons {690.4E}[45]
- ☐ Max voltage = sum of rated open-circuit voltage of series connected modules times correction factors for cold temp {690.7A}
- ☐ All power sources req disconnects {690.15}
- ☐ DC disconnect req'd for ungrounded conductors F65 {690.13}
- ☐ PV output circuits req in-sight disconnect {690.16B}
- ☐ Disconnect for ungrounded conductors must be readily accessible switch or breaker with no exposed live parts F65 {690.17}
- ☐ Warning req'd at DC disconnect if all terminals hot while open F65 {690.17}
- ☐ Rated max currents & voltages labeled on DC disconnect {690.53}
- ☐ No disconnect on grounded conductor if it would be left energized {690.13}
- ☐ PV disconnecting means req'd to be on outside or inside nearest point of entrance of conductors EXC {690.14(C1)}
- ☐ Source circuits through interior OK in metal conduit F66 {690.31E}
- ☐ AC disconnects energized from 2 directions req warning label F65 {690.17}
- ☐ Backfed breakers not req'd to be secured in place EXC {705.12D6}
  - • Stand-alone systems (non-utility-interactive) {690.10E}

Arrays & Inverters — 11 NEC

- ☐ Req'd markings on modules: polarity, max OCPD rating for module protection, open-circuit voltage, operating voltage, max system voltage, operating current, short-circuit current & max power {690.51}
- ☐ PV circuits may not share raceways with non-PV systems EXC {690.4B}
  - • OK with barriers, tagging & grouping {690.4B}
- ☐ DC ground-fault protection (DC GFP) req'd {690.5}
- ☐ Inverter listed as interactive if used in interactive system {690.60}
- ☐ DC arc-fault protection req'd systems > 80V {690.11}[46]
- ☐ Interactive systems to automatically disconnect in grid outage EXC {690.61}
  - • OK to feed subpanel isolated from service by transfer switch {690.61}

Grounding   11 NEC

- ☐ Module frames & all metal parts must be grounded _____ {690.43A}
- ☐ Size EGCs of PV output circuit per **T6** & min 14 AWG_____ {690.45}
- ☐ EGCs must be run in same raceway as PV array circuit conductors {690.43F}
- ☐ Bond ground-mounted array structures _____ {690.43C}
- ☐ DC 2-wire system > 50V must have grounded conductor _____ {690.41}
- ☐ Same conductor can perform DC grounding, AC grounding & bonding between AC & DC systems **F65,66** _____ {690.47C3}
- ☐ When grounded conductor bonded to EGC internal within DC GFP device, bond not to be duplicated with an external connection_____ {690.42X}

Overcurrent Protection & Wiring   11 NEC

- ☐ Single OCPD OK for series-connected string _____ {690.9E}
- ☐ Sum of PV & main breakers not > 120% of panel rating _____ {705.12D2}
- ☐ Source circuit currents = 125% × sum of parallel circuit currents ____ {690.8A1}
- ☐ Locate PV breaker opposite end of bus from main or feeder input ___ {705.12D7}
- ☐ Apply label warning against moving PV breaker _____ {705.12D7}
- ☐ Size conductors for 125% of max PV source short circuit currents {690.8B1}
- ☐ Max allowable voltage in SFD 600V _____ {690.7C}
- ☐ Consider high ambient temp (use 90°C wire)_____ {690.31}
- ☐ No multiwire or 240V circuits in panels with 120V supply _____ {690.10C}
- ☐ Single conductor cables type USE or L&L as PV wire in exposed outdoor source circuits (behind modules) _____ {690.31B}

FIG. 67	Voltage Correction Factors {NEC T690.7}

Degrees Fahrenheit

	−40	−4	32	68

Multiply by this amount

1.25
1.18
1.10
1.02

	−40	−20	0	20

Degrees Centigrade

SWIMMING POOL

Electricity and water can be a lethal mix. Precautions must be taken for shock hazard protection and to prevent corrosion of electrical equipment. Bonding is important to eliminate voltage gradients in the pool area. For GFCI requirements, see **p.199**. Installation of a pool might req relocating overhead service conductors.

Overhead Conductor Clearances   09 IRC   11 NEC

- ☐ 22½ ft. clearance in any direction from water _____ [T4203.5] {680.8A}
- ☐ 14½ ft. in any direction from diving platform _____ [T4203.5] {680.8A}

Underground Wiring   09 IRC   11 NEC

- ☐ Non-pool underground wiring min 5 ft. from pool EXC [4203.7] {680.10}
  - • If space limited, RMC, IMC, or PVC systems OK __ [4203.7] {680.10}
- ☐ Cover depth min 6 in. for RMC or IMC, 18 in. for PVC [4203.7] {680.10}

Feeders to Pool Panelboards   09 IRC   11 NEC

- ☐ New feeder req's RMC, IMC, LFNMC, or PVC EXC [T4202.1] {680.25A}
  - • EMT OK on or within buildings _____ [T4202.1] {680.25A}
- ☐ Raceway req's min 12 AWG insulated EGC EXC __ [T4202.1] {680.25B}
  - • Existing FMC or cable with EGC OK _____ [T4205.6] {680.25AX}

Pool Pump Motors   09 IRC   11 NEC

- ☐ RMC, IMC, PVC, or listed MC OK for branch circuit [T4202.1] {680.21A1}
- ☐ Branch circuits in AC, FMC, or NM only within building [T4202.1] {680.21A1}
- ☐ EMT branch circuit OK on or within building_____ [T4202.1] {680.21A2}
- ☐ Flexible connection OK in LFMC or LFNMC _____ [T4202.1] {680.21A3}
- ☐ Cord & plug connected motors OK with cord ≤3ft___ [4202.2] {680.21A4}
- ☐ Cords req EGC min 12 AWG & per **T6**_____ [4202.2] {680.7B}

Underwater Wet-Niche Lighting **F68**   09 IRC   11 NEC

- ☐ Min 18 in. below water level _____ [4206.4.2] {680.23A5}
- ☐ Luminaire bonded & secured to shell with locking device [4206.5] {680.23B5}
- ☐ Luminaire must req tool for removal _____ [4206.5] {680.23B5}

Underwater Wet-Niche Lighting (cont.) F68   09 IRC   11 NEC

- ☐ Low-voltage transformers req L&L for pool _____ [4206.1]   {680.23A2}
- ☐ Conductors from load side of GFCI or transformer not in same raceway or box as non-GFCI wires _____ [4206.3]   {680.23F3}
- ☐ Forming shell req's bonding terminal if PVC conduit _ [4206.5]   [680.23B1]
- ☐ Nonmetallic conduit req's 8 AWG bonding conductor [4205.3]   {680.23B2}
- ☐ Bonding conductor insulated & potted in forming shell[4205.3]   {680.23B2}
- ☐ Min 16 AWG EGC in cord to wet-niche fixture F69 _ [4205.4]   {680.23B3}
- ☐ EGC connections on terminals only—no splices _____ [4205.2]   {680.23F2}

Equipotential Bonding F68   09 IRC   11 NEC

- ☐ Purpose of bonding is to reduce voltage gradients __ [4204.1]   {680.26A}
- ☐ Bond metal parts of pool structure, ladders, equipment, fences & screens or structures < 5 ft. from pool EXC _____ [4204.2]   {680.26B}[47]
  - • Small isolated parts < 4 in. or < 1 in. into pool structure _____ [4204.2]   {680.26B5}
- ☐ Bond motors except listed & double-insulated type __ [4204.2]   {680.26B6X}
- ☐ Provide bond wire to area of double-insulated motor [4204.2]   {680.26B6}
- ☐ Bonding conductor min #8 solid Cu _____ [4204.4]   {680.26B}
- ☐ Unencapsulated steel shell req'd to be bonded _____ [4204.2]   {680.26B1}
- ☐ Cu conductor grid req'd if pool shell steel encapsulated in nonconductive compounds (coated rebar) _____[4204.2][48]   {680.26B1}
- ☐ Cu conductor grid req's 8 AWG Cu in 12 ×12 in. pattern, conforming to contour of pool & deck, ≤ 6 in. from outer contour of pool shell, all conductors bonded at crossings___[4204.2][48]   {680.26B1}
- ☐ Perimeter surfaces for 3 ft. beyond pool req equipotential bonding with steel wire or reinforcement _____[4204.2][48]   {680.26B2b}
- ☐ Connect perimeter to unencapsulated steel pool shell or Cu conductor grid at min 4 points _____ [4204.2][48]   {680.26B2}
- ☐ Min 9 sq. in. bonded metal contacting pool water __ [4204.3][49]   {680.26C}

Receptacles (see p.199 for GFCI requirements) 09 IRC   11 NEC

- ☐ Min 1 receptacle from pool walls _____ [4203.1.2]   {680.22A3}
- ☐ Pump motor receptacles not < 10 ft. from pool wall EXC
  - • 6 ft. OK for single-receptacle twist-lock types ___ [4203.1.1]   {680.22A1}
- ☐ Dimensions include distance around barriers without penetrating a floor, wall, doorway, or window opening_____ [4203.1]   {680.22A5}

Lighting Outlets & Luminaires   09 IRC   11 NEC

- ☐ Outdoors ≥ 5 ft. from pool edge unless 12 ft. above [4203.4.1]   {680.22C1}
- ☐ Indoors ≥7 ft. 6 in. above water if enclosed & GFCI [4203.4.2]   {680.22C2}
- ☐ Existing lighting OK if GFCI & ≥ 5 ft. from pool edge & ≥ 5 ft. high_____ [4203.4.3]   {680.22C3}
- ☐ Switches min 5 ft. from pool edge or separated by barrier _____ [4203.2]   {680.22D}

FIG. 68

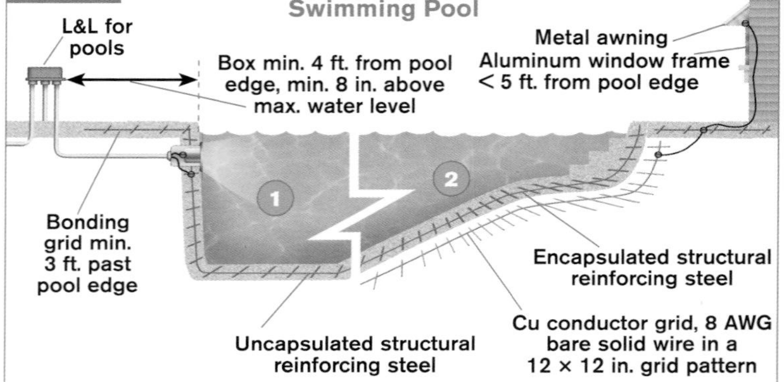

Swimming Pool

L&L for pools

Box min. 4 ft. from pool edge, min. 8 in. above max. water level

Metal awning Aluminum window frame < 5 ft. from pool edge

Bonding grid min. 3 ft. past pool edge

Encapsulated structural reinforcing steel

Uncapsulated structural reinforcing steel

Cu conductor grid, 8 AWG bare solid wire in a 12 × 12 in. grid pattern

❶ or ❷ are options for creating an equipotential bonding grid.

HOT TUB/SPA

Outdoor hot tubs and spas follow the same rules as swimming pools in addition to the general rules below. A hydromassage tub (**p.209**) is not a spa because it is emptied after each use.

General   09 IRC   11 NEC
- ☐ LFMC or LFNMC up to 6 ft. OK for package unit __ [T4202.1]   {680.42A1}
- ☐ Cord up to 15 ft. OK for GFCl-protected package unit [4202.2]   {680.42A2}
- ☐ Bands to secure hot tub staves exempt from bonding [4204.4]   {680.42B}

Indoor Spas   09 IRC   11 NEC
- ☐ Indoor packaged units ≤20A OK for cord & plug ____ [4202.2]   {680.43X}
- ☐ Min 1 receptacle 6–10 ft. from inside wall of spa__ [4203.1.4]   {680.43A1}
- ☐ Wall switches min 5 ft. from inside wall of spa _____ [4203.2]   {680.43C}1

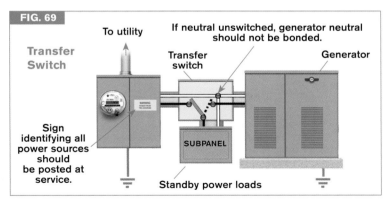

FIG. 69

Transfer Switch

To utility

If neutral unswitched, generator neutral should not be bonded.

Transfer switch

Generator

SUBPANEL

Sign identifying all power sources should be posted at service.

Standby power loads

GENERATORS

Generators provide a source of emergency power during a utility outage. Care must be taken to ensure that the 2 sources of power—utility and generator—cannot be connected simultaneously. This dangerous condition results from failure to install proper transfer switches and improper use of portable generators.

Generators   11 NEC
- ☐ Must be suitable for environment, rainproof if outdoors _____ {445.10}
- ☐ Rainproof generators not OK enclosed indoors _____ {110.3B}
- ☐ Conductors sized 115% of nameplate current rating_____{445.13}
- ☐ Live or moving parts guarded against accidental contact _____{445.14}
- ☐ GEC req'd for permanently installed generators _____{250.30A3}
- ☐ Remove bonding jumper if transfer switch does not switch neutral **F69** _____ {250.24A5}

Transfer Switches F69   11 NEC
- ☐ Sign req'd at service indicating generator location_____{702.8A}
- ☐ Transfer equipment must prevent simultaneous connection of generator & utility service _____ {702.6}

Electric Vehicle (EV) Charging Systems   11 NEC
- ☐ Systems >20A 125V no exposed live parts _____ [625.13]
- ☐ Coupler L&L for EV _____ {625.16}
- ☐ Interlock must de-energize connector when uncoupled from EV ____{625.18}
- ☐ Electric vehicle OK as standby power source through listed utility interactive connection _____ {625.26}

OLD WIRING

A high percentage of residential electrical fires occur in older homes. Proper over-current protection helps prevent insulation failure, though in some cases time and exposure take too great a toll on wiring, and it must be replaced with new materials. Fuses provide overcurrent protection only if they are the right size. Too often, they are altered or bypassed (a penny behind the fuse). Older ceramic fuse panels and panels with cartridge fuses also pose a risk of electrocution because of exposed electrical contacts. For these reasons, many insurance companies req upgrading of fuse systems. The references below are from the NEC. The IRC is a code for new construction and does not address old wiring.

Fuses	11 NEC
☐ No exposed contact fuseholders (must be dead front) F70	{240.50D}
☐ Edison base (plug fuses) not OK for 240V circuits	{240.51A}
☐ Type S fuse req'd if tampering or overfusing exists F70	{240.51B}
☐ Type S fuse adapter must be proper size for wire	{240.4D}
☐ No fuses in neutral conductor F70	{240.22}

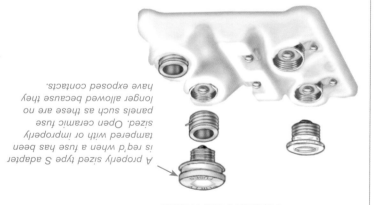

FIG. 70

Ceramic Fuse Holder

A properly sized type S adapter is req'd when a fuse has been tampered with or improperly sized. Open ceramic fuse panels such as these are no longer allowed because they have exposed contacts.

OLD WIRING

KNOB & TUBE (K&T)

K&T wiring is the oldest wiring method found in American homes. When left in its original state, it can be reliable; safety was inherent in its design. As a wiring method in uninsulated joist and stud cavities it is protected from damage and provided with air circulation, which prevents heat buildup. Unfortunately, when these systems are modified by unqualified persons, the inherent safety of K&T is often compromised. Adding new loads to an old system is tricky and seldom done correctly. Rubber insulation on K&T wiring becomes brittle over time and is prone to mechanical damage, especially when thermal insulation is added to an attic. Older rubber insulation has only a 60°C rating.

General   **11 NEC**

☐ No new K&T _____ {394.10}

☐ Additions to existing K&T OK if properly protected _____ {394.10}

☐ Must enter plastic boxes through separate holes _____ {314.17C}

☐ Must be protected with loom where entering box_____ {314.17B&C}

☐ Loom must extend from last insulator to ¼ in. inside box **F73** __{314.17B&C}

☐ Do not envelop with thermal insulation _____ {394.12}

☐ Wires must be kept out of direct contact with wood framing _____ {394.17}

☐ Tubes req'd where passing through framing members **F71**_____ {394.17}

☐ 3 in. min between wires, 1 in. to surfaces **F72** _____ {394.19A1}

☐ Conductors on sides (not face) of exposed joists & rafters EXC {394.23A&B}
   • OK on edges or faces of rafters or joists in attics < 3 ft. high __ {394.23BX}

☐ Protect with running boards up to 7 ft. high in attic with stairs ____ {394.23A}

☐ Provide protection where exposed < 7 ft. above floor _____ {398.15C}

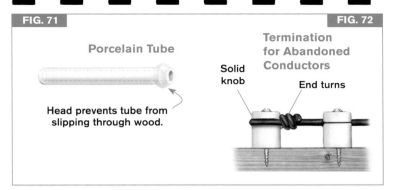

FIG. 71

Porcelain Tube

Head prevents tube from slipping through wood.

FIG. 72

Termination for Abandoned Conductors

Solid knob

End turns

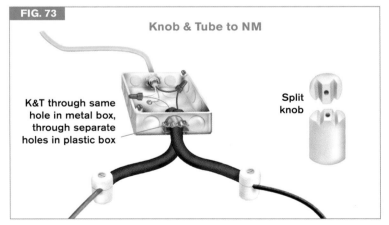

FIG. 73

Knob & Tube to NM

K&T through same hole in metal box, through separate holes in plastic box

Split knob

REPLACEMENT RECEPTACLES

Houses built before adoption of the 1962 NEC will not have 3-hole receptacles in all locations. Appliances with 3-prong cords are designed to be used with only grounded 3-hole receptacles. A GFCI can provide shock hazard protection for 2-conductor circuits; though without an EGC, it may not protect equipment.

General　11 NEC
☐ AFCI protection req'd for replacements in areas where circuit req's AFCI protection effective 1/1/2014 {406.4D4}so {p.194}
☐ Protection can be breaker, AFCI outlet device, or upstream AFCI outlet {406.4D4}so
☐ Replacement receptacles must be tamper-resistant {406.4D5}so
☐ Outdoor wet location replacement receptacles must be WR {406.4D6}so

Replacements When No Grounding Present　11 NEC
☐ 2-hole receptacle OK if in area where GFCI not req'd {406.4D2a}
☐ Must have GFCI protection in area that now req's GFCI {406.4D3}
☐ OK to install GFCI even if no ground present. {406.4D2b&c}
☐ Non-grounded GFCI or GFCI-protected receptacles req label stating "No Equipment Ground". {406.4D2b}
☐ Ungrounded 3-hole receptacle supplied through a GFCI also req label stating "GFCI Protected" {406.4D2d}
☐ Separate EGC can be added from receptacle box & connect to service enclosure, GEC, or ground bar of panel at circuit origin {250.130C}
☐ OK to run EGC separately from circuit conductors {300.3B2}
☐ Not OK to jumper neutral & EGC. {250.142B}

Replacements When Grounding Present in Box　11 NEC
☐ Replacements must be 3-hole if EGC present. {406.4D1}
☐ Bond 3-hole receptacle to grounded box with wire OR {250.146}
• Use grounding-type receptacle (captive metal screw from yoke) {250.146B}

OLD NM

Pre-1984 nonmetallic sheathed cable contained conductors with insulation rated 60°C. When installed in a hot attic, the ampacity of this old wire is easily exceeded. Precautions must also be taken to isolate this old low-temperature wiring from luminaires that req high-temperature rated connections F36. Much of this old wire was used in houses with problematic electrical equipment. Replacement circuit breakers for older panels can be very expensive—providing one more incentive to replace such systems. For further information on old wiring, refer to the Code Check website, www.codecheck.com.

Aluminum Wiring　11 NEC
☐ Snap switches with direct Al connection req L&L as "CO/ALR" {404.14C}
☐ Receptacles ≤ 20A with direct Al connection req L&L as "CO/ALR" {406.2C}
☐ Al to Cu splicing devices must be listed for same {110.14}
☐ Terminals (including breaker terminals) for Al req L&L {110.14A}

Pre-1984 NM　11 NEC
☐ Derate for ambient temp {310.10, 310.16}
☐ No 60°C conductors in attics > 131°F {310.16}
☐ No direct connection to luminaires that req > 60°C conductors {410.117A}
☐ Isolate old wiring from high-temp wiring F36 {410.117A}
☐ Box for tap conductors min 1 ft. from luminaire, max 6 ft. wire {410.117C}

TABLE 23

SIGNIFICANT CHANGES IN THE 2011 NEC & THE 2009 IRC / 2008 NEC

#	Page	Code & Year	Description	#	Page	Code & year	Description
1	177	11 NEC	Exemption for guarded/isolated roof areas	16	194	08 NEC & 09 IRC	Expanded AFCI other rooms than bedrooms
2	180	11 NEC	Seal underground conduits entering buildings	17	194	11 NEC	AFCI req'd for replacement or extension circuits
3	180	11 NEC	Max 1 branch circuit back to source building	18	195	09 IRC	Conformed to NEC luminaire box ratings
4	181	11 NEC	Clarified when cord-set GFCIs req'd	19	199	11 NEC	GFCI controls req'd to be readily accessible
5	186	08 NEC & 09 IRC	Allows isolated foundation pier rebar as Ufer	20	199	08 NEC & 09 IRC	GFCI exceptions eliminated for garages & unfinished basements
6	189	08 NEC & 09 IRC	List of specific acceptable grounding connection methods replaced simple prohibition against sheet metal screws	21	199	08 NEC & 09 IRC	GFCI within 6 ft. of laundry, utility, or bar sinks
7	189	11 NEC	Bonding req'd at line-side reducing washers	22	199	11 NEC	GFCI within 6 ft. all sinks in addition to those req'd for kitchen countertops
8	190	08 NEC & 09 IRC	Intersystem bonding method specified	23	199	08 NEC & 09 IRC	GFCI protection req'd for 240V boat hoists
9	191	08 NEC & 09 IRC	OCPDs not allowed over steps of a stairway	24	200	08 NEC & 09 IRC	Receptacles req'd by NEC 210.52 or IRC 3901.1 also req'd to be TR
10	191	08 NEC & 09 IRC	Labeling independent of transient conditions	25	200	11 NEC	Exceptions for tamper-resistant receptacles
11	191	08 NEC & 09 IRC	Labeling of spare breakers	26	201	08 NEC & 09 IRC	Clarification that switched receptacles do not count as part of req'd receptacles
12	191	08 NEC & 09 IRC	Neutral current not allowed through enclosure	27	201	11 NEC	New requirement for receptacles in foyers
13	191	08 NEC & 09 IRC	All multiwire circuits handle tie or single-handle 2-pole	28	202	11 NEC	Countermounted bath receptacles allowed
14	191	08 NEC & 09 IRC	Multiwire circuits req'd to be grouped in panel	29	202	08 NEC & 09 IRC	Receptacles req'd for balconies > 20 sq. ft.
15	191	11 NEC	Warning label to identify source of feed-through circuits in panel	30	202	11 NEC	Removed exemption for balconies < 20 sq. ft.
				31	202	08 NEC & 09 IRC	Weather-resistant receptacles req'd

TABLE 23	SIGNIFICANT CHANGES IN THE 2011 NEC & THE 2009 IRC / 2008 NEC (CONT.)		
#	Page	Code & year	Description
32	203	08 NEC & 09 IRC	Clarified when range or sink divides island or peninsula countertop into separate spaces
33	206	11 NEC	Neutral req'd in switch box
34	206	11 NEC	Exceptions if neutral can be added at later time
35	207	08 NEC & 09 IRC	L&L LEDs allowed in closet storage areas
36	209	08 NEC & 09 IRC	Hydromassage tub req'd's individual circuit
37	209	08 NEC & 09 IRC	Hydromassage area piping bonded to motor
38	211	09 IRC	NFPA 72 systems must become permanent part of property to replace req'd alarms
39	211	09 IRC	CO alarms req'd—hardwired like smoke alarms
40	215 212	08 NEC & 09 IRC	Derate > 2 NM cables in contact with thermal insulation
41	214	11 NEC	No cables or raceways other than RMC or IMC allowed < 1½ in. from sheet steel roof deck
#	Page	Code & year	Description
42	216	08 NEC & 09 IRC	SE cable no longer exempt from 60° limitation that applies to NM
43	218	08 NEC & 09 IRC	Wiring in raceways above grade in wet locations req'd to be wet-rated
44	219	08 NEC & 09 IRC	FMC no longer allowed in wet locations
45	224	11 NEC	PV installations req qualified personnel
46	224	11 NEC	DC AFCI protection req'd
47	226	11 NEC	Screens & metal windows < 5 ft. from pool edge included in equipotential bonding
48	226	08 NEC & 09 IRC	Specific methods for creating equipotential bonding grid for nonconductive pool shells & for extending past pool edge
49	226	08 NEC & 09 IRC	Contact with pool water to equipotential bond
50	230	11 NEC	Replacement receptacles req type & protection of new receptacles for the area in which they are installed

TABLE 24	COMMON NUMBERING SYSTEM FOR WIRE, CABLE & RACEWAY ARTICLES (BASED ON NEC CHAPTER 3)		
I. GENERAL	**II. INSTALLATION**		**III. CONSTRUCTION SPECIFICATIONS**
xxx.1 Scope	xxx.10 Uses Permitted	xxx.26 Bends: Number in 1 Run xxx.44 Expansion Fittings	xxx.100 Construction
xxx.2 Definitions	xxx.12 Uses Not Permitted	xxx.28 Reaming & Threading xxx.46 Bushings	xxx.104 Conductors
xxx.6 Listing Requirements	xxx.14 Dissimilar Metals	xxx.28 Trimming xxx.48 Joints	xxx.108 Equipment grounding
	xxx.16 Temperature Limits	xxx.30 Securing & Supporting xxx.50 Conductor Terminations	xxx.120 Marking
	xxx.20 Size	xxx.40 Boxes & Fittings xxx.56 Splices & Taps	
	xxx.22 Number of Conductors	xxx.42 Couplings & Connectors xxx.60 Grounding	
	xxx.24 Bending radius	xxx.80 Ampacity	

In 1752, Benjamin Franklin, aided by his son, William, conducted the famous, but highly dangerous kite experiment. For an animated explanation, visit: www. codecheck.com/cc/BenAndTheKite.html.